T0345253

QUANTUM MECHANICS
for THINKERS

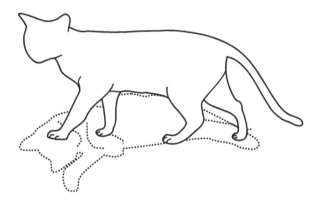

QUANTUM MECHANICS for THINKERS

Gennaro Auletta
Shang-Yung Wang

PAN STANFORD PUBLISHING

Published by

Pan Stanford Publishing Pte. Ltd.
Penthouse Level, Suntec Tower 3
8 Temasek Boulevard
Singapore 038988

Email: editorial@panstanford.com
Web: www.panstanford.com

British Library Cataloguing-in-Publication Data
A catalogue record for this book is available from the British Library.

Quantum Mechanics for Thinkers

ISBN 978-981-4411-71-4 (Hardcover)
ISBN 978-981-4411-72-1 (eBook)

Printed in the USA

To our families

Contents

PART III ONTOLOGICAL ISSUES: PROPERTIES

Foreword

The discovery of quantum mechanics and its comprehension are at the basis of the foundations of modern technology. This fact is not widely recognized. I believe that if one asks the layman which are the most important technological applications of quantum mechanics, he would mostly select nuclear power. After some reflections he could mention lasers, but he would not think of the most important one, i.e., the transistor that is at the basis not only of computers but of practically any device we commonly use (with some notable exceptions like bicycles, wind surfs, and skis).

People who are not trained in quantum mechanics can use a transistor without difficulties, and with some minor technical training they can understand the specifications and use transistors to build simple devices like a wireless radio: transistors behave in a way that is not very different from the old thermionic tubes. However, quantum mechanics has been crucial in the design of transistors, which, when finally constructed, worked exactly as predicted by quantum mechanics.

In spite of the ubiquitousness of quantum mechanics applications, quantum mechanics remains some kind of mystery not only for learned people with a humanistic background, but also for most of the scientists, with the exception of physicists and chemists. The intrinsic difficulty in understanding the principles of quantum mechanics certainly contributes to this deplorable situation. However, this situation is worsened by an aura of incomprehensibility that derives from most of the presentations of quantum mechanics that one find in the literature. Indeed, books that describe quantum mechanics may be divided into two main categories:

- Those that require an advanced knowledge of mathematical analysis (differential and integral calculus), thus casting away most of the people. Such books are perfect for people interested in getting a working knowledge of quantum mechanics, but are of no use for those interested in knowing only what quantum mechanics is and in understanding its implications.

- Those that are directed toward the general public. Although some of these books are excellent, their presentation is limited to a qualitative description. By the time one gets ready to see how all extraordinary properties of quantum mechanics could be implemented in a quantitative description of the system, the presentation, in most cases, stops, usually adding something like "More details would be too technical; they need too much mathematics and therefore cannot be described here." At the end, quantum mechanics seems to be something like magic that can be understood only by fifth-level wizards.

On the contrary, this book makes a strong effort to arrive to a quantitative formulation of quantum mechanical for very simple systems, a formulation that is constructed using minimal mathematical requirements. In this way the reader can easily arrive at the conceptual core of quantum mechanics in its precise mathematical formulation without having to know analysis and calculus. This can be done only if the authors are very careful in choosing the model systems that one uses for the presentation: the choice made in this book is very appropriate so that the reader becomes acquainted with the formalism of quantum mechanics in the simplest possible way.

Only in the second part of the book, after a minimal description of the analytic mathematical tools needed, the reader finds the extension (to a generic system) of the formalism that he or she has learned in the first part. In this way the reader arrives at an understanding of the usual formalism of quantum mechanics separating the conceptual steps, described in the first part, from the technical issues, described in the second part.

In the last part of the book, Ontological Issues, the authors discuss the general implications of quantum mechanics that have been discussed in many places, including popularization articles: the measurement problem, non-locality and non-separability, quantum information, and finally the interpretation of quantum mechanics. The authors' viewpoint on these highly debated subjects is deep and original: the presentation is quite concise, although it does not shy away from giving technical details where needed.

The book is well written and is very readable. It fulfills at its best the premise of the title *Quantum Mechanics for Thinkers*.

Giorgio Parisi

Introduction

Reasons for Studying Quantum Mechanics

Quantum mechanics represents one of the great conceptual revolutions of the 20th century. It has raised a huge number of fundamental questions of both physical and philosophical kind.

- What does matter mean at all?
- What are the main properties or characteristics of matter?
- Can matter be reduced to information?
- Is our universe probabilistic at the most fundamental level?
- Are there non-local correlations in nature?
- Are non-causal interconnections between physical systems possible?
- Is the bound on the speed of information propagation set by the theory of relativity violated?
- What do terms like state, observable, and property mean at all?
- Can physical reality exist without observers?
- Are observers necessary for having a macroscopic world?
- What are the general features of information processing and exchange in our universe?

These questions (and there are also many others) give a first feeling about the depth of the conceptual turn represented by quantum mechanics. Even those classical hypotheses or laws that have passed the quantum mechanical check have somehow been transformed or at least been corrected. It is important for people who desire to deal with fundamental problems in science, especially in quantum theory or in those fields (like chemistry, mathematics, and informatics) that are closely related to quantum theory, to have a deep and clear understanding of this kind of problems. This book provides such an opportunity. We think that undergraduate students in physics could also take advantage of this book, and then transition to more difficult stuff. This book could also be of some use in the last years of the high school. Indeed, one of the major problems we find for these classes is that most of our students go out of the school without having ever heard a single word about quantum mechanics, that is, about the basic physical theory that we have, and it is likely that most of them will never have the opportunity to come back to these issues.

The book is also addressed to people interested in the philosophy of science or in problems at the interface between science and philosophy. As a matter of fact, one of the biggest problems of modern thought is a fracture between science and philosophy causing severe alienation to both fields. Indeed, science without philosophy can become a pure technique, where finally ad hoc solutions and pure simulations dominate, whereas philosophy without science can shift toward esotericism and aestheticism. As a matter of fact, the issues that have been raised within natural sciences, and especially in physics, have always implied a deep shift of the philosophical paradigms. The affirmation of Galilean and Newtonian classical mechanics, which is an important part of the first scientific revolution, has led to a radical rearrangement of the theory of knowledge, first making of the physical science a privileged reference and then, with Kant's doctrine of the *a priori* synthetic judgments, as the unique and authentic form of knowledge.

Quantum mechanics implies, or should imply, even a more radical change of the philosophical modules. However, this has happened in an incomplete and partial form. This is because the discussion on the foundations of this theory is not yet accomplished and so far has not even been dealt with at a sufficiently deep level. Theoretically dealing with the foundations of quantum mechanics is an urgent task, especially considering its huge predictive power and the wide domain of applicability. Its practical consequences already determine many aspects of our modern society (atomic bombs and atomic energy, semiconductors, transistors, and photovoltaic cells, lasers and light-emitting diodes, applications to technology of new states of matter like Bose–Einstein condensates, etc.) and many other may be determined in the near future (quantum cryptography, quantum teleportation, quantum computation, photography without light, etc.).

Aim of the Book

We find that most of the problems that we have stated above are often treated by public opinion and even by cultivated laypersons with superficiality and without a true understanding of the physical and conceptual foundations of quantum theory. It is very often heard or read that quantum mechanics allows telepathy or that reality does not exist. Statements like these show a deep misunderstanding about the true meaning of quantum theory. Therefore, the main aim of the book is to allow the students, the scholars, the philosophers, and even the laypersons interested in these issues to have a quick access to quantum mechanics without dealing with a true textbook that demands proper specialized studies in physics (and related mathematics) for about a couple of years. The phrase "quick access" does not mean that this is a popular science book. It is in fact a scientific book, but addressed to people who do not already posses the prerequisite for dealing with such a sophisticated scientific stuff.

In order to understand the theoretic and philosophical problems in quantum mechanics, it is indeed necessary to master certain formal instruments. In other words, this book does contain quite a few equations. However, we have tried to reduce the formalism to the minimum extent required for understanding the basis of the theory. Moreover, we have also explained from scratch mathematical tools like vector and matrix algebra, probability (in the first part of the book), as well as integration and differentiation (in the second part of the book). This is the reason why the first part is confined to an algebraic approach. In this way, the book is somehow self-contained and only presupposes some high-school background in mathematics.

What Is Required of the Reader?

Although we shall try to do things as simply as possible, this does not mean that the reader shall not meet some difficulties and should not make some efforts to understand the mathematics and the underlying physics. However, our basic assumption is that the study of this book is in the range of university students and scholars of any faculty, or of any cultivated layperson, who are motivated and interested in deepening their knowledge of the subject. Where the reader should meet some particular difficulties in mathematics, we recommend to make use of some online resources where many mathematical concepts are explained with different degrees of difficulty. In particular, we suggest the online mathematics reference MATHWORLD.[a] As an alternative, the reader can also take into account the Mathematics Portal of WIKIPEDIA[b] and the ENCYCLOPEDIA OF MATHEMATICS.[c] Finally, for a first introduction to this type of mathematics we strongly recommend the textbook by Heller.[d]

To minimize the mathematics and to emphasize the underlying physics, we have chosen to present many of the technical details in the form of in-section boxes and end-of-section problems. There are 30 boxes and 130 problems altogether, with the solutions to most of the problems provided at the end of the book. However, the reader is encouraged to try to work out the problems by him- or herself before resorting to the solutions provided. There are also many resuming tables that help the reader quickly find the information that he or she desires. Moreover, we have included 70 figures which not only provide a kind of graphical help but often can even be understood as an integral part of the explanation. In order to help the reader better organize the concepts developed in the book, we have composed a summary of the main concepts at the end of each chapter. Finally, the book contains an extensive bibliography of about 150 entries, and two full, accurate, and comprehensive subject and author indexes for assisting the reader's quick search.

While this book could be an excellent starting point for self-study of quantum mechanics, it is obviously better if the reader is helped by someone with a physics background in dealing with this study. This could happen through an introductory course to quantum mechanics but also through a tutorial. A word of caution is also necessary. The present book does *not* substitute a complete course in quantum mechanics as taught in any physics department and taking advantage of more advanced textbooks.[e] With the help of

---

[a] http://mathworld.wolfram.com
[b] http://en.wikipedia.org/wiki/Portal:Mathematics
[c] http://www.encyclopediaofmath.org
[d] (Heller, 2006).
[e] (Le Bellac, 2006), (Auletta *et al.*, 2009).

this book, a careful reader can understand what quantum mechanics is (and this is *the* aim of the book) but cannot learn to make use of it. In other words, this book helps the reader understand what quantum mechanics is, what are its conceptual foundations, and how its basic formalism works, but does not make of him or her an expert in quantum mechanics.

Outline of the Book

The book is divided into three major parts and is organized as follows:

I. Basic Issues
   The first part contains five fundamental chapters. Chapter 1 provides a short review of classical mechanical concepts. Chapters 2–5 represent the foundation block that deals with the basic notions of quantum theory. However, already Chapter 5 raises many conceptual problems that may give a taste of what follows. Some readers may be satisfied to study this part. It is relatively easy but also needs some time, especially if the reader has never been engaged with mathematics or he or she was, but many years ago. Our suggestion in this case is to read each chapter repeatedly before going further in order to become fully familiar with this language.

II. Formal Issues
   The second part consists of three technical chapters, Chapters 6–8. In this part we introduce some of the most important quantum mechanical observables: position and momentum, energy, and angular momentum and spin. Arguably, this is probably the most difficult part for people not acquainted with physics or mathematics. It is, however, necessary if one really wants to understand the deep meaning of the philosophical conclusions drawn in the subsequent chapters. We stress that the main difficulty is not in the equations themselves, since each step is explained and we presume that a patient reader who will follow those steps shall also be able to consistently progress. The main problem is rather in the large quantity of information packed together. Then we suggest to proceed by taking time in order to assimilate each step and eventually read each

section again and again. If, having tried several times, the reader does not understand the meaning of certain steps, we suggest to take them as facts since it is plausible that some developments can become clearer afterwards. If the reader encounters insurmountable difficulties, he or she may initially skip the latter sections of these chapters and try to come back to specific aspects when the third part of the book demands the knowledge of some previous notions. Sooner or later, however, if the reader wants a deeper understanding of the theory, it becomes necessary to study the whole of it. Indeed, only to have assimilated the Schrödinger equation or the model of the hydrogen atom can bring the reader even to really appreciate not only the usefulness but even the beauty of this theory.

III. Ontological Issues

The third part is composed of four advanced chapters. The measurement problem is dealt with in Chapter 9, the issue of quantum non-locality in Chapter 10, and quantum information in Chapter 11. Finally, the interpretation of the theory that puts together the previous three subjects (and all the main issues raised in the book) is dealt with in Chapter 12. As a matter of fact, Chapters 9–12 represent the block that will turn out to be the most satisfactory one for people searching for a true understanding of quantum mechanics. Here, the reader can appreciate how worthwhile was the previous study for arriving at such a point!

It is likely that the first part could be very useful for students in the last years of the high school or for laypersons who intend to understand the very basic notions of quantum mechanics. Scholars in mathematics and chemistry will especially like the second part, while scholars in informatics and philosophy will perhaps find the third part more interesting. Undergraduate students in physics should study the whole book carefully as a kind of fore-preparation for the more technical studies.

Last but not least, the reader can visit the book's website[a] for communications about the book and errata of the book.

---

[a]http://www.gennaroauletta.net/qmftbook

PART I

BASIC ISSUES: STATES

Chapter 1

Classical Mechanics

The aim of this chapter is to present some very basic principles of classical mechanics. They can help us understand the novelty of quantum theory.

1.1 Classical-Mechanical Description

Classical mechanics was founded by Newton and contemporary 17th-century natural philosophers, and further developed in the 18th and 19th centuries by many physicists and mathematicians, as a result of the search for a rational, causal, and scientific explanation of the universe. In classical mechanics, real-world objects are modeled as a collection of point particles, i.e., objects with negligible size. The motion of a point particle is characterized by a small number of parameters such as its position, mass, velocity, and the forces applied to it. The position of a point particle is defined with respect to an arbitrary fixed reference point in the three-dimensional Euclidean space, which is usually chosen as the origin of some coordinate system. The velocity of a point particle is the rate of change of its position with time. The force applied to a point particle (called the external force) is the cause for the point particle

Quantum Mechanics for Thinkers
Gennaro Auletta and Shang-Yung Wang
Copyright © 2014 Pan Stanford Publishing Pte. Ltd.
ISBN 978-981-4411-71-4 (Hardcover), 978-981-4411-72-1 (eBook)
www.panstanford.com

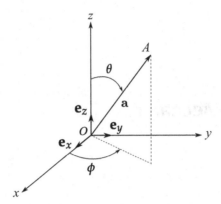

Figure 1.1 A vector **a** represented by the arrow connecting the initial point O (the origin) with the terminal point A. See Box 1.1 for details.

to undergo a certain change in its velocity, i.e., an acceleration (Newton's second law of motion). For the sake of simplicity, here we shall refer to a point particle or a collection of point particles simply as an object. Newton's first law of motion states that if an object experiences no net external force, then its velocity is constant, i.e., the object is either at rest or it moves in a straight line with a constant speed. The first law of motion postulates the existence of a certain set of frames of reference called inertial reference frames, relative to which the motion of an object not subject to external forces is a straight line at a constant speed. For this reason, Newton's first law is also referred to as the law of inertia.

Mechanical forces act along a given direction. Indeed, a force is ascertained by considering its effects, for instance pushing or pulling along a certain direction. The same is true for position, velocity, momentum (velocity at which an object travels multiplied by its mass), and also for acceleration (the variation of velocity in time). Then, all these quantities (and also other ones) can be represented by Euclidean vectors (sometimes called spatial vectors or simply vectors). A vector is a geometric object that has both a magnitude (or length), representing for instance the strength of the force, and a direction [see Box 1.1]. A vector is graphically represented as an arrow, connecting an initial point with a terminal point, as shown in Fig. 1.1.

Box 1.1 Euclidean vectors

In mathematics, a Euclidean vector is a geometric object that has a magnitude (or length) and a direction. Geometrically, a vector **a** is represented by the arrow connecting the initial point O (taken to be the origin) with the terminal point A [see Fig. 1.1]. The *magnitude* of the vector, symbolized as $\|\mathbf{a}\|$, is the distance between the initial and terminal points; its *direction* is specified by two angles θ and ϕ, where θ is called the polar angle of the vector measured from the z direction (with $0 \leq \theta \leq 180°$), and ϕ the azimuthal angle of the vector measured from the x axis to its orthogonal projection (showed by the dashed line) in the xy plane (with $0 \leq \phi < 360°$). A vector of unit magnitude is called a unit vector. Conventionally, a vector **a** is represented in Cartesian coordinates as

$$\mathbf{a} = a_x \mathbf{e}_x + a_y \mathbf{e}_y + a_z \mathbf{e}_z, \tag{1.1}$$

where $\mathbf{e}_x$, $\mathbf{e}_y$, and $\mathbf{e}_z$ are the unit vectors in the x, y, and z directions, respectively.

Vectors can be added and multiplied by a real number. Here, we only deal with addiction and multiplication of a vector by a constant. Given two vectors **a** and **b**, their sum **a** + **b** is also a vector. The addition method is given by the so-called parallelogram rule, which states that **a** + **b** is the diagonal of the parallelogram, where **a** and **b** are adjacent sides [see Fig. 1.2]. Let k be a real number. The vector $k\mathbf{a}$, obtained by multiplying **a** by k, is a vector with magnitude $|k|\|\mathbf{a}\|$ and pointing in the same direction of **a** if $k > 0$, or in the opposite direction of **a** if $k < 0$. (In the next chapter we shall learn the exact meaning of the expression $|k|$; by now, consider it as positive number.)

In classical mechanics forces are clearly local (examples are spring force, friction, etc.). Although we know today that all of the fundamental forces can be treated in terms of fields (which are characterized by specific interdependences between the involved systems and therefore may involve non-local aspects), classical mechanics does not deal with this subject and even assumes (or at least assumed until the late 19th century) that also the effect of potentials (connected with the fields) could be explained through kinds of local interactions that somehow "propagate" though a physical medium, like the ether that was still assumed in the 19th century for explaining the propagation of electromagnetic waves. Then, classical mechanics satisfies separability[a] in that when there is no causal influence of local and ultimately mechanical type between two systems, they can be considered separated with respect to each other. By *separated* it is meant here that every operation performed locally in one of the two systems has no influence on, nor can be influenced by, the other system.

The most important quantity in classical mechanics is *energy*, which is the ability of a physical system to do work on other physical systems (e.g., a motor bringing a lift to a certain height). Energy is thought to be a function of both the position and the momentum. As mentioned, position tells us the place in the space that a certain object occupies while momentum is the velocity at which the object travels multiplied by its mass. In other words, momentum tells us the impact that a certain object may have in the collision with other objects. It is indeed intuitively clear that, in the case of an accident, a car can have a bigger impact on other cars if the speed is greater, but also that at equal speed a large vehicle (like a truck), which possesses therefore more mass, can have a bigger impact than a car. Position and momentum are sufficient to describe a mechanical system as far as the energy can be determined.

1.2 Basic Principle of Classical Mechanics

In order to understand classical mechanics, it is useful to grasp its basic principles,[b] a compact theoretical building established in the 18th and 19th centuries. This will turn out to be relevant also for understanding the conceptual foundations of quantum mechanics.

Classical mechanics is a deterministic theory. Determinism consists in the idea that given the laws of classical mechanics and any current state of a mechanical system as well as an appropriate

---

[a]Einstein *et al.* (1935).
[b](Auletta, 2004a).

knowledge of the forces acting on it, it is possible to predict or retrodict any past or future state.[a] This means, that given any current state, both past and future are univocally determined. In turn, determinism is grounded on two assumptions:

(i) The *omnimoda determinatio*[b] assumption that any state of a classical system represents a complete collection of properties. In other words, given any property that can be meaningfully attributed to the system, it is possible to decide whether or not the current state of the system instantiates it. With a *property* we understand the value of a parameter (like time, energy, speed, position, etc.) used to describe the system.

(ii) The assumption of the existence of a set of dynamical laws allowing to derive any future trajectory or state from the knowledge of the current state. In the case of classical mechanics, it was supposed that the connection between different states of the same system across the time was continuous.[c] For this reason, the parameters describing the state of a given system are expressed, in mathematical terms, as functions of the continuous time variable (we have already seen that the energy is a function of position and momentum).

Classical mechanics is also methodologically reductionist. The basic assumption of classical mechanics here is that every physical system can be reduced to its elementary components. To this purpose, we need two basic assumptions: linearity and separability. We have already considered separability. Let us consider linearity. Classical mechanics is indeed *a priori* a linear theory. By a *linear* theory, we mean the sum of the forces acting on a system results in an effect that is the sum of the effects of the forces taken separately. In other words, two different forces acting on an object along two different directions can be summed and the resulting effect is proportional to this sum. Indeed, the acceleration **a** is proportional to the net (external) force **F** (and inversely proportional to the mass m of the object) according to the famous Newton's second law of

---

[a] (Laplace, 1796).
[b] (Baumgarten, 1739, Par 148).
[c] (Leibniz MS, Vol. VII, p. 25), (Boscovich, 1754).

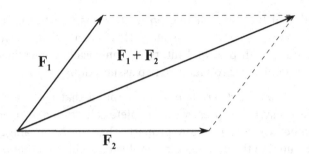

Figure 1.2 The sum of forces, also called classical superposition of forces. This addition method is sometimes called the parallelogram rule because $\mathbf{F}_1$ and $\mathbf{F}_2$ form the sides of a parallelogram and $\mathbf{F}_1 + \mathbf{F}_2$ is the diagonal of the parallelogram, where $\mathbf{F}_1$ and $\mathbf{F}_2$ are adjacent sides.

motion

$$\mathbf{F} = m\mathbf{a}, \tag{1.2}$$

where we stress that vectors are represented by bold letters in opposition to pure (or scalar) quantities like the mass. In other words, the mass offers a resistance (called inertia) to any attempt at moving the body, as it is clear by rewriting the previous equation as

$$\mathbf{a} = \frac{\mathbf{F}}{m}, \tag{1.3}$$

which tells us that the acceleration is proportional to the net force acting on the body and inversely proportional to its mass. This is intuitively clear when we consider that is more difficult to move an object the more mass it has (on the Earth surface the mass of a body, which is the quantity of matter it has, is connected with its weight, which is the gravitational pull exerted by the Earth). Therefore, two forces sum, for instance, as shown in Fig. 1.2. The physics of the 20th century has introduced many non-linear effects as occurring in complex and chaotic systems. However, the core of classical mechanics was and still is linear, otherwise we could not speak of a mechanics at all.

These two basic postulates (determinism and reductionism) implied the idea that in classical mechanics it is possible, at least in principle, to reduce the measurement error below any arbitrary threshold, so that it can become inessential. Therefore, possible measurement errors should be rather considered consequences of the unavoidable technological limitations that characterize any civilization we know and not as denoting an impossibility as such. As we shall see in the following chapters, quantum mechanics is very different under this point of view.

1.3 Summary

In this chapter we have dealt with the basic principles of classical mechanics:

- Determinism, which is ground on the assumptions of perfect determination and continuity.
- Reductionism, which presupposes linearity and separability.
- The possibility, at least in principle, to measure perfectly the properties of a system.

Chapter 2

Superposition Principle

The aim of this chapter is to present the very peculiar undulatory character of quantum systems that allows something unknown to classical physics called self-interference and therefore a superposition of the states in which the system under observation can be. We shall also become acquainted with the basic formalism of the theory, especially by learning to compute probabilities.

2.1 Origin and Foundations of Quantum Mechanics

Quantum mechanics was built as a physical theory between 1900 and 1927. Its founding fathers, among whom we recall Bohr, Schrödinger, Born, and Heisenberg, started this enterprise by developing the work of the giants like Planck, Einstein, and de Broglie. It is also true that the last three physicists, even if playing a crucial role in laying the foundation of the theory, never accepted it as a new fundamental explanation of the physical world. The basic principles of quantum mechanics are essentially the superposition principle, the complementarity principle, and the uncertainty principle. The latter, as we shall see, should be rather considered a corollary of other further principles. For historical

Quantum Mechanics for Thinkers
Gennaro Auletta and Shang-Yung Wang
Copyright © 2014 Pan Stanford Publishing Pte. Ltd.
ISBN 978-981-4411-71-4 (Hardcover), 978-981-4411-72-1 (eBook)
www.panstanford.com

reasons as well as because it facilitates the basic understanding of quantum theory, we shall consider it as a true principle.

In the early days of quantum mechanics, the incredible insight of de Broglie was that matter can present undulatory features.[a] After Maxwell's treatise on electromagnetism[b] and Hertz's experimental verification,[c] nobody doubted about the undulatory nature of light but everybody understood matter in corpuscular terms (in the previous centuries many physicists were influenced by the basic principles of classical mechanics and supported the idea that also light was corpuscular). De Broglie's new insight could eventually lead to a unification of the phenomena of matter and light. However, as we shall see, in the quantum mechanical framework the undulatory nature of matter (as well as of light) presents aspects that are totally inexplicable classically. Although in the following sections we shall very often use photons (light quanta) as a paradigm for explaining several aspects of quantum mechanics,[d] the following results apply as well as to matter for the reason indicated here.

2.2 Classical and Quantum Superposition

The quantum superposition principle constitutes an extension of the linearity and superposition principles of classical waves, but in a surprising form. In the previous chapter, we have already seen the classical superposition of forces. Here, we shall consider a slightly different case of superposition. Indeed, classically it is well known that waves are disturbances that propagate through space along certain paths. Waves are characterized by the wavelength (distance between adjacent peaks) and the amplitude (height of the peak) [see Fig. 2.1]. A distinct property of waves is the phenomenon of *interference*. For instance, if two different sea waves enter simultaneously a port from two different entrances, they will combine in the port area summing their effects in certain points (giving rise to constructive interference) and erasing their effects in

---

[a](De Broglie, 1924).
[b](Maxwell, 1873).
[c](Hertz, 1887).
[d]We obviously never take into consideration relativistic complications.

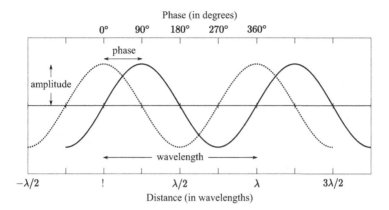

Figure 2.1 Snapshot of two waves of the same wavelength and amplitude at a particular moment in time. The amplitude of the wave is the distance between a peak and the baseline. Its wavelength (denoted by λ) is the distance between adjacent peaks and is inversely proportional to the frequency, which is the number of peaks that go through a given point in a unit time interval (the larger is the frequency, the shorter the wavelength). The phase of a wave (solid line) relative to a reference wave (dotted line), whose peak corresponds to $0°$ phase, is the distance between their respective peaks (modulo the wavelength). The two waves are $90°$ out of phase.

other points (giving rise to destructive interference) [see Fig. 2.2]. What determines wave interference is their phase difference ϕ (the so-called *relative phase*, given by the distance between their respective peaks modulo the wavelength[a]) [see Fig. 2.1]. If the phase difference is $0°$ (or $360°$), the waves are *in phase* (the peaks and valleys coincide), and in this case the interference is constructive. If it is $180°$ they are *completely out of phase* (a peak of a wave corresponds to a valley of the other wave and vice versa), and the resulting interference is destructive. Clearly, we also have a continuum of intermediate cases. It is very important to understand that, in a classical framework, such an interference phenomenon concerns the spatial behavior of waves, i.e., to the way different waves diffuse in space. Something analogous but inherently different occurs in quantum mechanics.

---

[a] The term *modulo* means the remainder of a quotient [see Box 11.2 for details].

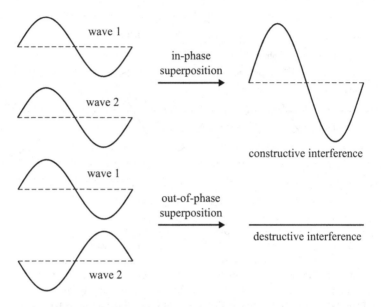

Figure 2.2 Constructive and destructive interference depending on the phases of the waves. In the former case they are in phase, in the latter case they are completely out of phase (intermediate cases result in intermediate interferences). Adapted from (Auletta, 2011a, p. 12).

2.3 A Photon in an Interferometer

Let us consider a quantum system, for instance, a photon. The photon is the quantum of light, that is, the minimum entity constituting light. By now, we will abstract from what this could precisely mean and confine our attention to its wave-like properties. Let us assume that such a photon is in a certain physical state, for instance, that it travels along a certain path. To make clear the reasoning, have a look at Fig. 2.3, in which an apparatus called the *interferometer* is depicted schematically (here, the so-called Mach–Zehnder version is shown). As we can see, a beam of photons is pumped into the apparatus by the laser located below on the left. After a short horizontal path, the photons meet a device, called the *beam splitter*, here indicated by BS1, with the function to split the incoming beam into two beams. In specific, a beam splitter is a half-silvered mirror with the property to partly transmit (here the horizontal component) and partly reflect (here the vertical

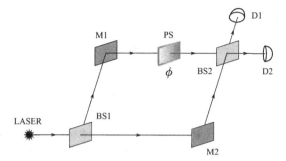

Figure 2.3 Schematic setup of the Mach–Zehnder interferometer (top-lateral view). The source beam coming from the laser is split at the first beam splitter BS1. After reflections at the two mirrors M1 and M2, the upper and lower paths are recombined at the second beam splitter BS2 and then detected at the photodetectors D1 and D2. The phase shifter PS causes a relative phase difference ϕ of the upper beam.

component) a given light beam. It is very important to stress the following two points:

(i) The laser can be regulated in such a way that it sends one photon at a time into the interferometer (in other words, there is always a single photon between the input at BS1 and the final detectors).

(ii) As we have mentioned the photon is the minimum entity of light, for reasons that will be clear below, it must be considered an indivisible entity. Therefore, rigorously speaking, it cannot be physically split into two parts by the beam splitter. As we shall see, such a splitting has to do with the *probabilities* for the photon to be transmitted or reflected.

In the simplest case, these two probabilities are equal, that is, that the photon has 50% probability to be transmitted and 50% probability to be reflected. Since it is in general assumed that probability varies between 0 (when we have 0% probability) and 1 (when we have 100% probability), then the case that we are considering here corresponds to a probability of $\frac{1}{2}$ for the photon to be either transmitted or reflected. A short review of the concept and basic properties of probability can be found in Box 2.1.

Let us now indicate with a specific symbol the component of the photon that has been transmitted. By now we do not need to care

Box 2.1 Probability

Probability theory plays an important role in the study of mathematics and physics. It deals with the relation of certain outcomes (like to get a head) of an random experiment (like to flip a coin). A possible outcome of an random experiment (e.g., head or tail in a coin flipping) is called a sample point. The set of all possible outcomes of an experiment is called the sample space, usually denoted by Ω. A subset of the sample space to which a probability can be assigned is called an event. In other words, an event is a set of outcomes of the experiment (different outcomes may be grouped in a single event). Each time the experiment is run, a given event A either occurs, if the outcome of the experiment is an element of A, or does not occur, if the outcome of the experiment is not an element of A. In particular, the sample space Ω itself is an event; by definition it always occurs (because it comprehends all possible outcomes). At the other extreme, the empty set $\emptyset$ is also an event; by definition it never occurs.

Let A and B be two events, then A implies B (denoted in the set theory by $A \subset B$), if A occurs, B must also occur. The union of A and B (denoted by $A \cup B$) is the event obtained by combining the elements of A and B. The intersection of A and B (denoted by $A \cap B$) is the event whose elements are common to both A and B. If the intersection of events A and B is empty, (denoted by $A \cap B = \emptyset$), then A and B are said to be mutually exclusive (or disjoint) events.

It is an experimental fact that if an ordinary coin is flipped N times, coming up heads N_h times, the ratio of N_h to N is nearly $\frac{1}{2}$, and the large we make N the closer the ratio approaches $\frac{1}{2}$. We express the results by the following statement: The probability of a coin coming up heads is $\frac{1}{2}$. This illustrates the intuitive frequency concept of probability. Let us generalize this as follows. If a random experiment has N possible outcomes, all mutually exclusive and equally likely, and the number N_A of these outcomes lead to the event A, then the probability of A is defined by

$$\wp(A) = \frac{N_A}{N}, \tag{2.1}$$

which is the classical definition of probability. An intuitive and easy (yet classical) example of a set of mutually exclusive events is represented by the six possible outcomes when throwing a dice. If we throw a single dice (so, we do not consider possible combinations given by throwing different dice), the probability to get any of the faces is $\frac{1}{6}$ since all six outcomes are equally probable. Moreover, probability satisfies the following basic rules:

$$0 \leq \wp(A) \leq 1 \text{ for all } A \subset \Omega, \tag{2.2a}$$

$$\wp(\Omega) = 1, \tag{2.2b}$$

$$\wp(A \cup B) = \wp(A) + \wp(B) - \wp(A, B), \tag{2.2c}$$

$$\wp(A \cup B) = \wp(A) + \wp(B) \text{ if } A \cap B = \emptyset, \tag{2.2d}$$

where $\wp(A, B) = \wp(A \cap B)$ is the *joint probability* of events A and B, i.e., the probability that A and B both occur. It is noted that Eqs. (2.2c) and (2.2d) are referred to as the addition rule for probability and that the latter is a special case of the former.

about the exact meaning of the term *component*, but do not forget what previously said above about the indivisibility of the photon. For the sake of simplicity, we call the transmitted component *down* since it takes the lower path. It can be denoted by the symbol

$$|d\rangle, \tag{2.3}$$

where we have used d as a shorthand for *down*. We are totally free in the choice of the symbol, provided that it is univocal. Instead, the symbol $|\ \rangle$ is not arbitrary (though being conventional) since in quantum mechanics it denotes the state of a quantum system. For reasons that will be clear below, it is called a *ket*. Similarly, we can call *up* the other component (the reflected one) of the photon since it takes the upper path. It can then be denoted by the symbol

$$|u\rangle. \tag{2.4}$$

Now, we shall try to describe the action of BS1 on the photon. Suppose that its initial state (before meeting BS1) is described again by $|d\rangle$. This is justified by the fact that the photon going to BS1 is along the path that is parallel to the *down* component. Now, the action of BS1 can be mathematically described by

$$|d\rangle \xrightarrow{\text{BS1}} c_d|d\rangle + c_u|u\rangle. \tag{2.5}$$

Here the transformation induced by BS1 on the photon is indicated by an arrow with the superscript BS1. The initial state of the photon is written on the left-hand side of the arrow, while the final transformed state of the photon, with the two components down and up, is written on the right-hand side of the arrow.

2.4 Probability Amplitudes

To understand the meaning of the expression (2.5), we first note that c_d and c_u are some coefficients (numbers) representing the "amounts" in which the two components are present in the state resulting from the action of BS1, i.e., the degrees to which the photon is transmitted and reflected. A short consideration will tell us that these coefficients must somehow be negatively related, namely, the more likely the photon is reflected, the less likely it is transmitted and vice versa. As will be discussed below, the coefficients c_d and c_u are related to the probabilities $\wp_d$ and $\wp_u$ of finding the photon in the lower and upper path, respectively. In fact, the probability in quantum mechanics is computed as the square modulus of the corresponding coefficient. We recall that the modulus of a real number a is denoted by $|a|$ and is the absolute value of a. Hence, $|a|$ is always positive irrespective of the fact whether a is positive or negative. The coefficients c_d and c_u whose square moduli give probabilities are called the *probability amplitudes*. As a consequence of the wave-like character of quantum systems, the probability amplitudes are in general complex numbers. For a short review of the basic properties of complex numbers, see Box 2.2.

Box 2.2 Complex numbers

Complex numbers are an extension of real numbers and can be written in the form $a + bi$, where a and b are two real numbers and $i = \sqrt{-1}$ is called the imaginary unit. In this form, a is called the real part and b the imaginary part of the complex number $a + bi$. They can be geometrically represented in the Cartesian coordinates, with the real part represented by a displacement along the x axis, and the imaginary part by a displacement along the *y* axis. The complex number *a*+*b*i is then identified with the point (*a*, *b*) in a plane, and the resultant plane is called the complex plane [see Fig. 2.4].

The complex conjugate of a complex number $c = a + b$i is the complex number $c^* = a - b$i, i.e., the number given by the real part of *c* plus the imaginary part with interchanged sign. The modulus (or absolute value) of a complex number *c* is denoted by $|c|$ and defined by

$$|c| = \sqrt{c\,c^*} = \sqrt{a^2 + b^2}. \qquad (2.6)$$

For instance, we have $|1 + i| = \sqrt{(1+i)(1-i)} = \sqrt{2}$ and the square modulus of $1 + i$ is given by $|1 + i|^2 = 2$. If *c* is a real number (i.e., $b = 0$), then $|c| = \sqrt{a^2} = |a|$, which is the usual absolute value of the real number *a*, that is, the numerical value of *a* without regard to its sign. Geometrically, $|c|$ is the distance of the point (*a*, *b*) to the origin of the complex plane. The modulus has the following fundamental properties:

$$|c| \geq 0 \quad \text{(equality holds if and only if } c = 0\text{)}, \qquad (2.7a)$$

$$|c_1 c_2| = |c_1||c_2|, \qquad (2.7b)$$

$$|c_1 + c_2| \leq |c_1| + |c_2|, \qquad (2.7c)$$

where *c*, c_1, and c_2 are arbitrary complex numbers. In particular, the inequality (2.7c) is called the triangle inequality.

Since the square moduli of the coefficients in the expression (2.5) stand for probabilities, they satisfy the *normalization* condition that the sum of all probabilities of a set of mutually exclusive events be equal to one [see Box 2.1]. Therefore, the coefficients c_d and c_u in the expression (2.5) have to satisfy the normalization condition

$$|c_d|^2 + |c_u|^2 = 1. \qquad (2.8)$$

It in turn implies that the two probabilities (and also the respective transmission and reflection coefficients) are not independent since

$$|c_d|^2 = 1 - |c_u|^2, \qquad (2.9)$$

in accordance with the fact that both the coefficients and the corresponding probabilities are negatively related. In the case in

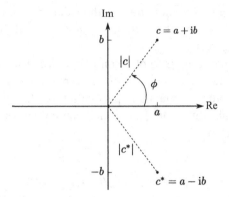

Figure 2.4 Complex plane and the geometric representation of $c = a + ib$ and its complex conjugate $c^* = a - ib$.

which they are *real* numbers, they also satisfy the requirement to vary between -1 and 1, i.e., $-1 \leq c_d, c_u \leq 1$.

Problem 2.1 Find the complex conjugates and square moduli of the following complex numbers: (a) $2 + i$ and (b) $1 + 3i$.

2.5 Formulation of the Superposition Principle

Coming back to the expression (2.5), we can see that the state of the photon after it has passed through BS1 is classically unknown. Indeed, the photon is now *delocalized*, that is, we cannot say that it is exclusively in the state $|d\rangle$, nor can we say that it is exclusively in the state $|u\rangle$. In other words, the photon now is in a state that does not show the deterministic character that is peculiar to classical mechanics [see Section 1.2]. Indeed, the parameter *path* of the photon does not have a determined value since, according to the expression (2.5), any single photon going through the interferometer is *both* in the lower and in the upper paths of the interferometer. This is the fundamental content of the quantum superposition principle, which can be formulated in its generality as follows.

Principle 2.1 (Superposition Principle) *If a quantum system can be either in a state $|\psi_1\rangle$ or in a state $|\psi_2\rangle$, it can also be in any linear* *combination of these two states, i.e., it can also be in the superposition state $|\psi\rangle$ defined by*

$$|\psi\rangle = c_1|\psi_1\rangle + c_2|\psi_2\rangle, \qquad (2.10)$$

where c_1 and c_2 are complex coefficients satisfying the normalization condition $|c_1|^2 + |c_2|^2 = 1$.

As a matter of fact, the state $|\psi\rangle = c_d|d\rangle + c_u|u\rangle$ describes the most general state of a photon after it has passed through a generic beam splitter. We stress here two essential features of this principle:

(i) We do not deal here with a lack of knowledge, namely, with some ignorance of a certain situation such that to acquire further information could fill a knowledge gap and let us know for sure whether the photon is in fact in the state $|d\rangle$ or $|u\rangle$. The indetermination of the quantum state must be taken as an objective and irreducible state of affairs. As a matter of fact, it does not depend on some statistics since it is true of single quantum systems. Indeed, the laser has been regulated to send one photon at a time into the interferometer. In other words, quantum mechanics does not satisfy the *omnimoda determinatio* requirement of classical mechanics [see Section 1.2].

(ii) While classical systems show a superposition of spatial waves, here it is the state itself of the system to be delocalized. For this reason, one has spoken of the *self–interference* of quantum systems.[a] This means that we do not have two different waves interfering with each other, but a *single* physically indivisible photon being in the state $|\psi\rangle$ of which $|d\rangle$ and $|u\rangle$ are two component states. This is the specific sense in which we should understand these components. We shall consider below some further consequences of this fact.

2.6 Transmission, Reflection, and Phase Shift

Let the probabilities that after having passing through BS1 the photon is in the down path and in the upper path be demoted

---

[a](Dirac, 1958, p. 9).

by $\wp_d$ and $\wp_u$, respectively. We assume that in our example the probabilities of transmission and reflection are equal (i.e., a 50–50 beam splitter), hence $\wp_d = \wp_u = \frac{1}{2}$. Then, from the fact that $\wp_d = |c_d|^2$ and $\wp_u = |c_u|^2$, it is not difficult to see that the coefficients (if, for the sake of simplicity, real numbers) could be $c_d = c_u = \frac{1}{\sqrt{2}}$. Hence we have

$$|d\rangle \xrightarrow{\text{BS1}} \frac{1}{\sqrt{2}} (|d\rangle + |u\rangle), \tag{2.11}$$

where the factor $\frac{1}{\sqrt{2}}$ on the right-hand side of the arrow denotes that BS1 is a 50–50 beam splitter.

We now consider how this state evolves further. First, note that the mirrors M1 and M2 do not change the state of the photon. After M1, the component $|u\rangle$ of the photon meets a device called *phase shifter* (indicated by PS in Fig. 2.3), which produces a phase shift (the phase difference between lower and upper components) of the amount ϕ. Recall that we have considered waves (or state components in the quantum mechanical case) as being in phase (when the phase difference is of $0°$, $360°$, or multiples) or completely out of phase (when the phase difference is of $180°$ or odd multiples). By inspection of Fig. 2.1, we can see that there are also intermediate cases in which the components of the photon are partly out of the phase or partly in phase. Therefore, it is necessary that the formulation of the superposition principle allows for a description of all the possible relative phases between the component states. This is exactly where the complex coefficients come to the rescue. Since the relative phase ϕ varies periodically, going from $0°$ to $360°$ (or 0 to 2π in radians) and starting a new cycle, the suitable choice to represent the relative phase would be the exponential function with a purely imaginary exponent [see Boxes 2.3 and 2.4]

$$e^{i\phi} = \cos \phi + i \sin \phi. \tag{2.12}$$

From the profiles of the sine and cosine functions depicted in Fig. 2.6, we see that these two functions are sinusoidals that behave precisely as waves. Since they are periodic functions with period 2π, the function $e^{i\phi}$ allows us to describe arbitrary phase differences. As a result, the action of PS on the component $|u\rangle$ can be mathematically described by

$$|u\rangle \xrightarrow{\text{PS}} e^{i\phi}|u\rangle. \tag{2.13}$$

Box 2.3 Trigonometric functions

Trigonometric functions (also called the circular functions) play an important role in mathematics and physics. They are functions of an angle ϕ and can be constructed geometrically in terms of a circle. Consider the right triangle OPQ in Fig. 2.5. The sine and cosine of the angle ϕ are defined in terms of the segments OP, PQ, and OQ by

$$\sin\phi = \frac{PQ}{OP}, \quad \cos\phi = \frac{OQ}{OP}. \tag{2.14}$$

Note that OP represents the radius of the circumference while OQ and OR are its projections on the x and y axes, respectively. Note also that $PQ = OR$. The other four trigonometric functions, called the tangent, cotangent, secant, and cosecant of the angle ϕ, can then be defined in terms of the sine and cosine by

$$\tan\phi = \frac{\sin\phi}{\cos\phi}, \quad \cot\phi = \frac{\cos\phi}{\sin\phi}, \quad \sec\phi = \frac{1}{\cos\phi}, \quad \csc\phi = \frac{1}{\sin\phi}. \tag{2.15}$$

Note that cotangent and cosecant interchange sine and cosine relative to tangent and secant, respectively. While the angle ϕ is usually measured in degrees, in the context of trigonometry it is more convenient to measure the angle ϕ in radians, which is the ratio of the length of a circular arc to the radius of the arc. A full angle of 360 degrees is therefore 2π radians. Similarly, a straight angle (i.e., 180 degrees) is π radians and a right angle is $\frac{\pi}{2}$ radians. The graphs of the sine and cosine functions are displayed in Fig. 2.6. The values of the sine, cosine, and tangent of certain special angles can be calculated easily and are listed in Table 2.1.

The inverse trigonometric functions arcsin, arccos, arctan, etc., are the inverse functions of the trigonometric functions (they map one of these to their angle). As can be seen from Fig. 2.6, the inverse trigonometric functions are multivalued, namely, every input is associated with two or more distinct outputs. For instance, we find from Table 2.1 that $\arctan 1 = \frac{\pi}{4}, \frac{5\pi}{4}$, etc. Hence, the inverse trigonometric functions are not uniquely defined unless the respective ranges are suitably restricted.

Table 2.1 Values of the sine, cosine, and tangent of certain special angles. The symbols $+\infty$ and $-\infty$ denote plus infinity and minus infinity, respectively

ϕ (in degrees)	ϕ (in radians)	$\sin\phi$	$\cos\phi$	$\tan\phi$
0°	0	0	1	0
45°	$\dfrac{\pi}{4}$	$\dfrac{1}{\sqrt{2}}$	$\dfrac{1}{\sqrt{2}}$	1
90°	$\dfrac{\pi}{2}$	1	0	$+\infty$
135°	$\dfrac{3\pi}{4}$	$\dfrac{1}{\sqrt{2}}$	$-\dfrac{1}{\sqrt{2}}$	-1
180°	π	0	-1	0
225°	$\dfrac{5\pi}{4}$	$-\dfrac{1}{\sqrt{2}}$	$-\dfrac{1}{\sqrt{2}}$	1
270°	$\dfrac{3\pi}{2}$	-1	0	$-\infty$
315°	$\dfrac{7\pi}{4}$	$-\dfrac{1}{\sqrt{2}}$	$\dfrac{1}{\sqrt{2}}$	-1
360°	2π	0	1	0

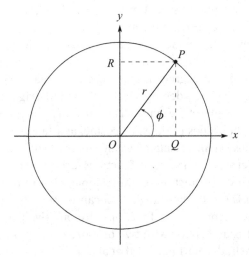

Figure 2.5 Geometric representation of sine and cosine on a circle. See Box 2.3 for details.

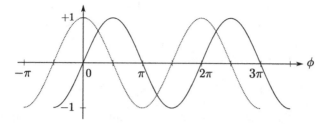

Figure 2.6 Plot of the sine (solid line) and cosine (dotted line) functions. They can be understood as mathematical waves being out of phase by 90° (or $\frac{\pi}{2}$ in radians).

Box 2.4 Euler formula

Another way of representing complex numbers is by utilizing the polar coordinates (on which we will say more in Chapter 8) in the complex plane and the Euler formula

$$e^{i\phi} = \cos\phi + i\sin\phi, \qquad (2.16)$$

where e is a number approximately equal to 2.718281 (as we shall see, it is the base of the natural logarithm) and the real argument ϕ of the cosine and sine functions is given in radians. A complex number $c = a + bi$ may be written in the so-called polar form as [see Box 2.2 and Fig. 2.4]

$$c = |c|(\cos\phi + i\sin\phi) = |c|e^{i\phi}. \qquad (2.17)$$

Here, ϕ is known as the complex argument or phase, i.e., the angle between the x axis and the Cartesian vector (a, b) measured counterclockwise in radians. As can be seen from Fig. 2.4, the complex conjugate of $e^{i\phi}$ is given by

$$e^{-i\phi} = \cos\phi - i\sin\phi. \qquad (2.18)$$

In the particular case in which $\phi = 180°$ (or π in radians), the Euler formula becomes

$$e^{i\pi} + 1 = 0, \qquad (2.19)$$

which connects the constants π (fundamental for the circumference) and e with the numbers 0, 1, and i.

The Euler formula is one the most beautiful in the history of mathematics and allows to express sines and cosines in terms of exponentials and vice versa. Indeed, summing and subtracting Eqs. (2.16) and (2.18) we have

$$\cos\phi = \frac{1}{2}\left(e^{i\phi} + e^{-i\phi}\right), \quad \sin\phi = \frac{1}{2i}\left(e^{i\phi} - e^{-i\phi}\right). \quad (2.20)$$

Using the Euler formula, one can derive important trigonometric identities. For instance, from the following identity

$$\begin{aligned} e^{i(\alpha+\beta)} &= \cos(\alpha + \beta) + i\sin(\alpha + \beta) \\ &= (\cos\alpha + i\sin\alpha)(\cos\beta + i\sin\beta) = e^{i\alpha}e^{i\beta}, \quad (2.21) \end{aligned}$$

we obtain the so-called addition formulas

$$\sin(\alpha + \beta) = \sin\alpha\cos\beta + \cos\alpha\sin\beta, \quad (2.22a)$$
$$\cos(\alpha + \beta) = \cos\alpha\cos\beta - \sin\alpha\sin\beta. \quad (2.22b)$$

Moreover, for $\alpha = -\beta = \phi$ we have

$$\left|e^{i\phi}\right|^2 = e^{i\phi}e^{-i\phi} = e^0 = 1, \quad (2.23)$$

which implies

$$\cos^2\phi + \sin^2\phi = 1. \quad (2.24)$$

This is the well-known Pythagorean trigonometric identity.

Combining Eqs. (2.11) and (2.13), we can now mathematically write the transformations induced by the beam splitter BS1 and the phase shifter PS as

$$|d\rangle \xrightarrow{\text{BS1}} \frac{1}{\sqrt{2}}\left(|d\rangle + |u\rangle\right)$$

$$\xrightarrow{\text{PS}} \frac{1}{\sqrt{2}}(|d\rangle + e^{i\phi}|u\rangle). \quad (2.25)$$

Hence, the state of the photon *before* it reaches the beam splitter BS2 is a superposition of components $|d\rangle$ and $|u\rangle$, and the relative phase between the two components is given by ϕ.

Problem 2.2 Calculate $e^{i\phi}$ for $\phi = 0°$ (or 0 in radians), $\phi = 45°$ (or $\frac{\pi}{4}$ in radians), $\phi = 90°$ (or $\frac{\pi}{2}$ in radians), $\phi = 135°$ (or $\frac{3\pi}{4}$ in radians), and $\phi = 180°$ (or π in radians) by making use of the Euler formula (2.12) and Table 2.1.

Problem 2.3 Check the derivation of Eqs. (2.20) and (2.22).

Problem 2.4 Have you understood this section? If not, your task is to read it again. Then try to write down a complete list of the concepts that you have not understood (you may read it again for writing down such a list). Once that you have the list, try to find out where is the piece of information that you have missed in order to understand any item of the list. If you think that there are terms that you find strange or weird, make use of the web resources we have indicated in the introduction. Once you have finished, read it again and verify if you understand it fully. If not, start again the procedure until you have absolute clarity about these issues. This work will take you some time but is very important to the extent to which it will teach you the basic requirement of scientific thought: to systematically and methodically reflect on a problem by singling out its components. For this reason, we strongly suggest to follow it also for the next difficulties that you may meet in this book but also for problems in your professional career. Consider that great scientists are not gods or prophets but only persons like you and us. However, one of the reasons of their success is that they have applied this method with more determination and coherence than other people.

2.7 Action of the Second Beam Splitter

The two components of the photon finally meet at the second beam splitter BS2 [see Fig. 2.3]. We note that the geometry of the apparatus can be set in such a way that they arrive simultaneously. It is very important to understand that the transformation induced by BS2 is to merge the two components and then to split again the whole, here we again assume that the two outgoing components have equal probability. Let us first consider the action of BS2 on the two components separately. The transformation on $|d\rangle$ is similar to that induced by BS1 (indeed the input states are so far the same and the beam splitters are parallel), that is,

$$|d\rangle \xrightarrow{\text{BS2}} \frac{1}{\sqrt{2}} (|1\rangle + |2\rangle), \tag{2.26}$$

where with $|1\rangle$ and $|2\rangle$ we indicate here the two paths leading to the detectors D1 and D2, respectively. We could have still used the symbols $|d\rangle$ and $|u\rangle$, in which case we would have written precisely Eq. (2.11). However, the new symbols are more useful for understanding what happens at BS2 and also for calculating the final detection probabilities. The transformation on $|u\rangle$ has, instead, a different form

$$|u\rangle \xrightarrow{\text{BS2}} \frac{1}{\sqrt{2}} (|1\rangle - |2\rangle). \tag{2.27}$$

The reason for this expression will be clearly explained in Section 4.6. By now, let us take it as an experimental fact. Since the total input to BS2 is represented by the output state (i.e., the last line) in Eq. (2.25), we need to combine the latter two equations and obtain

$$\frac{1}{\sqrt{2}} \left(|d\rangle + e^{i\phi}|u\rangle\right) \xrightarrow{\text{BS2}} \frac{1}{\sqrt{2}} \left[\frac{1}{\sqrt{2}} (|1\rangle + |2\rangle) + e^{i\phi}\frac{1}{\sqrt{2}} (|1\rangle - |2\rangle) \right]$$

$$= \frac{1}{2} \left(|1\rangle + |2\rangle + e^{i\phi}|1\rangle - e^{i\phi}|2\rangle\right). \tag{2.28}$$

Collecting the terms in last line, we obtain the final state $|f\rangle$ of the photon *after* it leaves the beam splitter BS2, but *before* it is set to be detected at D1 or D2

$$|f\rangle = \frac{1}{2} \left[(1 + e^{i\phi}) |1\rangle + (1 - e^{i\phi}) |2\rangle \right]. \tag{2.29}$$

It is important to note that the state $|f\rangle$ is precisely of the same form as that in Eq. (2.10), with the coefficients of components $|1\rangle$ and $|2\rangle$ given by

$$c_1 = \frac{1}{2} (1 + e^{i\phi}) \quad \text{and} \quad c_2 = \frac{1}{2} (1 - e^{i\phi}), \tag{2.30}$$

respectively. Moreover, Eq. (2.29) reveals clearly that in order to account for the relative phase between the components, the coefficients of the components of a quantum state in general are complex numbers. This is also the reason why we need to calculate the square moduli (and not simply the squares) of the coefficients to obtain the corresponding probabilities.

Problem 2.5 Explain why a simple square of the probability amplitude (which, as we have seen above, is in general a complex number) would not give the correct probabilities.

2.8 Computing the Detection Probabilities

As we have mentioned, to find the probabilities of certain measurement outcomes, we need to calculate the square moduli of the probability amplitudes associated with the components describing the outcomes. In the case of the Mach–Zehnder interferometer, our problem is how to compute the probability that either the detector D1 or the detector D2 clicks when a photon is in the interferometer. We see now how useful has been to indicate the outputs after BS2 as states $|1\rangle$ and $|2\rangle$. Since D1 clicking is associated with the component $|1\rangle$ while D2 clicking is associated with $|2\rangle$, we calculate the square moduli of the two respective coefficients given by Eqs. (2.30). Therefore, the probability that the photon will be detected by D1 is given by

$$\wp_1 = |c_1|^2 = \left| \frac{1}{2} \left(1 + e^{i\phi} \right) \right|^2 = \frac{1}{4} \left(1 + e^{i\phi} \right) \left(1 + e^{-i\phi} \right), \quad (2.31)$$

where we recall that according to Box 2.2 the square modulus of a complex number c is computed by multiplying c by its complex conjugate c^* (and the complex conjugate of $e^{i\phi}$ is $e^{-i\phi}$). Since we have [see Eq. (2.23)]

$$\left| e^{i\phi} \right|^2 = 1, \quad (2.32)$$

the above expression for $\wp_1$ can be simplified to

$$\wp_1 = \frac{1}{4} \left(2 + e^{i\phi} + e^{-i\phi} \right) = \frac{1}{2} \left(1 + \cos\phi \right), \quad (2.33)$$

where we have used the first of Eqs. (2.20).

Let us now compute the probability that D2 clicks. It is given by

$$\wp_2 = |c_2|^2 = \left| \frac{1}{2} \left(1 - e^{i\phi} \right) \right|^2 = \frac{1}{2} \left(1 - \cos\phi \right), \quad (2.34)$$

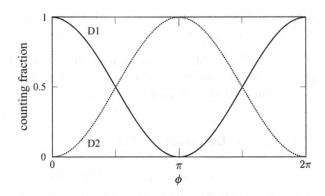

Figure 2.7 Statistical results of photon counting at detectors D1 and D2. The number fractions of photons counted at detectors D1 (solid line) and D2 (dotted line) are plotted as functions of the phase shift ϕ. It should be noted that for each value of the phase shift ϕ the sum of the two fractions is always unity.

It is straightforward to verify that the normalization condition is indeed satisfied, i.e.,

$$\wp_1 + \wp_2 = \frac{1}{2}\left(1 + \cos\phi + 1 - \cos\phi\right) = 1. \qquad (2.35)$$

For $\phi = 90°$ (or $\frac{\pi}{2}$ in radians), both D1 and D2, when several experimental runs are performed, will click with a probability of $\frac{1}{2}$ (since $\cos\frac{\pi}{2} = 0$). In fact, when we perform many experimental runs with different value of ϕ, we obtain a typical interference profile like the ones shown in Fig. 2.7. What is also interesting is that there are limiting values of ϕ for which either D1 or D2 never clicks [see Problem 2.8].

Problem 2.6 Verify the formula for the probability $\wp_2$ that D2 clicks in Eq. (2.34).

Problem 2.7 Show that the squares of the coefficients given by Eqs. (2.30) do not express true probabilities.

Problem 2.8 Given the state $|f\rangle$ in Eq. (2.29), what are the limiting values of the phase shift ϕ for which one of the two detectors never clicks?

2.9 Summary

In this chapter we have

- Followed the path of a photon in an interferometer.
- Learned that the photon in an interferometer is delocalized.
- Understood the phenomena of constructive and destructive interference.
- Seen that a photon in an interferometer can do self–interference.
- Formulated the principle of superposition, i.e., if two states are allowed then an arbitrary linear combination of them is also allowed.
- Learned the action of the beam splitter and phase shifter in an interferometer.
- Understood that the coefficients of the superposition represents the probability amplitudes.
- Learned to find the probabilities of measurement outcomes by computing the square moduli of the corresponding probability amplitudes.

Chapter 3

Quantum States as Vectors

The aim of this chapter is to present the vector nature of quantum states. We shall also become acquainted with the concepts of projectors and the mathematical tool of matrix calculation.

3.1 Photon Polarization

In this chapter we shall elaborate the concepts presented in the previous chapter. To this end, let us consider another important property of the photon: polarization. Light is an electromagnetic wave and ordinarily it oscillates in all directions orthogonal to its propagation direction, as it happens for the light coming from the sun or spreading out of a normal electric bulb. However, when passing through a polarization filter (e.g., polarized sunglasses) light may acquire a privileged oscillation direction, and in this case we say that it is polarized in that direction (in this way polarized sunglasses are able to eliminate reflected glare).

Let us consider two polarization directions in particular: horizontal polarization, which we denote by the ket $|h\rangle$, and vertical polarization, which we denote by the ket $|v\rangle$. As we know from the superposition principle [see Principle 2.1], this also means that a

Quantum Mechanics for Thinkers
Gennaro Auletta and Shang-Yung Wang
Copyright © 2014 Pan Stanford Publishing Pte. Ltd.
ISBN 978-981-4411-71-4 (Hardcover), 978-981-4411-72-1 (eBook)
www.panstanford.com

photon can be in any superposition of these two states. Therefore, an arbitrary photon polarization state $|\psi\rangle$ can be written as

$$|\psi\rangle = c_h|h\rangle + c_v|v\rangle, \tag{3.1}$$

where c_h and c_v are complex coefficients satisfying the normalization condition

$$|c_h|^2 + |c_v|^2 = 1. \tag{3.2}$$

Since the resulting state $|\psi\rangle$ is a combination of two independent polarization directions, it will in general represent itself a given polarization direction.[a] In the easiest (symmetric) case it represents a polarization at $45°$ (or $\frac{\pi}{4}$ in radians) relative to the horizontal polarization when

$$c_h = c_v = \frac{1}{\sqrt{2}}. \tag{3.3}$$

3.2 Action of the Polarization Filter

Let us suppose to insert a polarization filter along an arbitrary direction **a** on the path of an incoming photon in the state $|\psi\rangle$ given by Eq. (3.1) [see Section 2.4]. The bold letter expresses the fact that this direction is an Euclidean vector [see Fig. 1.1 and Box 1.1]. Experimentally, if we send many photons in the same state $|\psi\rangle$, it is observed that only a portion of the photons will pass through the filter. It is found that the fraction of photons passing through is the same as the probability that a single photon will pass through the filter. The latter is given by

$$\wp_{\text{pass}}(\theta) = \cos^2\theta, \tag{3.4}$$

where θ is the angle between the filter direction **a** and the photon polarization direction given by $|\psi\rangle$ [see Fig. 3.1]. The probability for the photon to be blocked (or absorbed) by the filter is given by

$$\wp_{\text{block}}(\theta) = 1 - \wp_{\text{pass}}(\theta) = \sin^2\theta, \tag{3.5}$$

where we have used the trigonometric relation $\cos^2\theta + \sin^2\theta = 1$ [see Box 2.4] and the normalization condition for probabilities.

---

[a]For the sake of presentational simplicity, we consider here only the so-called linear polarization.

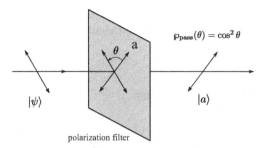

Figure 3.1 The photon that is initially in a polarization state $|\psi\rangle$ pass will pass through a polarization filter in the direction **a** with a probability given by $\cos^2 \theta$, where θ is the angle between the filter direction **a** and the photon polarization direction given by $|\psi\rangle$. As will be explain below, the resulting state is $|a\rangle$.

Below we shall understand the exact meaning of these expressions. For the time being, let us take them as experimental facts and only remark that we again face the puzzling nature of quantum mechanics: it is indeed classically impossible that systems prepared in the same state (like the state $|\psi\rangle$) behave in different ways, that is, a part will pass through the filter while a part will not.

In the easiest case when **a** is the horizontal direction and $|\psi\rangle$ is the polarization state with $\theta = 45°$ (or $\frac{\pi}{4}$ in radians), after having sent many photons in the same state $|\psi\rangle$, we would observe that half of the photons will pass through the filter and half of the photons will be blocked by the filter (in fact, a polarization at an angle $\theta = 45°$ is halfway between horizontal and vertical polarization directions). Obviously, all photons that have passed through the filter are now polarized along the direction **a**. In other words, the filter acts as a *projector* that projects the state $|\psi\rangle$ onto the state $|a\rangle$ represented by the polarization direction **a**. Consequently, we can write the state $|\psi\rangle$ as a combination of the alternatives: the "passing through the filter" state, $|a\rangle$, and the "blocked by the filter" state, $|a'\rangle$. Therefore, taking into account the expressions (3.4) and (3.5), we can write

$$|\psi\rangle = \cos\theta |a\rangle + \sin\theta |a'\rangle. \tag{3.6}$$

The validity of this formula is guaranteed by the fact that the probabilities of measurement outcomes in quantum mechanics are the square moduli of the corresponding probability amplitudes.

In our case, the probabilities (3.4) and (3.5) are precisely the square moduli of the respective coefficients $\cos\theta$ and $\sin\theta$ given by Eq. (3.6).

3.3 Vector Spaces and Bases

The most evident justification of Eq. (3.6) is given by geometric considerations [see Fig. 3.2]. Since polarization directions can be represented by vectors like **a** as well as by polarization states like $|\psi\rangle$, it is quite natural to consider also quantum states as vectors. In particular, Eq. (3.6) can be interpreted as the vector $|\psi\rangle$ being the linear combination (or superposition) of the vectors $|a\rangle$ and $|a'\rangle$. The example we have introduced can be quite useful because it makes things very intuitive. Moreover, it seems to be a quite natural extension of the classical superposition that we have considered in Section 1.2. However, if the reader has really acquired this result, the following two provisos become immediately necessary:

(i) The ket $|a\rangle$, together with other quantum states like $|a'\rangle$, $|\psi\rangle$, etc., is a *state vector*. It is not an ordinary vector such as **a**, which are those generally considered in classical mechanics [see Box 1.1]. Indeed, classical vectors (like positions, velocities, forces, and so on) are represented by ordinary spatial vectors.

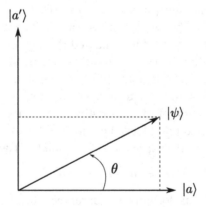

Figure 3.2 The vectors $|a\rangle$ and $|a'\rangle$ constitute a basis for geometrically expressing the vector $|\psi\rangle$ in a two-dimensional Hilbert space.

They are geometric objects in a three-dimensional vector space, called the Euclidean space. On the other hand, quantum states are geometric objects in an abstract vector space called the Hilbert space, whose dimensions depend on the system under consideration (until now we have only considered the two-dimensional case, but there can be more than two and in some cases even infinite dimensions, as we shall see later). In short, a vector space is a mathematical structure formed by a collection of vectors, i.e., objects that may be added together (called addition) and multiplied by numbers (called scalar multiplication).

(ii) Quantum state vectors can describe many different physical phenomena and parameters, not only polarization. In the previous chapter, we have seen that they describe the path taken by a photon in an interferometer. In the specific case of the polarization vector **a**, we have also associated with it a state vector $|a\rangle$, but such an association between a spatial vector and a state vector is rather an exception. In other words, quantum state vectors do not need to correspond to any ordinary vectors. Moreover, we should carefully distinguish the polarization direction **a** from the polarization state $|a\rangle$ in which the photon is when having a polarization in the direction **a**. So, although **a** and $|a\rangle$ are associated, since the latter represents photons in a polarization state along the **a** direction, they are nevertheless two different kinds of vectors that are only contingently connected here due to the specificity of the example.

An inspection of Fig. 3.2 shows that the state vectors $|a\rangle$ and $|a'\rangle$ constitute a *basis* in a two-dimensional Hilbert space. In other words, they are the basic direction vectors that allow us to express other state vectors (like $|\psi\rangle$ and any other state vector representing a polarization direction) as their linear combinations. We note that from this specific point of view the state vectors $|a\rangle$ and $|a'\rangle$ are not very different from the basis vectors $\mathbf{e}_x$ and $\mathbf{e}_y$ in a two-dimensional Euclidean space [see Box 1.1]. A basis constituted by the basis state vectors $|a\rangle$ and $|a'\rangle$ is usually indicated in the following way as

$$\{|a\rangle, |a'\rangle\}. \tag{3.7}$$

Moreover, just like the unit vectors in the x and y directions can be conveniently represented in component form as $\mathbf{e}_x = (1, 0)$ and $\mathbf{e}_y = (0, 1)$, respectively, so can the two vectors $|a\rangle$ and $|a'\rangle$. However, we write them in component form as

$$|a\rangle = \begin{pmatrix} 1 \\ 0 \end{pmatrix} \quad \text{and} \quad |a'\rangle = \begin{pmatrix} 0 \\ 1 \end{pmatrix}. \tag{3.8}$$

Here, we have preferred to used the column vectors to represent state vectors (or kets) and have reserved row vectors to another type of vectors that will be discussed in Section 3.4. This is for the sake of simplicity in the mathematical notion, as it will became clear very soon. Then, the state vector $|\psi\rangle$ that is defined by the superposition (3.6) can be written in component form as

$$|\psi\rangle = \cos \begin{pmatrix} 1 \\ 0 \end{pmatrix} + \sin \begin{pmatrix} 0 \\ 1 \end{pmatrix} = \begin{pmatrix} \cos\theta \\ \sin\theta \end{pmatrix}. \tag{3.9}$$

Problem 3.1 Express the following two state vectors in component form

$$|\phi\rangle = \frac{1}{\sqrt{2}}(|a\rangle + |a'\rangle) \quad \text{and} \quad |\phi'\rangle = \frac{1}{\sqrt{2}}(|a\rangle - |a'\rangle)$$

3.4 Scalar Products and Brackets

We are now in a position to discuss an important mathematical concept of the Hilbert space, namely, the scalar product of two vectors. To this end, we shall introduce to every ket (or state vector) $|\psi\rangle$ in a Hilbert space a dual counterpart, called a *bra* and denoted by $\langle\psi|$. The bra $\langle\psi|$ is the *complex conjugate transpose* (or conjugate transpose for short) of the corresponding ket $|\psi\rangle$:

- The term *transpose* means to interchange columns and rows.
- Since the coefficients in a state vector (or probability amplitudes) are in general complex numbers [see Sections 2.4–2.6], the coefficients in a bra dual to a ket are the complex conjugates of the corresponding coefficients in the ket.

The conjugate transpose of the kets $|a\rangle$ and $|a'\rangle$ defined by Eq. (3.8) are respectively given by

$$\langle a| = \begin{pmatrix} 1 & 0 \end{pmatrix} \quad \text{and} \quad \langle a'| = \begin{pmatrix} 0 & 1 \end{pmatrix}, \tag{3.10}$$

where 0 and 1 are real numbers and are therefore the complex conjugates of themselves. For the ket $|\psi\rangle$ given by Eq. (3.6), its dual bra is given by

$$\langle\psi| = \cos\theta\,\langle a| + \sin\theta\,\langle a'|, \qquad (3.11)$$

which is then written in component form as

$$\langle\psi| = \cos\theta\,(1\ 0) + \sin\theta\,(0\ 1) = (\cos\theta\ \sin\theta). \qquad (3.12)$$

This allows us to formulate the general rules for addition and scalar multiplication of vectors. Given two arbitrary state vectors

$$|a\rangle = \begin{pmatrix} a_1 \\ a_2 \end{pmatrix} \quad \text{and} \quad |b\rangle = \begin{pmatrix} b_1 \\ b_2 \end{pmatrix}, \qquad (3.13)$$

where the components a's and b's are complex numbers, we have

$$\alpha|a\rangle = \begin{pmatrix} \alpha\,a_1 \\ \alpha\,a_2 \end{pmatrix}, \qquad (3.14a)$$

$$\alpha\langle a| = \begin{pmatrix} \alpha\,a_1^* & \alpha\,a_2^* \end{pmatrix}, \qquad (3.14b)$$

$$|a\rangle + |b\rangle = \begin{pmatrix} a_1 \\ a_2 \end{pmatrix} + \begin{pmatrix} b_1 \\ b_2 \end{pmatrix} = \begin{pmatrix} a_1 + b_1 \\ a_2 + b_2 \end{pmatrix}, \qquad (3.14c)$$

$$\langle a| + \langle b| = \begin{pmatrix} a_1^* & a_2^* \end{pmatrix} + \begin{pmatrix} b_1^* & b_2^* \end{pmatrix} = \begin{pmatrix} a_1^* + b_1^* & a_2^* + b_2^* \end{pmatrix}, \qquad (3.14d)$$

where α is a complex number. Note that the sum $|a\rangle + \langle b|$ is never allowed.

The scalar product of two state vectors is a generalization of the dot product of two Euclidean vectors. A brief discussion of the latter can be found in Box 3.1. In specific, the scalar product of two state vectors $|a\rangle$ and $|b\rangle$ is a complex number and is defined by

$$\langle a|b\rangle = \begin{pmatrix} a_1^* & b_1^* \end{pmatrix} \begin{pmatrix} b_1 \\ b_2 \end{pmatrix} = a_1^* b_1 + b_1^* b_2. \qquad (3.15)$$

The general rule for calculating the scalar product is then to multiply the first and second columns in the row vector by the corresponding rows in the column vector and then add the results together. This is a special case of the so-called matrix multiplication, which will be discussed in more detail in Section 3.6. Since the expression $\langle a|b\rangle$ can be thought of as a bra acting on a ket, it is called a *bracket* (note that the two central vertical bars resulting by writing the

---

Box 3.1 Dot product of vectors

The dot product (or scalar product) is an important operation between two Euclidean vectors. For two vectors **a** and **b**, their dot product **a** · **b** is a real number and is defined by

$$\mathbf{a} \cdot \mathbf{b} = \|\mathbf{a}\| \, \|\mathbf{b}\| \, \cos\theta, \qquad (3.16)$$

where $\|\mathbf{a}\|$ and $\|\mathbf{b}\|$ are the magnitudes of the vectors **a** and **b**, respectively, and θ is the angle between **a** and **b**. The dot product **a** · **b** is a number that expresses how much the vectors **a** and **b** aligned. When two vectors are aligned (i.e., $\mathbf{a} = k\mathbf{b}$ with $k \neq 0$), their dot product is obviously different from zero. When the dot product of two vectors is zero, the vectors are said to be orthogonal since they are pointing in perpendicular directions.

The magnitude of a vector $\|\mathbf{a}\|$ can be expressed in terms of the dot product as

$$\|\mathbf{a}\| = \sqrt{\mathbf{a} \cdot \mathbf{a}}. \qquad (3.17)$$

It is suitable to introduce Cartesian coordinates. By denoting with $\mathbf{e}_x$, $\mathbf{e}_y$, and $\mathbf{e}_z$ the unit vectors in the x, y, and z directions (in Cartesian coordinates), respectively, we can decompose any vector in their components and write the dot product of the vectors $\mathbf{a} = a_x\mathbf{e}_x + a_y\mathbf{e}_y + a_z\mathbf{e}_z$ and $\mathbf{b} = b_x\mathbf{e}_x + b_y\mathbf{e}_y + b_z\mathbf{e}_z$ as

$$\mathbf{a} \cdot \mathbf{b} = a_x b_x + a_y b_y + a_z b_z. \qquad (3.18)$$

For the sake of notational simplicity, the magnitude squared of a vector **a** is usually written as $\mathbf{a}^2$, which in Cartesian coordinates takes the form

$$\mathbf{a}^2 \equiv \|\mathbf{a}\|^2 = a_x^2 + a_y^2 + a_z^2. \qquad (3.19)$$

Moreover, the dot product satisfies the following properties:

$$\mathbf{a} \cdot \mathbf{b} = \mathbf{b} \cdot \mathbf{a}, \qquad (3.20a)$$

$$\mathbf{a} \cdot \mathbf{a} \geq 0 \quad \text{(equality holds if and only if } \mathbf{a} = 0\text{)}, \qquad (3.20b)$$

$$\mathbf{a} \cdot (\mathbf{b} + \mathbf{c}) = (\mathbf{a} \cdot \mathbf{b}) + (\mathbf{a} \cdot \mathbf{c}), \qquad (3.20c)$$

$$(k\mathbf{a}) \cdot \mathbf{b} = \mathbf{a} \cdot (k\mathbf{b}) = k(\mathbf{a} \cdot \mathbf{b}), \qquad (3.20d)$$

where k is an arbitrary real number.

scalar product have been contracted into a single one for the sake of notational simplicity). In addition, the scalar product satisfies the following properties [see Eqs. (3.20)]:

$$\langle b|a \rangle = \langle a|b \rangle^*, \tag{3.21a}$$

$$\langle a|a \rangle \geq 0 \quad \text{(equality holds if and only if } |a\rangle = 0), \tag{3.21b}$$

$$(\langle b| + \langle c|) |a\rangle = \langle b|a\rangle + \langle c|a\rangle, \tag{3.21c}$$

$$\langle a| (|b\rangle + |c\rangle) = \langle a|b\rangle + \langle a|c\rangle, \tag{3.21d}$$

$$\langle b| (\alpha|a\rangle) = \alpha \langle b|a\rangle, \tag{3.21e}$$

$$(\alpha \langle b|) |a\rangle = \alpha \langle b|a\rangle, \tag{3.21f}$$

where α is a complex number.

The scalar product $\langle a|b \rangle$ is a number that expresses how much the state vectors $|a\rangle$ and $|b\rangle$ overlap. When two state vectors coincide (i.e., $|a\rangle = |b\rangle \neq 0$), their scalar product is obviously different from zero. When the scalar product of two state vectors is zero, they are said to be orthogonal since they no longer have an overlap. For instance, the basis state vectors $|a\rangle$ and $|a'\rangle$ given by Eq. (3.8) are orthogonal because [see Problem 3.2]

$$\langle a|a' \rangle = 0. \tag{3.22}$$

The scalar product also helps defining the magnitude (or length) of a state vector, which is also called its *norm*. Similarly to Eq. (3.17) the norm of the state vector $|\psi\rangle$ is denoted by $\||\psi\rangle\|$ and is defined as the square root of the scalar product of $|\psi\rangle$ with itself, i.e.,

$$\||\psi\rangle\| = \sqrt{\langle \psi|\psi \rangle}. \tag{3.23}$$

We recall that in quantum mechanics a state vector has to fulfill the normalization condition [see Section 2.4], which is tantamount to having unit magnitude, i.e.,

$$\||\psi\rangle\| \equiv \sqrt{\langle \psi|\psi \rangle} = 1. \tag{3.24}$$

This in turn implies that

$$\langle \psi|\psi \rangle = 1. \tag{3.25}$$

A state vector satisfies the normalization condition (3.25) is said to be *normalized*. If a non-zero state vector $|\psi\rangle$ is not normalized, then the corresponding normalized state vector (in the finite-dimensional case) is given by

$$|\psi\rangle_{\text{normalized}} = \frac{|\psi\rangle}{\||\psi\rangle\|}. \tag{3.26}$$

The combination of orthogonality and normalization is called *orthonormality* (or the orthonormal condition). In other words, two state vectors are orthonormal if they are orthogonal and both of unit magnitude. A set of state vectors forms an orthonormal set if all state vectors in the set are mutually orthogonal and all of unit magnitude. An orthonormal set of state vectors which also forms a basis is called an orthonormal basis. It is straightforward to verify that the basis $\{|a\rangle, |a'\rangle\}$ is an orthonormal basis, i.e., the following orthonormal conditions hold:

$$\langle a|a\rangle = \langle a'|a'\rangle = 1 \quad \text{and} \quad \langle a|a'\rangle = \langle a'|a\rangle = 0. \tag{3.27}$$

The above orthonormal conditions can be written in a compact form as

$$\langle j|k\rangle = \delta_{jk}, \tag{3.28}$$

where $j, k = a, a'$ and δ_{jk} is the Kronecker delta defined by

$$\delta_{jk} = \begin{cases} 1 & \text{for } j = k, \\ 0 & \text{for } j \neq k. \end{cases} \tag{3.29}$$

To put it in another way, the Kronecker delta tells us that the two indexes j and k need to coincide otherwise its value is zero. As we will see below, the concept of orthonormal basis plays an important role in quantum mechanics.

Problem 3.2 Derive the result (3.22) by making use of the vector components.

Problem 3.3 Compute the scalar products $\langle\psi|a\rangle$ and $\langle\psi|a'\rangle$, where $|\psi\rangle$, $|a\rangle$, and $|a'\rangle$ are given by Eqs. (3.8) and (3.9).

Problem 3.4 Compute the scalar product of the state vectors

$$|\psi\rangle = c|a\rangle + c'|a'\rangle \quad \text{and} \quad |\varphi\rangle = d|b\rangle + d'|b'\rangle,$$

where

$$|b\rangle = \frac{1}{\sqrt{2}}(|a\rangle + |a'\rangle) \quad \text{and} \quad |b'\rangle = \frac{1}{\sqrt{2}}(|a\rangle - |a'\rangle).$$

(*Hint*: Expand first the ket $|\phi\rangle$ in the basis $\{|a\rangle, |a'\rangle\}$ by taking advantage of the latter two equations.)

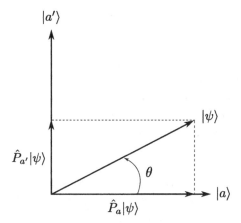

Figure 3.3 The action of a polarization filter along the direction **a** as a projection.

3.5 Polarization Filters as Projectors

We have mentioned the fact that a polarization filter along a given direction **a** let pass through only a portion of the photons sent in the state $|\psi\rangle$. We have also said that the filter acts as a selecting devise that projects this part of the photons along the direction **a** by eliminating the part along the orthogonal direction **a**′, precisely those that will not pass through but are absorbed by the filter. The reason for this circumstance is that we have here a situation of mutual exclusivity, namely, either a photon passes through the filter or not. We mathematically express this by saying that the scalar product between $|a\rangle$ (repressing photons passing through the filter) and $|a'\rangle$ (representing photons that are blocked by the filter) is zero, i.e., $\langle a|a'\rangle = 0$. We can represent this situation as in Fig. 3.3. The two entities shown in the figure as $\hat{P}_a$ and $\hat{P}_{a'}$ are called the *projectors* (or *projection operators*) into the states $|a\rangle$ and $|a'\rangle$, respectively. They mathematically describe the action of the filters in the direction **a** and **a**′, respectively.

To find the definition of the projector, let us make use of the general form of superposition and write

$$|\psi\rangle = c_a|a\rangle + c_{a'}|a'\rangle, \tag{3.30}$$

where c_a and $c_{a'}$ are complex coefficients satisfying the normalization condition

$$|c_a|^2 + |c_{a'}|^2 = 1. \tag{3.31}$$

Since $\{|a\rangle, |a'\rangle\}$ is an orthonormal basis, we can write down the overlaps of $|\psi\rangle$ with $|a\rangle$ and $|a'\rangle$, on the one hand, and on the other hand express the coefficients c_a and $c_{a'}$ in their terms as (they turn out to express precisely the same concept)

$$\langle a|\psi\rangle = c_a \quad \text{and} \quad \langle a'|\psi\rangle = c_{a'}. \tag{3.32}$$

In other words, probability amplitudes are themselves scalar products (or brackets). Upon substituting Eq. (3.32) into Eq. (3.30), we can express the state vector $|\psi\rangle$ as

$$|\psi\rangle = \langle a|\psi\rangle|a\rangle + \langle a'|\psi\rangle|a'\rangle. \tag{3.33}$$

By interchanging the order of the scalar products and the accompanying state vectors, we can rewrite the right-hand side of Eq. (3.33) as

$$|\psi\rangle = (|a\rangle\langle a|)|\psi\rangle + (|a'\rangle\langle a'|)|\psi\rangle. \tag{3.34}$$

This is absolutely allowed because the scalar products $\langle a|\psi\rangle$ and $\langle a'|\psi\rangle$, being complex numbers, can always be interchanged with vectors in a product. Equation (3.34) is very instructive and leads to the following definitions of the projectors

$$\hat{P}_a = |a\rangle\langle a| \quad \text{and} \quad \hat{P}_{a'} = |a'\rangle\langle a'|. \tag{3.35}$$

We note that if the initial state is given by the state $|\psi\rangle$, then the states projected by the projector $\hat{P}_a$ and $\hat{P}_{a'}$ are denoted by $\hat{P}_a|\psi\rangle$ and $\hat{P}_{a'}|\psi\rangle$, respectively, which allows us to write Eq. (3.34) as

$$|\psi\rangle = \hat{P}_a|\psi\rangle + \hat{P}_{a'}|\psi\rangle. \tag{3.36}$$

By comparing this equation with Eq. (3.6), we immediately see that

$$\hat{P}_a|\psi\rangle = \cos\theta|a\rangle \quad \text{and} \quad \hat{P}_{a'}|\psi\rangle = \sin\theta|a'\rangle, \tag{3.37}$$

as it should be in accordance with our examination in Section 3.2.

Problem 3.5 Making use of the orthonormal conditions (3.27), prove the identity

$$(\langle a| + \langle a'|)(|a\rangle + |a'\rangle) = \langle a|a\rangle + \langle a'|a'\rangle.$$

3.6 Projectors as Matrices

While the scalar product is a product of a bra (a row vector) and a ket (a column vector), the projector is a product of a ket (a column vector) and a bra (a row vector). Then, by taking into account the component form of the vectors $|a\rangle$ and $|a'\rangle$ given by Eq. (3.8), we can write the projectors $\hat{P}_a$ and $\hat{P}_{a'}$ respectively as [see Eq. (3.35)]

$$\hat{P}_a = |a\rangle\langle a| = \begin{pmatrix} 1 \\ 0 \end{pmatrix} \begin{pmatrix} 1 & 0 \end{pmatrix} \tag{3.38a}$$

and

$$\hat{P}_{a'} = |a'\rangle\langle a'| = \begin{pmatrix} 0 \\ 1 \end{pmatrix} \begin{pmatrix} 0 & 1 \end{pmatrix}. \tag{3.38b}$$

In order to compute these two projectors we need to introduce some mathematical notions. First, it is important to understand that a projector is not a number, like the result of a scalar product, which is indeed a probability amplitude. Instead, it describes the *action* of projection. As a matter of fact, the projector is the first example of another type of mathematical structure called *linear operators* that also play an important role in quantum mechanics. A linear operator $\hat{O}$ acts on a state vector $|\psi\rangle$, resulting in a new state vector $\hat{O}|\psi\rangle$, and satisfies the following properties:

$$\hat{O}(|\psi_1\rangle + |\psi_2\rangle) = \hat{O}|\psi_1\rangle + \hat{O}|\psi_2\rangle, \tag{3.39a}$$

$$\hat{O}(\alpha|\psi\rangle) = \alpha\,\hat{O}|\psi\rangle, \tag{3.39b}$$

where the $|\psi\rangle$'s are arbitrary state vectors and α is a complex number. Since in this book we deal only with linear operators, here and henceforth we shall refer to them simply as operators whenever no confusion may arise. Note that here and in what follows an operator is written with a hat so that it can be easily distinguished from a number. The easiest way in which linear operators are uniquely and concisely represented in a given orthonormal basis is by using *matrices*. A matrix is a generalization of the concept of vector having *both* columns and rows. An $m \times n$ matrix consists of m rows and n columns. A matrix is said to be square if $m = n$, and rectangular if $m \neq n$. In particular, an $m \times 1$ matrix is a column vector, and a $1 \times n$ matrix is a row vector. The individual entries in a matrix are called its elements. The element in the ith row and the jth column of a matrix $\hat{A}$ is referred to as the (i, j)th element of $\hat{A}$ and is commonly written as a_{ij}. An element in row j and column j of a square matrix is called a *diagonal* element of the square matrix, while an element that is not diagonal is called *off-diagonal*. The convention that we shall use is that a matrix will also be written with a hat.

We now discuss some basic operations of matrices. For the sake of presentational simplicity, we will consider here 2×2 square matrices. The generalization to rectangular matrices is straightforward. Addition and scalar multiplication of matrices are defined analogously to that of vectors. Given two arbitrary 2×2 matrices

$$\hat{A} = \begin{bmatrix} a_{11} & a_{12} \\ a_{21} & a_{22} \end{bmatrix} \quad \text{and} \quad \hat{B} = \begin{bmatrix} b_{11} & b_{12} \\ b_{21} & b_{22} \end{bmatrix}, \tag{3.40}$$

where the elements a's and b's are complex numbers, we have

$$\alpha \hat{A} = \begin{bmatrix} \alpha\, a_{11} & \alpha\, a_{12} \\ \alpha\, a_{21} & \alpha\, a_{22} \end{bmatrix} \tag{3.41a}$$

and

$$\hat{A} + \hat{B} = \begin{bmatrix} a_{11} + b_{11} & a_{12} + b_{12} \\ a_{21} + b_{21} & a_{22} + b_{22} \end{bmatrix}, \tag{3.41b}$$

where α is a complex number. In addition, matrix multiplication is an operation that, in the easiest case, takes a pair of square matrices, and produces another square matrix that is called their product $\hat{A}\hat{B}$. For instance, the product of matrices $\hat{A}$ and $\hat{B}$ is written as $\hat{A}\hat{B}$ and defined by

$$\hat{A}\hat{B} = \begin{bmatrix} a_{11}b_{11} + a_{12}b_{21} & a_{11}b_{12} + a_{12}b_{22} \\ a_{21}b_{11} + a_{22}b_{21} & a_{21}b_{12} + a_{22}b_{22} \end{bmatrix}. \tag{3.42}$$

In general, if $\hat{A}$ is an $n \times m$ matrix and $\hat{B}$ is an $m \times p$ matrix, the product $\hat{A}\hat{B}$ of their multiplication is an $n \times p$ matrix but the product $\hat{B}\hat{A}$ is not defined. In other words, in order to perform the product $\hat{A}\hat{B}$ we need that the number of columns of $\hat{A}$ is the same as the number of rows of $\hat{B}$.

The matrix $\hat{A}$ can be also viewed as a collection of two row vectors

$$\langle a_1 | = \begin{pmatrix} a_{11} & a_{12} \end{pmatrix}, \quad \langle a_2 | = \begin{pmatrix} a_{21} & a_{22} \end{pmatrix}, \tag{3.43}$$

while the matrix $\hat{B}$ as a collection of two column vectors

$$|b_1\rangle = \begin{pmatrix} b_{11} \\ b_{21} \end{pmatrix}, \quad |b_2\rangle = \begin{pmatrix} b_{12} \\ b_{22} \end{pmatrix}, \tag{3.44}$$

so that formally we can write $\hat{A}$ and $\hat{B}$ respectively as

$$\hat{A} = \begin{pmatrix} \langle a_1| \\ \langle a_2| \end{pmatrix} \quad \text{and} \quad \hat{B} = \left(|b_1\rangle \ |b_2\rangle \right). \tag{3.45}$$

Then, matrix multiplication can be calculated in analogy with the scalar product between vectors and the matrix $\hat{A}\hat{B}$ given by Eq. (3.42) can be expressed in a compact form as

$$\hat{A}\hat{B} = \begin{bmatrix} \langle a_1|b_1\rangle & \langle a_1|b_2\rangle \\ \langle a_2|b_1\rangle & \langle a_2|b_2\rangle \end{bmatrix}. \tag{3.46}$$

In terms of the elements, on the other hand, we can write the (i, j)th element of the matrix $\hat{A}\hat{B}$ as

$$(\hat{A}\hat{B})_{ij} = \sum_{k=1}^{2} a_{ik} \, b_{kj}, \tag{3.47}$$

where $i, j = 1, 2$ and use has been made of the summation notation [see Box 3.2]. Matrix multiplication satisfies the following properties:

$$\hat{A}(\hat{B}\hat{C}) = (\hat{A}\hat{B})\hat{C}, \tag{3.48a}$$

$$\hat{A}(\hat{B} + \hat{C}) = \hat{A}\hat{B} + \hat{A}\hat{C}, \tag{3.48b}$$

$$(\hat{A} + \hat{B})\hat{C} = \hat{A}\hat{C} + \hat{B}\hat{C}, \tag{3.48c}$$

$$\alpha(\hat{A}\hat{B}) = (\alpha\hat{A})\hat{B} = \hat{A}(\alpha\hat{B}), \tag{3.48d}$$

where $\hat{A}$, $\hat{B}$, and $\hat{C}$ are matrices such that the above operations are defined and α is a complex number. Therefore, in the basis $\{|a\rangle, |a'\rangle\}$ the projectors $\hat{P}_a$ and $\hat{P}_{a'}$ can be expressed in matrix form as [see Eqs. (3.38)]

$$\hat{P}_a = |a\rangle\langle a| = \begin{pmatrix} 1 \\ 0 \end{pmatrix} (1 \ 0) = \begin{bmatrix} 1 & 0 \\ 0 & 0 \end{bmatrix} \tag{3.49a}$$

and

$$\hat{P}_{a'} = |a'\rangle\langle a'| = \begin{pmatrix} 0 \\ 1 \end{pmatrix} (0 \ 1) = \begin{bmatrix} 0 & 0 \\ 0 & 1 \end{bmatrix}. \tag{3.49b}$$

Box 3.2 Summation and series

Summation is the operation of addition. The result of summation is called the sum. The summation operation can be conveniently indicated by using the summation symbol (the capital sigma, Σ). For instance, the sum of the square of the first four natural numbers can be written as

$$\sum_{k=1}^{4} k^2 = 1 + 2^2 + 3^2 + 4^2 = 30, \qquad (3.50)$$

where $k = 1$ in the subscript and 4 in the superscript denote that the summation index k takes integer values from 1 to 4.

Let $\{a_k\}_{k=1}^{n} = \{a_1, a_2, \ldots, a_n\}$ be a sequence of n numbers, in which each term a_k is given by a certain rule, whose character is here irrelevant. A series is the sum of a sequence and can be written in summation notation as

$$\sum_{k=1}^{n} a_k = a_1 + a_2 + \cdots + a_n. \qquad (3.51)$$

If the number of terms n is finite, the series is called a finite series; otherwise, it is called an infinite series. An example of an infinite series is the geometric series

$$\sum_{k=0}^{\infty} \frac{1}{2^k} = 1 + \frac{1}{2} + \frac{1}{4} + \frac{1}{8} + \frac{1}{16} + \cdots, \qquad (3.52)$$

where the symbol ∞ denotes infinity. It is not difficult to see that this series is built by taking the fraction $\frac{1}{2}$ and considering in succession infinite exponents, from zero on. Infinite series play an important role in mathematical analysis and require the notion of limits to be fully understood and manipulated.

It is noted that the matrix multiplication performed in obtaining the matrices given by Eqs. (3.49) is a particular instance of the square matrix multiplication discussed above (where we have column and row vectors instead of matrices).

As we shall see in the following chapters, in quantum mechanics every physical quantity can be in principle represented by an operator and is called an *observable*. Then, in general, if $\{|b_k\rangle\}$ is an orthonormal basis (where k is some discrete index labeling the basis vectors), the (i, j)th element of the matrix representing the observable $\hat{O}$ is given by

$$O_{ij} = \langle b_i|\hat{O}|b_j\rangle. \tag{3.53}$$

Here we note that $\langle b_i|\hat{O}|b_j\rangle$ is a legitimate mathematical object and is in fact a shorthand notation for the scalar product of the kets $|b_i\rangle$ and $\hat{O}|b_j\rangle$, i.e.,

$$\langle b_i|\hat{O}|b_j\rangle = \langle b_i|\hat{O}b_j\rangle, \tag{3.54}$$

where $|\hat{O}b_j\rangle = \hat{O}|b_j\rangle$.

Problem 3.6 Compute the following matrices:

$$\hat{A} = \begin{bmatrix} 1 & 2 \\ 3 & 4 \end{bmatrix} + \begin{bmatrix} 1 & 4 \\ 2 & 3 \end{bmatrix} + \begin{bmatrix} 1 & 3 \\ 4 & 2 \end{bmatrix} \quad \text{and} \quad \hat{B} = \begin{bmatrix} 1 & 2 & 3 \\ 4 & 5 & 6 \end{bmatrix} \begin{bmatrix} 1 & 2 \\ 3 & 4 \\ 5 & 6 \end{bmatrix}.$$

Problem 3.7 Using the matrix multiplication rule (3.47), check the derivation of the matrices given by Eqs. (3.49).

3.7 Action and Properties of Projectors

Let us now apply the projectors in Eqs. (3.38) to the polarization states encountered so far to show that they can fully describe the experimental outcomes that we are dealing with. First, let us remark that a projector applied to the same state vector into which it projects leaves this state vector unchanged. Indeed, we have

$$\hat{P}_a|a\rangle = \begin{bmatrix} 1 & 0 \\ 0 & 0 \end{bmatrix} \begin{pmatrix} 1 \\ 0 \end{pmatrix} = \begin{pmatrix} 1 \\ 0 \end{pmatrix} = |a\rangle \tag{3.55a}$$

and

$$\hat{P}_{a'}|a'\rangle = \begin{bmatrix} 0 & 0 \\ 0 & 1 \end{bmatrix} \begin{pmatrix} 0 \\ 1 \end{pmatrix} = \begin{pmatrix} 0 \\ 1 \end{pmatrix} = |a'\rangle. \tag{3.55b}$$

Physically, this means that photons *already* in the polarization state $|a\rangle$ will all pass through a filter that is aligned along the same direction. Likewise, photons already in the state $|a'\rangle$ will all pass through a filter aligned along the direction **a'** that is orthogonal to

a. We also expect that a projector applied to a state orthogonal to the state into which it projects gives zero as the result. Indeed, we find that

$$\hat{P}_a |a'\rangle = \begin{bmatrix} 1 & 0 \\ 0 & 0 \end{bmatrix} \begin{pmatrix} 0 \\ 1 \end{pmatrix} = 0 \tag{3.56a}$$

and

$$\hat{P}_{a'} |a\rangle = \begin{bmatrix} 0 & 0 \\ 0 & 1 \end{bmatrix} \begin{pmatrix} 1 \\ 0 \end{pmatrix} = 0. \tag{3.56b}$$

The physical meaning of this result is that photons in a polarization state (e.g., $|a'\rangle$) that is orthogonal to the direction of a given polarization filter (e.g., **a**) will all be blocked by the filter.

Much more interesting is the action of the projectors $\hat{P}_a$ and $\hat{P}_{a'}$ on an arbitrary polarization state $|\psi\rangle$ of the form given by Eq. (3.33) [see also (3.32)]:

$$\hat{P}_a |\psi\rangle = \langle a|\psi\rangle \hat{P}_a |a\rangle + \langle a'|\psi\rangle \hat{P}_a |a'\rangle$$

$$= \langle a|\psi\rangle |a\rangle \tag{3.57a}$$

and

$$\hat{P}_{a'} |\psi\rangle = \langle a|\psi\rangle \hat{P}_{a'} |a\rangle + \langle a'|\psi\rangle \hat{P}_{a'} |a'\rangle$$

$$= \langle a'|\psi\rangle |a'\rangle, \tag{3.57b}$$

which are in accordance with Eqs. (3.37). Being the outcomes of the action of projection, the projected states $\hat{P}_a |\psi\rangle$ and $\hat{P}_{a'} |\psi\rangle$ have a magnitude that is smaller than or equal to the state vector $|\psi\rangle$ (which is taken to be normalized and hence is of unit magnitude). In specific, from Eqs. (3.35) and (3.32) and, since for any normalized vector $|\psi\rangle$, $\||\psi\rangle\| = 1$, we have

$$\|\hat{P}_a |\psi\rangle\|^2 = \|\langle a|\psi\rangle |a\rangle\|^2$$

$$= |\langle a|\psi\rangle|^2 \||a\rangle\|$$

$$= |\langle a|\psi\rangle|^2$$

$$= |c_a|^2 \tag{3.58a}$$

and

$$\|\hat{P}_{a'} |\psi\rangle\|^2 = \|\langle a'|\psi\rangle |a'\rangle\|^2$$

$$= |\langle a'|\psi\rangle|^2 \||a'\rangle\|$$

$$= |\langle a'|\psi\rangle|^2$$

$$= |c_{a'}|^2. \tag{3.58b}$$

Because of the normalization condition (3.31), we conclude that $0 \leq |c_a|^2, |c_{a'}|^2 \leq 1$. The physical meaning of the above equations is our main result here. The action of a projector, say $\hat{P}_a$, on an arbitrary state $|\psi\rangle$ is precisely that it lets pass through the component of $|\psi\rangle$ that is parallel to the direction of the projector with a probability given by the square modulus of the corresponding probability amplitude $|\langle a|\psi\rangle|^2$. When $|\psi\rangle$ is not already a state along the direction of the projector, this probability is less then one, as expressed in Fig. 3.3.

We will now discuss three important properties of projectors. The first property is that the projectors are positive semidefinite operators, meaning that

$$\langle\psi|\hat{P}_a|\psi\rangle \geq 0 \quad \text{and} \quad \langle\psi|\hat{P}_{a'}|\psi\rangle \geq 0 \quad \text{for all } |\psi\rangle. \tag{3.59}$$

This property can be proved easily by rewriting the above expressions as

$$\langle\psi|\hat{P}_a|\psi\rangle = \langle\psi|a\rangle\langle a|\psi\rangle = |\langle a|\psi\rangle|^2 \geq 0, \tag{3.60a}$$

$$\langle\psi|\hat{P}_{a'}|\psi\rangle = \langle\psi|a'\rangle\langle a'|\psi\rangle = |\langle a'|\psi\rangle|^2 \geq 0, \tag{3.60b}$$

where use has been made of Eq. (3.21a). The second property is that the square of a projector is the projector itself, while the product of projectors onto orthogonal states vanishes. In particular, we have

$$\hat{P}_a^2 = \begin{bmatrix} 1 & 0 \\ 0 & 0 \end{bmatrix}\begin{bmatrix} 1 & 0 \\ 0 & 0 \end{bmatrix} = \begin{bmatrix} 1 & 0 \\ 0 & 0 \end{bmatrix} = \hat{P}_a, \tag{3.61a}$$

$$\hat{P}_{a'}^2 = \begin{bmatrix} 0 & 0 \\ 0 & 1 \end{bmatrix}\begin{bmatrix} 0 & 0 \\ 0 & 1 \end{bmatrix} = \begin{bmatrix} 0 & 0 \\ 0 & 1 \end{bmatrix} = \hat{P}_{a'}, \tag{3.61b}$$

$$\hat{P}_a\hat{P}_{a'} = \begin{bmatrix} 1 & 0 \\ 0 & 0 \end{bmatrix}\begin{bmatrix} 0 & 0 \\ 0 & 1 \end{bmatrix} = 0, \tag{3.61c}$$

$$\hat{P}_{a'}\hat{P}_a = \begin{bmatrix} 0 & 0 \\ 0 & 1 \end{bmatrix}\begin{bmatrix} 1 & 0 \\ 0 & 0 \end{bmatrix} = 0, \tag{3.61d}$$

which can be expressed in a compact form as

$$\hat{P}_j\hat{P}_k = \delta_{jk}\hat{P}_k, \tag{3.62}$$

where j and k can be either a or a'. The reason of the last two expressions in Eqs. (3.61) is that projectors onto orthogonal states represent mutually exclusive measurement outcomes. Mathematically, this is a consequence of the orthogonality of the corresponding state vectors. In general, a set of projectors satisfying the property (3.62) is said to be orthogonal.

As said, projectors represent mutually exclusive outcomes of a measurement. These outcomes occur with certain probabilities and the sum of all probabilities of a set of mutually exclusive outcomes must be equal to 1 (like all possible six results when throwing a dice), according to the normalization condition. Consequently, all possible mutually exclusive outcomes of a measurement constitute a complete set of orthogonal states. The third property states that the sum of all projectors onto a complete set of orthonormal states is equal to the identity operator. In our two-dimensional case, the basis $\{|a\rangle, |a'\rangle\}$ is a complete orthonormal basis as

$$\hat{P}_a + \hat{P}_{a'} = \begin{bmatrix} 1 & 0 \\ 0 & 0 \end{bmatrix} + \begin{bmatrix} 0 & 0 \\ 0 & 1 \end{bmatrix} = \begin{bmatrix} 1 & 0 \\ 0 & 1 \end{bmatrix} = \hat{I}, \qquad (3.63)$$

or equivalently,

$$\sum_{j=a,a'} \hat{P}_j = \hat{I}, \qquad (3.64)$$

where $\hat{I}$ is two-dimensional identity matrix representing the identity operator in a two-dimensional Hilbert space. It is noted that Eq. (3.63) is also called the *completeness relation* or the *resolution of the identity*.

These three properties are general in nature and hence valid in arbitrary dimensions. In particular, in arbitrary dimensions the identity operator is represented by the identity matrix with all diagonal elements equal to 1 and all off-diagonal elements equal to 0. The reason why it is called identity matrix is that it represents the operation that induces no change for all operators and states, i.e., we have

$$\hat{I}\hat{A} = \hat{A}\hat{I} = \hat{A} \quad \text{and} \quad \hat{I}|\psi\rangle = |\psi\rangle, \qquad (3.65)$$

where $\hat{A}$ is an arbitrary operator and $|\psi\rangle$ is an arbitrary state.

Problem 3.8 Using of the properties of the scalar product, prove the results (3.55) and (3.56).

Problem 3.9 Redo the calculations of Eqs. (3.57) in vector and matrix forms.

Problem 3.10 Check Eq. (3.61) by making use of the product of matrices.

Problem 3.11 Prove Eq. (3.61) by making use of the properties of the scalar product. (*Hint*: Use the expressions $\hat{P}_a = |a\rangle\langle a|$ and $\hat{P}_{a'} = |a'\rangle\langle a'|$.)

Problem 3.12 Prove Eqs. (3.65) by using the projectors $\hat{P}_a$ and $\hat{P}_{a'}$ as matrices and the polarization state $|\psi\rangle$.

3.8 Summary

In this chapter we have

- Learned that a quantum state is represented by a normalized vector in a Hilbert space, called the state vector.
- Seen that a state vector can be expanded in a complete orthonormal basis, with the expansion coefficients being the respective probability amplitudes.
- Learned the scalar product of two state vectors and its basic properties.
- Learned the matrix representation of linear operators and the basic properties of matrix multiplication.
- Introduced the projectors associated with a complete orthonormal basis and shown that they represent mutually exclusive outcomes of a measurement.

Chapter 4

Bases and Operations

In this chapter we shall present the dilemma between an undulatory and corpuscular understanding of quantum systems. We shall show the historical reasons that have led to the necessity to posit the quantization principle. Then, we shall focus on the concept of observables and show two crucial aspects of quantum observables: there are different possible bases to expand a quantum state (hence superposition is a relative concept) and observables do not necessarily commute. Finally, the concept of non-local features is introduced.

4.1 Corpuscular Nature of Light

In Chapter 2 we have followed the photon in the interferometer. We have seen that it is in a superposition state, producing self–interference. We have also seen that, after many experimental runs, the curves describing the detection events feature a typical interference pattern [see Fig. 2.7]. However, if we ask what happens to a *single* photon in the superposition state (2.29) when it is detected, we face a puzzling situation: the photon has been revealed *either* by D1 *or* by D2, exactly as we would expect for a classical

Quantum Mechanics for Thinkers
Gennaro Auletta and Shang-Yung Wang
Copyright © 2014 Pan Stanford Publishing Pte. Ltd.
ISBN 978-981-4411-71-4 (Hardcover), 978-981-4411-72-1 (eBook)
www.panstanford.com

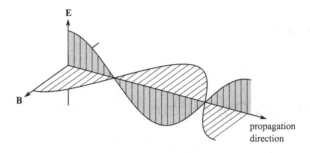

propagation
direction

Figure 4.1 Snapshot of an electromagnetic wave, which is a self-propagating transverse oscillating wave of electric and magnetic fields. Oscillation of the electric field **E** and magnetic field **B** associated with a linearly polarized electromagnetic wave is depicted here. The electric and magnetic fields oscillate in phase perpendicular to each other and perpendicular to the direction of energy and wave propagation. Adapted from (Auletta *et al.*, 2009, p. 20).

corpuscle. It is as if the superposition state were disappeared and we had a sudden localization of the photon in one of the two paths (either in the state $|1\rangle$ or in the state $|2\rangle$). How is this possible? It is timely to recall that, according to classical physics, light is a wave-like continuous phenomenon, whose electric and magnetic components oscillate in phase perpendicular to each other as well as perpendicular to the propagation direction [see Fig. 4.1]. On the other hand, matter was conceived classically as composed of corpuscles, i.e., of discontinuous elements.

Historically, the corpuscular and discontinuous nature of radiation was precisely the background on which quantum mechanics was born. In two seminal papers published in 1900, the German physicist Max Planck provided an extraordinary solution to the famous *black body problem* that had arisen in physics towards the end of the 19th century.[a] A black body is an idealized physical body that absorbs all incident electromagnetic radiation falling on it. Because of this perfect absorptivity at all wavelengths, a black body is also the best possible emitter of thermal radiation, which it radiates incandescently in a characteristic, continuous spectrum that depends on the body's temperature. Electromagnetic

[a](Planck, 1900a,b).

radiation absorbed by the black body eventually reaches a thermal equilibrium with the body, and then is reemitted as thermal radiation by the body. However, the classical theory of radiation cannot correctly predict the spectral properties of the black body. According to classical calculations, we should have the so-called ultraviolet catastrophe, a situation in which the intensity of the electromagnetic radiation trapped in the black body goes to infinity. The reason is that the energy intensity was calculated by assuming that there is a continuum in the energy of the electromagnetic radiation.

Planck was able to solve the theoretical problem by assuming that the energy of the electromagnetic radiation (which amounts to say the light itself) be discontinuous, that is, that the energy carried by the light always comes in units. It is interesting to remark that Planck did not consider the incident light as discontinuous but only the *emitted* one. In other words, he did not oppose the traditional continuous understanding of light but assumed that, being the structure of the black body composed of atoms, and therefore corpuscular and discontinuous, the radiation that was emitted by the black body was discontinuous as well. In this way, Planck postulated the formula

$$E = nh\nu \quad (n = 0, 1, 2, 3, \dots), \tag{4.1}$$

where E the energy carried by the light, ν is the frequency of the light, and h is a new constant of nature called the *Planck constant*, whose value in SI units (the International System of Units that is the modern form of the metric system) is

$$h = 6.626069 \times 10^{-34} \, \text{J} \cdot \text{s}. \tag{4.2}$$

In the above expression, we have

$$10^{-34} = \frac{1}{10^{34}},$$

while J · s means Joule times second and the Joule is the unit of energy. In order to understand how small the Planck constant is, it suffices to consider that a Joule roughly measures the energy that is required to elevate a weight of 1 kg by a vertical height of 10 cm on the surface of the Earth. We have already met the concept of frequency [see Fig. 2.1] and said that it is inversely proportional to the wavelength. Indeed, the frequency is the number of oscillations that a wave makes in a unit time interval, and it is obviously larger when the wavelength is shorter, which for light also demands more energy. In this context, we recall that visible light goes from red (longer wavelength with lower energy) to blue and violet (shorter wavelength with higher energy). In summary, Planck's postulate, now commonly known as the quantum postulate, states that energy may be emitted and absorbed in discrete units (expressed by $n = 0, 1, 2, 3, \ldots$). In turn, this means that in certain contexts energy takes the form of discrete packets called *quanta* (from which the term *quantum* mechanics), and that the indivisible light quanta of energy $h\nu$ are the single photons, as anticipated in Chapter 2.

Although the quantum postulate was initially formulated as a pure formal hypothesis in order to solve a specific problem (which could be taken to represent, therefore, what is called an *ad hoc* hypothesis), its consequences were truly revolutionary for physics. Indeed, for centuries it was postulated that Nature cannot do jumps (*Natura non facit saltus*). As Leibniz put it, continuity can be found in time, extension, quality, motion, and all nature transitions, since nothing happens jump-like (*numquam fit per saltum*).[a] On the contrary, here it was assumed for the first time that certain physical parameters like energy are not continuous [see Section 1.2].

It is very interesting to analyze a little bit the form of inference that led Planck to his conclusion. Planck was deeply committed to the validity of the classical laws. However, he could not explain why those laws pushed to a conclusion or an expectation that was in so strident contrast with experimental results. In such a situation, the best option is to perform an *abduction*,[b] that is, to try to show that there is some *new property* or behavior that was not taken into account by the previous applications of those laws and whose existence can solve the anomaly without rejecting the laws as such.[c] In the case considered here, Planck did not reject the assumption of the continuity of the natural laws [see Chapter 1], neither of the electromagnetic waves as such. He only postulated

---

[a](Leibniz MS, Vol. VII, p. 25). See also (Boscovich, 1754).
[b](Peirce, 1866), (Peirce, 1877), (Peirce, 1878).
[c](Auletta, 2009).

that, under special circumstances, when light is reemitted by a discontinuous matter structure, then it shows a new property, namely discontinuity. In the following we shall see that there is a deep insight in this result that remains valid even when the classical laws will be abandoned and a new theory, i.e., quantum mechanics, will be formulated.

4.2 Further Experimental Evidences

Some years after Planck's pathbreaking work, Einstein used the quantum postulate to solve another classical problem of physics, showing in this way that Planck's solution could not be considered a mere *ad hoc* solution.[a] This new problem was the *photoelectric effect*. It was experimentally known that when a certain metal is irradiated by ultraviolet light, electrons are emitted. Now, if light was continuous, as it was assumed at that time, one should need a significant amount of time (of irradiation) for cumulating energy sufficient to break the atomic bonds and expel the electrons. However, the photoelectric effect is almost instantaneous once the metal is irradiated by ultraviolet light. Einstein solved the problem by assuming that the *incident* (and no longer only the reemitted) light was composed of discrete energy packets relatively well localized that were called *photons* later on.

To assume that the incident light was discontinuous was a big conceptual step because it raised two fundamental questions: (i) whether or not light could be considered corpuscular as such, and (ii) if discontinuity could be generalized to other physical contexts. The majority of the physicists at that time had an abiding faith in the validity of classical mechanics (even Planck never took very seriously Einstein's work on the photoelectric effect). Things began to change when Bohr became aware that a discontinuity in the possible orbits at which the electrons are located in an atom could be a solution to the stability problem of the atom.[b] A couple of years earlier, Rutherford had proposed a planetary model

---

[a](Einstein, 1905).
[b](Bohr, 1913).

of atom, in which negatively charged particles (electrons) would revolve around a positively charged nucleus in a circular motion at all possible continuous distances. The model accounted for many experimental facts known at that time but presented the problem that an accelerating electric charge (due to the circular motion of the electrons) emits electromagnetic radiation, according to the classical theory of electromagnetism, with the consequence that the electrons should progressively lose energy and finally fall, with a spiral trajectory, into the nucleus in a very short time. However, this disagrees with the experience that shows matter is exceeding long-living (apart from radioactive matter). To solve this problem, Bohr proposed that the electrons in an atom can occupy only a certain number of orbits which correspond to certain discrete energy levels and that the electrons cannot jump from one energy level to another without emitting (when jumping to a lower energy level) or absorbing (when jumping to a higher energy level) a quantum of energy, i.e., a photon. This was a splendid confirmation of both Planck's and Einstein's results.

Einstein also applied Planck's formula to solve the problem of the specific heat of a solid (although the final formula is due to Debye). At the same time it was also discovered that the spectrum of the electromagnetic radiation emitted by diluted gases was not continuous. The accumulation of independent experimental evidence that, at least at microscopic scale, nature shows a discontinuous behavior became overwhelming when Compton discovered the remarkable effect that now bares his name. In the Compton effect experiment, an electron is scattered with a photon and results in a decrease in the energy of the photon.[a] This can be accounted for by assuming that during the scattering the electron first absorbs the photon and then emits another photon of lower energy. In this way, the reality of light quanta could be barely put into question.

In such a situation, an abductive solution is no longer sufficient. The accumulation of independent evidences showed that there can be a problem with the classical theory as such and not only with specific applications. In such situations, we face a terrible dilemma: We become growingly aware that the classical laws no longer

---

[a](Compton, 1923).

possess general validity and should be rather considered statistical regularities (for instance, valid for macroscopic bodies, for which some approximations can be valid but not for microscopical systems like atoms or electrons that demand new fundamental laws). This is the essence of the inferential procedure called *induction*.[a] However, how to find the new laws? Unfortunately, there is no inference that could provide a solution. In other words, there is no scientific method for finding solutions to current scientific problems. It is simply a matter of *insight* by some very creative people. The person who provided such a new insight was Heisenberg.

4.3 Quantization Principle

The big idea of Heisenberg was to assume that microscopic systems systematically behave in a discontinuous way (when interacting with other systems) and to introduce the appropriate mathematical tools for dealing with the discontinuities that were experimentally found.[b] The first point can be summarized in a quantization principle that puts together all the previous contributions.

Principle 4.1 (Quantization Principle) *Some relevant physical quantities of quantum systems can show discontinuous characteristics when the latter interact with other systems or are subject to fields or external forces. In those cases, the Planck constant h is associated with the minimum quantum of energy or action involved in those interactions.*

Therefore, Planck's idea that only emitted radiation is corpuscular is to a certain sense right, obviously not in the specific assumption, because light may show corpuscular properties in many different contexts and for various reasons, but because it is here implicit that, in order to show this discontinuous behavior, light need to *interact* with something else. The progress of science consists precisely in singling out the conditions and the extent to which certain regularities that were previously assumed need to

---

[a](Auletta, 2009).
[b](Heisenberg, 1925).

be understood, and therefore it always consists in error correction though which, when generalizing, previous statements are more specifically formulated.

About the mathematical tools, we have seen in Chapter 1 that being all of the physical quantities (time, energy, speed, position, etc.) continuous in a classical world, they are very well mathematically expressed by continuous variables and functions thereof. Heisenberg understood that these can no longer be the appropriate mathematical instruments and proposed to introduce *operators* as the mathematical representation of quantum mechanical quantities. Operators can indeed cover both the discontinuous and continuous cases. In the first case, they are represented by matrices [see Section 3.6]. In the previous chapter, we have already found two examples of operators: projectors and the identity operator [see Section 3.7]. An operator is much more a dynamical entity than a variable or a function. Indeed, it induces or represents a *transformation*. In the previous chapter, we have indeed seen that projectors can represent the action of a polarization filter on the photon, as it is evident by Eq. (3.57). To distinguish the quantum mechanical operator form of physical quantities from the classical variables and functions, we will refer to the former as physical *observables*. A projector can be understood as the easiest example of observables. For instance, $\hat{P}_a$ can be understood as a test addressed to the incoming photon by asking the question "Are you in the polarization state $|a\rangle$?" If the answer is yes, the photon will pass the test. If the answer is no, which means that the photon is in the orthogonal polarization state $|a'\rangle$, it will not pass it. In the intermediate cases, in which the photon is in a superposition state of $|a\rangle$ and $|a'\rangle$, it has a non-zero probability to pass the test, as again given by Eq. (3.57).

Let us now express the action of $\hat{P}_a$ described by Eqs. (3.55) and (3.56) as

$$\hat{P}_a|a\rangle = |a\rangle = +1\,(|a\rangle), \tag{4.3a}$$

$$\hat{P}_a|a'\rangle = 0 = 0\,(|a'\rangle). \tag{4.3b}$$

The two equations above are examples of the general *eigenvalue equation*

$$\hat{O}|o_j\rangle = o_j|o_j\rangle, \tag{4.4}$$

where $\hat{O}$ is an observable, o_j is called the *eigenvalue* of $\hat{O}$, and $|o_j\rangle$ is the *eigenstate* corresponding to the eigenvalue o_j (with j being some index labeling the eigenvalues). These notions will be clarified in the following sections. By now, let us remark that the eigenvalue o_j is the possible value that we read on a certain measurement device when we perform a measurement of the observable $\hat{O}$, a value that we obtain when the measured system is in fact in the state $|o_j\rangle$. In the polarization case, the states $|a\rangle$ and $|a'\rangle$ are eigenstates of the observable $\hat{P}_a$ with respective eigenvalues $+1$ and 0 [see Eqs. (4.3)], which express the possible results that we obtain when testing if the photon is in the state $|a\rangle$, namely, yes for $+1$ and no for 0. As mentioned, they can be thought of as measurement outcomes and therefore also as values shown on a measuring device (for instance, a counter or a graduated scale).

Problem 4.1 Make use again of Eqs. (3.55) and (3.56) to check the two eigenvalue equations of the observable $\hat{P}_{a'}$.

4.4 Quantum Observables in General

It is possible to generalize these results to observables that are combinations of projectors. Suppose that instead of the state vector (and the corresponding polarization direction) $|a\rangle$ or $|a'\rangle$, we consider a state vector (and a polarization direction) $|\psi\rangle$, given by Eq. (3.33), which for the sake of convenience we rewrite here as

$$|\psi\rangle = |a\rangle\langle a|\psi\rangle + |a'\rangle\langle a'|\psi\rangle. \tag{4.5}$$

Then, in this case we are tempted to mathematically write the corresponding observable as the operator

$$\hat{O} = \hat{P}_a + \hat{P}_{a'}, \tag{4.6}$$

whose eigenstates should be *both* $|a\rangle$ and $|a'\rangle$. However, we see immediately that $\hat{O}$ would coincide with the identity operator $\hat{I}$ [see Eq. (3.63)], which, by definition, is not an observable (since every state is its eigenstate with the same eigenvalue $+1$). Therefore, our equation (4.6) does not work. In order to distinguish the two *distinct* eigenstates $|a\rangle$ and $|a'\rangle$, we can assign to them different eigenvalues and construct an observable of the form

$$\hat{O}' = -\hat{P}_a + \hat{P}_{a'} = -|a\rangle\langle a| + |a'\rangle\langle a'|, \qquad (4.7)$$

or in terms of the matrices,

$$\hat{O}' = -\begin{bmatrix} 1 & 0 \\ 0 & 0 \end{bmatrix} + \begin{bmatrix} 0 & 0 \\ 0 & 1 \end{bmatrix} = \begin{bmatrix} -1 & 0 \\ 0 & +1 \end{bmatrix}. \qquad (4.8)$$

The operator $\hat{O}'$ is obviously neither the identity operator nor a projector, but is precisely the form of the observable we look for. Indeed, when applied to the states $|a\rangle$ and $|a'\rangle$, the resulting eigenvalues will be -1 and $+1$ (the two possible values that we obtain when measuring $\hat{O}'$), respectively, and therefore we shall be able to discriminate between the two different outcomes. It is true that the derivation that we have performed is very particular. But this does not matter since it leads us to a result of general validity, provided that we abandon the specific assumptions that were used to derive this conclusion (this kind of inference is called conditional reasoning). Indeed, the main conclusion here is that at least in one case it is possible to construct an appropriate observable that distinguishes between the two different eigenvalues -1 and $+1$ associated respectively with the two different projectors $\hat{P}_a$ and $\hat{P}_{a'}$. The two outcomes represented by the two eigenvalues are the two different possible *results* that we could obtain by performing a certain operation associated with the observable $\hat{O}'$. But which kind of operation is it? We have mentioned that it should be a measurement. This is a very important issue and we need to be careful by addressing it step by step. By now, let us deal with some formal aspects of observables and operators.

In order to be the mathematical representation of physical observables, operators (or matrices) need to obey some constraints. The most important one is that the eigenvalues be real (and not imaginary or complex) numbers. As we have seen, eigenvalues of an observable are indeed associated with the possible results of a certain test or experiment and therefore should even be displayed by some measuring devices. But imaginary or complex numbers cannot be suitable to this purpose. Mathematically this requirement of real eigenvalues is tantamount to imposing that the operator (or matrix) representing a physical observable will be equal to its conjugate transpose (a mathematical operation that we have already seen for state vectors in Section 3.4).

The *conjugate transpose* (or Hermitian conjugate) of an $m \times n$ matrix $\hat{A}$ with complex elements is the $n \times m$ matrix $\hat{A}^\dagger$ obtained from $\hat{A}$ by taking the transpose and then taking the complex conjugate of each element, i.e., the conjugate transpose is formally defined by

$$(A^\dagger)_{ij} = (A_{ji})^*, \tag{4.9}$$

where the subscripts denote the (i, j)th element with $1 \leq i \leq n$ and $1 \leq j \leq m$. For a ket (i.e., a 2 × 1 matrix)

$$|\psi\rangle = \begin{pmatrix} a \\ b \end{pmatrix}, \tag{4.10}$$

its transposed conjugate is its dual bra (i.e., a 1 × 2 matrix):

$$|\psi\rangle^\dagger = \begin{pmatrix} a^* & b^* \end{pmatrix} = \langle\psi|. \tag{4.11}$$

Then, for a 2 × 2 matrix

$$\hat{A} = \begin{bmatrix} a & b \\ c & d \end{bmatrix}, \tag{4.12}$$

we have

$$\hat{A}^\dagger = \begin{bmatrix} a^* & c^* \\ b^* & d^* \end{bmatrix}. \tag{4.13}$$

In addition, the conjugate transpose satisfies the following properties:

$(\hat{A} + \hat{B})^\dagger = \hat{A}^\dagger + \hat{B}^\dagger$ for any matrices $\hat{A}$ and $\hat{B}$ of the same dimensions, $\tag{4.14a}$

$(\alpha\hat{A})^\dagger = \alpha^*\hat{A}^\dagger$ for any complex number α and any matrix $\hat{A}$, $\tag{4.14b}$

$(\hat{A}\hat{B})^\dagger = \hat{B}^\dagger\hat{A}^\dagger$ for any $m \times n$ matrix $\hat{A}$ and any $n \times p$ matrix $\hat{B}$, $\tag{4.14c}$

$(\hat{A}^\dagger)^\dagger = \hat{A}$ for any matrix $\hat{A}$, $\tag{4.14d}$

Eigenvalues of $\hat{A}^\dagger$ are the complex conjugates of those of $\hat{A}$. $\tag{4.14e}$

In technical terms the above requirement amounts to saying that the operator (or matrix) representing a physical observable needs to be *Hermitian*. We may express this requirement in operator form as

$$\hat{O} = \hat{O}^{\dagger}, \tag{4.15}$$

where $\hat{O}^{\dagger}$ denotes precisely the conjugate transpose of the observable $\hat{O}$. From the property (4.14e), it follows immediately that the eigenvalues of a Hermitian operator (or matrix) are real numbers. Moreover, Hermitian operators have orthogonal eigenstates and the eigenstates form a complete orthonormal basis. It will be important to note that while physical observables are represented by Hermitian operators, the converse is not true in general. In other words, not all Hermitian operators correspond to physical observables: in the same way in which we have built in this section an observable by combining projectors, we can always build abstract operators by combining physical observables, but it is not *a priori* guaranteed that the result will correspond to a physical quantity.

Problem 4.2 Find the action of the observable $\hat{O}'$ in Eq. (4.7) on both the states $|a\rangle$ and $|a'\rangle$ in analogy with that of the projectors $\hat{P}_a$ and $\hat{P}_{a'}$ shown in Eq. (4.3). Check whether or not the eigenvalues are different.

Problem 4.3 Check that the projectors $\hat{P}_a$ and $\hat{P}_{a'}$ given by Eq. (3.49) are Hermitian.

4.5 Different Bases and Superposition

Having chosen Hermitian operators (or matrices) as the representation of quantum observables has many surprising consequences. We shall now begin to explore them. We have already seen that observables can be single projectors and the combinations thereof. This in turn reflects a general circumstance of quantum theory: quantum states can be expanded in different bases, each of which, at least in principle, represents the eigenbasis of a different observable. Here by *eigenbasis* we mean the complete set of eigenstates of a given observable, that is, those states that we could obtain if we measured that observable. To be less abstract, let us consider the usual basis $\{|a\rangle, |a'\rangle\}$ and the alternative polarization basis $\{|h\rangle, |v\rangle\}$, where $|h\rangle$ denotes the state of horizontal polarization and $|v\rangle$ the state of vertical polarization. Instead, the state $|\psi\rangle$ is as usual a state of arbitrary polarization. We first note that $|a\rangle$ and $|a'\rangle$ are eigenstates of the observable

$$\hat{O} = \hat{P}_a - \hat{P}_{a'}, \tag{4.16}$$

where for the sake of convenience we have reversed the signs of the eigenvalues in the previous example [see Eq. (4.7)]. On the other hand, it can be shown that $|v\rangle$ and $|h\rangle$ are eigenstates of the observable

$$\hat{O}' = \hat{P}_h - \hat{P}_v, \tag{4.17}$$

where $\hat{P}_h = |h\rangle\langle h|$ and $\hat{P}_v = |v\rangle\langle v|$ are the projectors onto the states $|h\rangle$ and $|v\rangle$, respectively. Since $|h\rangle$ and $|v\rangle$ are orthogonal and exhaust the two possible results when measuring the polarization of a photon, they also constitute a complete orthonormal basis [see Section 3.4]. Therefore, they satisfy a completeness relation of the form [see Eq. (3.63)]

$$\hat{P}_h + \hat{P}_v = \hat{I}. \tag{4.18}$$

Then, the state $|\psi\rangle$ can be written as [see also Eq. (3.1)]

$$|\psi\rangle = \hat{I}|\psi\rangle$$
$$= (\hat{P}_h + \hat{P}_v)|\psi\rangle$$
$$= |h\rangle\langle h|\psi\rangle + |v\rangle\langle v|\psi\rangle$$
$$= \langle h|\psi\rangle|h\rangle + \langle v|\psi\rangle|v\rangle. \tag{4.19}$$

Upon comparing Eq. (3.33) or (4.5) with Eq. (4.19), we find that the *same* state $|\psi\rangle$ can be expanded *both* in the basis $\{|a\rangle, |a'\rangle\}$ and in the basis $\{|h\rangle, |v\rangle\}$ [see Figs. 3.2 and 4.2]. Therefore, if we send a photon in the state $|\psi\rangle$ to a polarization filter along the horizontal direction, the photon will have a probability $|\langle h|\psi\rangle|^2$ to pass through and a probability $|\langle v|\psi\rangle|^2$ to be blocked. The meaning of this result is that given a photon in the state $|\psi\rangle$, we are totally free to set the polarization filter along any direction (represented by a basis state vector of a certain basis). Each time we have a certain probability that the photon will pass this test and a certain probability that it will

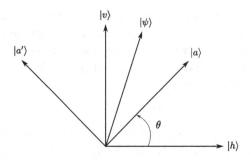

Figure 4.2 Change of basis. The basis $\{|a\rangle, |a'\rangle\}$ is obtained from the original basis $\{|h\rangle, |v\rangle\}$ by a counterclockwise rotation of 45°.

not. These two *outcomes* are associated with the two different basis states of a given basis. However, choosing a certain *test* amounts to choosing a certain observable (and hence to choosing a basis that is the eigenbasis of that observable). In other words, we are totally free to *measure* any observable we may desire (although, as we shall see later, we have no control on the possible outcomes themselves). The consequences of this circumstance are very deep and surprising.

It is very important to understand that in the specific example considered above, the basis state vectors $|a\rangle$ and $|a'\rangle$ can be respectively written as superpositions of $|h\rangle$ and $|v\rangle$:

$$|a\rangle = \frac{1}{\sqrt{2}}\left(|h\rangle + |v\rangle\right) \quad \text{and} \quad |a'\rangle = \frac{1}{\sqrt{2}}\left(|v\rangle - |h\rangle\right), \qquad (4.20)$$

as is evident by an inspection of Fig. 4.2. We recall the requirement that $|a'\rangle$ is orthogonal to $|a\rangle$, and one way to fulfill that is to take the state $|h\rangle$ in $|a'\rangle$ with an opposite sign (or a relative phase of π as $e^{i\pi} = -1$). Moreover, as shown in Fig. 4.2 both superpositions are symmetric (for the angle $\theta = 45°$). We can generalize this conclusion as follows. Apart from the limiting and uninteresting case in which $|v\rangle$ or $|h\rangle$ are orthogonal either to $|a\rangle$ or to $|a'\rangle$ (or, in other words, each one coincides with $|a'\rangle$ or $|a\rangle$)), we must conclude that $|a\rangle$ and $|a'\rangle$ both are superposition of the states $|h\rangle$ and $|v\rangle$ (which are then the basis state vectors).

Therefore, the main conclusion we may draw is that the concept of superposition is not absolute but *depends on the basis used*. Indeed, in the basis $\{|a\rangle, |a'\rangle\}$ neither $|a\rangle$ nor $|a'\rangle$ are superpositions.

It will be important to note that all bases are equally good for expressing the state $|\psi\rangle$ and that the choice of basis is only a matter of convenience.

Problem 4.4 Find the action of the projectors $\hat{P}_h$ and $\hat{P}_v$ on the state $|\psi\rangle$ as expanded in Eq. (4.19), in analogy with that of the projectors $\hat{P}_a$ and $\hat{P}_{a'}$ shown in Eq. (4.3).

Problem 4.5 Why, although in the expansion of $|a'\rangle$ in Eq. (4.20), the coefficient of $|h\rangle$ is negative, the probability of obtaining the outcome associated with $|h\rangle$ is equal to the probability of obtaining the outcome associated with $|v\rangle$?

4.6 Change of Basis as a Unitary Transformation

Before proceeding further, let us summarize the two main results in the previous section:

(1) The same state can be expanded in different bases. The choice of basis is only a matter of convenience (or of choice for selecting certain measurement contexts).
(2) The basis state vectors of one basis are in general superpositions of those of another basis.

We are now interested in understanding how to formulate the relations between two different bases. Let us first rewrite, in all its generality, the expansion of $|\psi\rangle$ in the basis $\{|a\rangle, |a'\rangle\}$ as [see Eq. (3.32)]

$$|\psi\rangle = \langle a|\psi\rangle|a\rangle + \langle a'|\psi\rangle|a'\rangle$$
$$= c_a|a\rangle + c_{a'}|a'\rangle, \qquad (4.21)$$

and similarly the expansion of the same $|\psi\rangle$ in the basis $\{|v\rangle, |h\rangle\}$ as

$$|\psi\rangle = \langle h|\psi\rangle|h\rangle + \langle v|\psi\rangle|v\rangle$$
$$= c_h|h\rangle + c_v|v\rangle. \qquad (4.22)$$

Our task is to find the relation between the two sets of coefficients $\{c_a, c_{a'}\}$ and $\{c_h, c_v\}$, and therefore the relation between the two different bases, a transformation known as the *change of basis*.

We shall consider here the change of basis from the basis $\{|a\rangle, |a'\rangle\}$ to the basis $\{|h\rangle, |v\rangle\}$. The reverse can be obtained in a similar manner [see Problem 4.6]. Taking into account the completeness relation for the projectors $\hat{P}_a$ and $\hat{P}_{a'}$ [see Eq. (4.18)], we have

$$\langle h|\psi\rangle = \langle h|\hat{I}|\psi\rangle$$
$$= \langle h| \left(|a\rangle\langle a| + |a'\rangle\langle a'|\right) |\psi\rangle \qquad (4.23a)$$

and

$$\langle v|\psi\rangle = \langle v|\hat{I}|\psi\rangle$$
$$= \langle v| \left(|a\rangle\langle a| + |a'\rangle\langle a'|\right) |\psi\rangle, \qquad (4.23b)$$

which in turn allows us to write down the relation between the coefficients $\{c_a, c_{a'}\}$ and $\{c_h, c_v\}$ as

$$c_h = \langle h|\psi\rangle$$
$$= \langle h|a\rangle\langle a|\psi\rangle + \langle h|a'\rangle\langle a'|\psi\rangle$$
$$= \langle h|a\rangle c_a + \langle h|a'\rangle c_{a'} \qquad (4.24a)$$

and

$$c_v = \langle v|\psi\rangle$$
$$= \langle v|a\rangle\langle a|\psi\rangle + \langle v|a'\rangle\langle a'|\psi\rangle$$
$$= \langle v|a\rangle c_a + \langle v|a'\rangle c_{a'}. \qquad (4.24b)$$

In matrix notation, the above expressions can be cast into a compact form as [see Section 3.6]

$$\begin{pmatrix} c_h \\ c_v \end{pmatrix} = \begin{bmatrix} \langle h|a\rangle & \langle h|a'\rangle \\ \langle v|a\rangle & \langle v|a'\rangle \end{bmatrix} \begin{pmatrix} c_a \\ c_{a'} \end{pmatrix}, \qquad (4.25)$$

or equivalently,

$$\begin{pmatrix} \langle h|\psi\rangle \\ \langle v|\psi\rangle \end{pmatrix} = \begin{bmatrix} \langle h|a\rangle & \langle h|a'\rangle \\ \langle v|a\rangle & \langle v|a'\rangle \end{bmatrix} \begin{pmatrix} \langle a|\psi\rangle \\ \langle a'|\psi\rangle \end{pmatrix}. \qquad (4.26)$$

Let us denote the transformation matrix in Eqs. (4.25) and (4.26) by $\hat{U}$, i.e.,

$$\hat{U} = \begin{bmatrix} \langle h|a\rangle & \langle h|a'\rangle \\ \langle v|a\rangle & \langle v|a'\rangle \end{bmatrix}. \qquad (4.27)$$

By expanding the basis vectors $|a\rangle$ and $|a'\rangle$ in the basis $\{|h\rangle, |v\rangle\}$ as

$$|a\rangle = (|h\rangle\langle h| + |v\rangle\langle v|)|a\rangle = \langle h|a\rangle|h\rangle + \langle v|a\rangle|v\rangle, \qquad (4.28a)$$

$$|a'\rangle = (|h\rangle\langle h| + |v\rangle\langle v|)|a'\rangle = \langle h|a'\rangle|h\rangle + \langle v|a'\rangle|v\rangle, \qquad (4.28b)$$

we find that the matrix elements of $\hat{U}$ are nothing but the expansion coefficients of the *original* basis vectors $|a\rangle$ and $|a'\rangle$ in the *new* basis $\{|v\rangle, |h\rangle\}$. The operator $\hat{U}$ is our first example of the so-called *unitary* matrices, which satisfy the condition

$$\hat{U}\hat{U}^\dagger = \hat{U}^\dagger\hat{U} = \hat{I}, \qquad (4.29)$$

where we recall that the matrix $\hat{U}^\dagger$ is the conjugate transpose of the matrix $\hat{U}$ [see Section 4.4]. Indeed, using the expression (4.27), we have [see also Problem 4.7]

$$
\begin{aligned}
\hat{U}\hat{U}^\dagger &= \begin{bmatrix} \langle h|a\rangle & \langle h|a'\rangle \\ \langle v|a\rangle & \langle v|a'\rangle \end{bmatrix} \begin{bmatrix} \langle h|a\rangle^* & \langle v|a\rangle^* \\ \langle h|a'\rangle^* & \langle v|a'\rangle^* \end{bmatrix} \\[2mm]
&= \begin{bmatrix} \langle h|a\rangle\langle a|h\rangle + \langle h|a'\rangle\langle a'|h\rangle & \langle h|a\rangle\langle a|v\rangle + \langle h|a'\rangle\langle a'|v\rangle \\ \langle v|a\rangle\langle a|h\rangle + \langle v|a'\rangle\langle a'|h\rangle & \langle v|a\rangle\langle a|v\rangle + \langle v|a'\rangle\langle a'|v\rangle \end{bmatrix} \\[2mm]
&= \begin{bmatrix} \langle h| (|a\rangle\langle a| + |a'\rangle\langle a'|) |h\rangle & \langle h| (|a\rangle\langle a| + |a'\rangle\langle a'|) |v\rangle \\ \langle v| (|a\rangle\langle a| + |a'\rangle\langle a'|) |h\rangle & \langle v| (|v\rangle\langle v| + |a'\rangle\langle a'|) |v\rangle \end{bmatrix} \\[2mm]
&= \begin{bmatrix} \langle h|h\rangle & \langle h|v\rangle \\ \langle v|h\rangle & \langle v|v\rangle \end{bmatrix} \\[2mm]
&= \begin{bmatrix} 1 & 0 \\ 0 & 1 \end{bmatrix} \\[2mm]
&= \hat{I}. \qquad\qquad\qquad\qquad\qquad\qquad\qquad (4.30)
\end{aligned}
$$

Therefore, the change of basis is a unitary transformation, i.e., a linear transformation that satisfies the property (4.29). We note that a square matrix $\hat{A}$ is called invertible if there exists a square matrix $\hat{B}$ such that

$$\hat{A}\hat{B} = \hat{B}\hat{A} = \hat{I}. \qquad (4.31)$$

If this is the case, then the matrix $\hat{B}$ that is uniquely determined by $\hat{A}$ is called the inverse of $\hat{A}$ and is denoted by $\hat{A}^{-1}$. It follows from Eq. (4.29) that a unitary operator $\hat{U}$ is always invertible and its inverse is given by $\hat{U}^{-1} = \hat{U}^\dagger$. As a result, a unitary operator represents in quantum mechanics the idea of a *reversible* transformation, since the inverse of any unitary transformation will bring back the transformed state to the original state.

It is important to note that any unitary operator can be interpreted as a rotation in the Hilbert space. In particular, the transformation from the basis $\{|a\rangle, |a'\rangle\}$ to the basis $\{|h\rangle, |v\rangle\}$ can be thought of as a rotation of the basis vectors as shown in Fig. 4.2. The second important point is closely related to the first one in that being a rotation in the Hilbert space, the unitary operator represents in quantum mechanics the transformation that preserves the scalar products of state vectors (hence also preserves both the probability amplitudes and probabilities). To understand this point, let us consider two arbitrary state vectors $|\psi\rangle$ and $|\phi\rangle$ and their respective transformed state vectors $|\psi'\rangle$ and $|\phi'\rangle$ under a unitary transformation represented by $\hat{U}$. Namely, we have

$$|\psi\rangle \xrightarrow{\hat{U}} |\psi'\rangle = \hat{U}|\psi\rangle \quad \text{and} \quad |\phi\rangle \xrightarrow{\hat{U}} |\phi'\rangle = \hat{U}|\phi\rangle. \tag{4.32}$$

The conjugate transpose of the ket $|\psi'\rangle$ is given by $\langle\psi'| = \langle\psi|\hat{U}^\dagger$, so the scalar product $\langle\psi|\phi\rangle$ transforms under $\hat{U}$ as

$$\langle\psi|\phi\rangle \xrightarrow{\hat{U}} \langle\psi'|\phi'\rangle = \langle\psi|\hat{U}^\dagger\hat{U}|\phi\rangle = \langle\psi|\hat{I}|\phi\rangle = \langle\psi|\phi\rangle. \tag{4.33}$$

Hence, the scalar product of the state vectors *after* the unitary transformation $\hat{U}$ is indeed the same as that of the state vectors *before* the transformation.

In Chapter 2 we have considered the action of both BS1 and BS2 on the components $|d\rangle$ and $|u\rangle$ [see Eqs. (2.26) and (2.27)]. In the bases $\{|d\rangle, |u\rangle\}$, we may write

$$|d\rangle = \begin{pmatrix} 1 \\ 0 \end{pmatrix} \quad \text{and} \quad |u\rangle = \begin{pmatrix} 0 \\ 1 \end{pmatrix}, \tag{4.34}$$

and express the action of the beam splitter by means of a unitary operator (also called the Hadamard operator) of the form[a]

$$\hat{U}_{\text{BS}} = \frac{1}{\sqrt{2}} \begin{bmatrix} 1 & 1 \\ 1 & -1 \end{bmatrix}. \tag{4.35}$$

---

[a](Cerf *et al.*, 1998).

Indeed, we can rewrite Eq. (2.26) as

$$\hat{U}_{BS}|d\rangle = \frac{1}{\sqrt{2}} \begin{bmatrix} 1 & 1 \\ 1 & -1 \end{bmatrix} \begin{pmatrix} 1 \\ 0 \end{pmatrix}$$

$$= \frac{1}{\sqrt{2}} \begin{pmatrix} 1 \\ 1 \end{pmatrix}$$

$$= \frac{1}{\sqrt{2}}(|d\rangle + |u\rangle). \tag{4.36}$$

We are now in a position to fully understand also the result (2.27). Indeed, we have

$$\hat{U}_{BS}|u\rangle = \frac{1}{\sqrt{2}} \begin{bmatrix} 1 & 1 \\ 1 & -1 \end{bmatrix} \begin{pmatrix} 0 \\ 1 \end{pmatrix}$$

$$= \frac{1}{\sqrt{2}} \begin{pmatrix} 1 \\ -1 \end{pmatrix}$$

$$= \frac{1}{\sqrt{2}}(|d\rangle - |u\rangle). \tag{4.37}$$

The unitarity of $\hat{U}_{BS}$ can be easily checked and is left as an exercise for the reader [see Problem 4.8]. It is also noted that $\hat{U}_{BS}$ is also Hermitian, which in turn means that $\hat{U}_{BS}$ is the reverse of itself, i.e. $\hat{U}_{BS} = \hat{U}_{BS}^{\dagger}$.

Problem 4.6 Derive the inverse change of basis by expressing the coefficients c_a and $c_{a'}$ in terms of the coefficients c_v and c_h.

Problem 4.7 Prove that $\hat{U}^{\dagger}\hat{U} = \hat{I}$ in analogy with the proof of $\hat{U}\hat{U}^{\dagger} = \hat{I}$ given by Eq. (4.30).

Problem 4.8 Check that the operator $\hat{U}_{BS}$ in Eq. (4.35) is both Hermitian and unitary.

Problem 4.9 Obtain the state $|f\rangle$ in Eq. (2.29) by applying the unitary operator $\hat{U}_{BS}$ in Eq. (4.35) to the state

$$\frac{1}{\sqrt{2}}(|d\rangle + e^{i\phi}|u\rangle).$$

Problem 4.10 Find the operator $\hat{U}_{\phi}$ representing the phase shifter PS on the *upper* path that produces a phase shift ϕ to the state $|u\rangle$, and show that the operator $\hat{U}_{\phi}$ is unitary.

4.7 Not all Operations Commute

An immediate consequence of the use of operators to describe quantum mechanical observables is that some of them do not commute with each other. Elementary algebra tells us that all (real and imaginary) numbers commute, i.e., for arbitrary two numbers a and b, we have $ab = ba$ (the order of factors in a product is irrelevant). This is not true of operators in general. Since, we have suggested to interpret operators as describing operations, we would like to introduce a very simple (classical) example of operations for understanding this property. Suppose that we are driving and executing following subsequent operations. First, we turn left and then turn right. A quick reflection will show that inverting the order of the operations (that is, turning first right and then left) will bring us to a different place (we would therefore obtain a different result). We can mathematically represent that by writing $\hat{R}\hat{L} \neq \hat{L}\hat{R}$, where we have made use of the hat, as usual, to indicate operators. Here, we also follow the usual convention to describe a succession of operations by use of multiplication (where the second operation is written to the left of the first one, and similarly for any subsequent operation).

Let us now consider a quantum mechanical example. Suppose that we have some photons all prepared in the following state along an arbitrary polarization direction **a** [see Fig. 4.3(a)]:

$$|a\rangle = c_h|h\rangle + c_v|v\rangle, \tag{4.38}$$

and we let them pass through a vertical polarization filter described by the projector

$$\hat{P}_v = |v\rangle\langle v|. \tag{4.39}$$

The action of the filter on the state $|a\rangle$ is given by

$$\hat{P}_v|a\rangle = |v\rangle\langle v| (c_h|h\rangle + c_v|v\rangle) = c_v|v\rangle, \tag{4.40}$$

since $\langle v|h\rangle = 0$ (the horizontal and vertical polarization directions are orthogonal). Therefore, only a fraction of the photons will pass through the filter (because by the assumption $c_h, c_v \neq 0$ and the normalization condition, we have $0 < |c_h|, |c_v| < 1$), and those passed through the filter will be precisely in the state of vertical

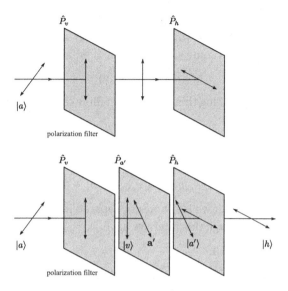

Figure 4.3 (a) An light beam in the polarization state $|a\rangle$ along an arbitrary polarization direction **a** is sent to the filter $\hat{P}_v$ with vertical polarization direction. After passing through the filter, the light beam is in the state $|v\rangle$ (with a reduced intensity, since some photons have been blocked). Then, no photon can pass through the filter $\hat{P}_h$ whose horizontal polarization direction is orthogonal to that of the photons in state $|v\rangle$. (b) If a third polarization filter $\hat{P}_{a'}$ is placed between the first and the last one, with an arbitrary polarization direction **a′** (different from **a**, **v**, and **h**), then some photons will pass through the last filter. Adapted from (Auletta *et al.*, 2009, p. 62).

polarization $|v\rangle$. Now, suppose also that we let those photons pass through a subsequent filter with horizontal polarization, which is described by the projector

$$\hat{P}_h = |h\rangle\langle h|. \tag{4.41}$$

It is easy to see that no photon will pass through the subsequent filter since

$$\hat{P}_h \left(c_v |v\rangle \right) = c_v \langle h|v\rangle |h\rangle = 0, \tag{4.42}$$

which follows from the fact that $\langle h|v\rangle = 0$ (the horizontal and vertical polarization directions are indeed orthogonal).

Let us now consider an alternative experimental setup, in which we insert a third filter between the vertical polarization and the horizontal polarization ones [see Fig. 4.3(b)]. The inserted filter is along the direction **a′** (different from **a**, **v**, and **h**) and selects polarization states in some superposition of $|h\rangle$ and $|v\rangle$ given by

$$|a'\rangle = c'_h|h\rangle + c'_v|v\rangle. \tag{4.43}$$

The projector describing this polarization filter can be written in the basis $\{|h\rangle, |v\rangle\}$ as

$$
\begin{aligned}
\hat{P}_{a'} &= |a'\rangle\langle a'| \\
&= \left(c'_h|h\rangle + c'_v|v\rangle\right)\left(c'^*_h\langle h| + c'^*_v\langle v|\right) \\
&= |c'_h|^2|h\rangle\langle h| + c'^*_vc'_h|h\rangle\langle v| + c'_vc'^*_h|v\rangle\langle h| + |c'_v|^2|v\rangle\langle v|.
\end{aligned} \tag{4.44}
$$

It is easy to see that if we let photons prepared in the state given by Eq. (4.40) pass through this filter, we have

$$
\begin{aligned}
\hat{P}_{a'}(c_v|v\rangle) &= c_v|a'\rangle\langle a'|v\rangle \\
&= c_v\left(|c'_h|^2|h\rangle\langle h| + c'^*_vc'_h|h\rangle\langle v| + c'_vc'^*_h|v\rangle\langle h| + |c'_v|^2|v\rangle\langle v|\right)|v\rangle \\
&= c_v\left(c'^*_vc'_h|h\rangle + |c'_v|^2|v\rangle\right),
\end{aligned} \tag{4.45}
$$

which is a very remarkable expression. It reveals that the action of the inserted filter on the state selected by the vertical polarization filter is again a superposition of horizontal and vertical polarization. Indeed, the numbers $c_v\left|c'_v\right|^2$ and $c_vc'^*_vc'_h$ are to be regarded simply as coefficients and the outgoing photons will be in the polarization state along the **a′** direction. This can be understood as follows. Taking into account that $c_v\left|c'_v\right|^2 = c_vc'_vc'^*_v$, we can rewrite the last line of the previous equation as

$$\hat{P}_{a'}\left(c_v|v\rangle\right) = c_vc'^*_v\left(c'_h|h\rangle + c'_v|v\rangle\right) = c_vc'^*_v|a'\rangle, \tag{4.46}$$

where in the last step we have made use of the expansion (4.43). The above result means that we have photons in the state $|a'\rangle$ as output but reduced by a factor $c_vc'^*_v$, in accordance with the analysis developed in Section 3.5. Now, if we let those photons pass through the horizontal polarization filter (represented by $\hat{P}_h$), we shall obtain

$$\hat{P}_h\left(c_vc'^*_v|a'\rangle\right) = c_vc'^*_v|h\rangle\langle h|\left(c'_h|h\rangle + c'_v|v\rangle\right) = c_vc'_hc'^*_v|h\rangle. \tag{4.47}$$

Hence, the probability that the photons will pass through the horizontal polarization filter is now given by

$$\left|c_vc'_hc'^*_v\right|^2 = \left(c_vc'_hc'^*_v\right)\left(c^*_vc'^*_hc'_v\right) = |c'_v|^2|c'_h|^2|c_v|^2, \tag{4.48}$$

which is certainly different from zero. Note that putting the filter $\hat{P}_{a'}$ between the filters $\hat{P}_v$ and $\hat{P}_h$ actually means to have interchanged the order of operations between $\hat{P}_{a'}$ and $\hat{P}_h$. Indeed, if we put $\hat{P}_{a'}$ after $\hat{P}_h$, we would not change the situation. This is because already after $\hat{P}_h$ (in the absence of a previous $\hat{P}_{a'}$) the output is zero (no photon passes the test, as shown in Fig. 4.3(a)). Thus, putting the filter $\hat{P}_{a'}$ *before* the filter $\hat{P}_h$ is equivalent to interchanging their positions. Therefore, we have proved that

$$\hat{P}_{a'}\hat{P}_h \neq \hat{P}_h\hat{P}_{a'}. \tag{4.49}$$

Consequently, we have also proved that quantum mechanical observables do not necessarily commute. This property is in sharp contrast with that of the classical quantities, which always commute. As a result, we shall refer to quantities that always commute as the c–numbers, while those not necessarily commute as the q–numbers.

We end this section by noting that non-commutativity of observables can also be obtained in terms of matrices. Since we are working in the basis $\{|h\rangle, |v\rangle\}$, we may write

$$|h\rangle = \begin{pmatrix} 1 \\ 0 \end{pmatrix} \quad \text{and} \quad |v\rangle = \begin{pmatrix} 0 \\ 1 \end{pmatrix}, \tag{4.50}$$

and express the state vectors $|a\rangle$ and $|a'\rangle$ in this basis as

$$|a\rangle = \begin{pmatrix} c_h \\ c_v \end{pmatrix} \quad \text{and} \quad |a'\rangle = \begin{pmatrix} c_h' \\ c_v' \end{pmatrix}. \tag{4.51}$$

The projectors $\hat{P}_h$, $\hat{P}_v$, and $\hat{P}_{a'}$ in the basis $\{|h\rangle, |v\rangle\}$ are respectively given by

$$\hat{P}_h = \begin{pmatrix} 1 \\ 0 \end{pmatrix} (1\ 0) = \begin{bmatrix} 1 & 0 \\ 0 & 0 \end{bmatrix}, \tag{4.52}$$

$$\hat{P}_v = \begin{pmatrix} 0 \\ 1 \end{pmatrix} (0\ 1) = \begin{bmatrix} 0 & 0 \\ 0 & 1 \end{bmatrix}, \tag{4.53}$$

and

$$\hat{P}_{a'} = \begin{pmatrix} c_h' \\ c_v' \end{pmatrix} (c_h'^*\ c_v'^*) = \begin{bmatrix} |c_h'|^2 & c_h' c_v'^* \\ c_h'^* c_v' & |c_v'|^2 \end{bmatrix}. \tag{4.54}$$

Now, we can obtain the same result by exchanging the order of the projectors $\hat{P}_h$ and $\hat{P}_{a'}$ acting on the state vector $|v\rangle$. In particular, we have

$$\hat{P}_{a'}\hat{P}_h = \begin{bmatrix} |c_h'|^2 & 0 \\ c_h'^* c_v' & 0 \end{bmatrix} \neq \begin{bmatrix} |c_h'|^2 & c_h' c_v'^* \\ 0 & 0 \end{bmatrix} = \hat{P}_h\hat{P}_{a'}. \tag{4.55}$$

Problem 4.11 Redo the above calculations in ket form by assuming that $|a'\rangle = c_v|v\rangle - c_h|h\rangle$.

Problem 4.12 Use the vector and matrix forms given by Eq. (4.50)–(4.54) to prove that

$$\hat{P}_h\left(\hat{P}_v\left(\hat{P}_{a'}|a\rangle\right)\right) = 0 \quad \text{and} \quad \hat{P}_h\left(\hat{P}_{a'}\left(\hat{P}_v|a\rangle\right)\right) \neq 0.$$

4.8 Features vs Properties

Now we face this amazing fact, which is classically inconceivable: To add an *additional* filter (represented here by $\hat{P}_{a'}$) may enhance the probability of having a detection. This is inconceivable because, classically, to insert an obstacle will *diminish*, not increase the probability of having a detection. Why do we have this extraordinary result? Because the projector $\hat{P}_{a'}$ in Eq. (4.44) has some weird cross terms, $|h\rangle\langle v|$ and $|v\rangle\langle h|$, which allow to restore a superposition of horizontal and vertical polarizations like that displayed by the state vector $|a'\rangle$ in Eq. (4.43). The consequence is that the photons that pass through the inserted filter have a non-zero probability to pass through the final horizontal polarization filter.

These cross terms do *not* exist classically. They express the circumstances that quantum components (of a superposition, for instance) are not independent of each other but allow a single non-local state (we may recall here what has been said in Section 2.5). Indeed, these cross terms describe characteristics of a state that even contribute to determine the final detection probability. But they are not properties of a quantum system. This is because any property is by definition local and is a specific value that we attribute to an observable in a determinate context, therefore it can even be associated with an event and represented by a projector. To distinguish them from properties, we shall refer to the specific quantum mechanical characteristics that generate this kind of cross terms as *features*.[a]

---

[a](Auletta/Torcal, 2010). In (Olivier/Zurek, 2001) the concept of discord and quantumness which hint at the same issue as feature have been introduced. We shall consider the relation between these concepts in Chapter 11.

4.9 Summary

In this chapter we have

- Seen the historical reasons that led to introduce the quantization principle.
- Introduced the notion of quantum observables.
- Considered the mathematical representation of observables as Hermitian operators.
- Seen that a state vector can be expanded in different orthonormal bases.
- Computed the change of basis transformation from one orthonormal basis to another.
- Learned that quantum transformations are represented by unitary operators.
- Understood that unitary transformations are reversible and preserve probabilities.
- Considered beam splitting as a unitary transformation.
- Learned that quantum observables may not commute.
- Introduced the notion of quantum features, which are non-local in nature and therefore are not properties.

Chapter 5

Complementarity Principle

In this chapter we shall come back to the issue of the undulatory nature of matter. Interferometer experiments with a blocked path show that this character cannot be taken in a classical sense. Indeed, quantum probabilities are deeply affected by the presence of features. The possibility of measurements without interaction shows that such probabilities have in quantum mechanics an ontological import. We then state a complementarity between local events and non-local features. Delayed choice experiments sets these aspects in new and enlightening terms.

5.1 Undulatory Nature of Matter

In Section 2.1 we have recalled the path-breaking developments due to de Broglie, showing that matter can assume undulatory nature. This prediction was confirmed when it was experimentally verified that electrons display a diffraction behavior[a] that is typically wave-like. Moreover, in the previous chapter we have also discovered that not only matter particles but also photons have a discontinuous

---

[a](Davisson/Germer, 1927).

Quantum Mechanics for Thinkers
Gennaro Auletta and Shang-Yung Wang
Copyright © 2014 Pan Stanford Publishing Pte. Ltd.
ISBN 978-981-4411-71-4 (Hardcover), 978-981-4411-72-1 (eBook)
www.panstanford.com

nature like corpuscles. Even more stunning is the so-called Kapitza–Dirac effect showing a characteristic reversal of the roles of matter and light: wave-like electrons are here diffracted by laser beams, where light is in a coherent state.[a] These results raise important questions about the nature of quantum entities. How to account for such a puzzling situation? The two understanding (wave-like or particle-like) of quantum systems seem to contradict each other. Indeed, classically they describe two completely different domains of physics: The corpuscular treatment is specific to mechanical systems and therefore to everything that is composed of matter. Instead, the undulatory interpretation was confined to the treatment of radiation (at least since the 19th century). This is both an interesting situation and a puzzling one, since, on the one hand, it seems to allow a unified treatment of any physical phenomenon by wiping out the sharp classical segregation of radiation from matter, and, on the other hand, it reintroduces such a dualism in the treatment of the same entity (whether matter or light).

The first consequence is that these entrenched undulatory and corpuscular characteristics of quantum systems cannot be interpreted in any classical sense. As a matter of fact, any quantum entity can produce a self–interference, which no classical wave-like entity can perform. Moreover, the discontinuities due to quantization are very different from the classical treatment of matter, since they affect the nature of the involved physical parameters (which can no longer be described by continuous variables and functions thereof, instead we need to introduce operators [see Section 4.3]).

5.2 Interferometry with a Blocked Path

To understand a little more, let us come back to the Mach–Zehnder interferometer experiment [see Sections 2.3 and 2.6–2.8], but with a very different situation.[b] We have purposedly inserted a screen in the lower path, blocking in this way the lower component of the incoming photon [see Fig. 5.1]. Let us consider how the state of the

---

[a](Freimund *et al.*, 2001).
[b](Elitzur/Vaidman, 1993).

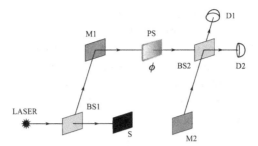

Figure 5.1 Mach–Zehnder interferometer with the lower path blocked by the screen S. The apparatus is shown in three dimensions with a top-lateral view.

photon will evolve. When the photon passes BS1, we have precisely the same situation of the experiment presented in Chapter 2. We have therefore an analogue of the transformation (2.11), which we rewrite here for the sake of convenience

$$|d\rangle \xrightarrow{\text{BS1}} \frac{1}{\sqrt{2}} (|d\rangle + |u\rangle). \tag{5.1}$$

Because of the presence of the screen S, it is evident that further evolution of the state is reduced to the component $|u\rangle$, so that we have

$$\frac{1}{\sqrt{2}} (|d\rangle + |u\rangle) \xrightarrow{\text{S}} \frac{1}{\sqrt{2}} |u\rangle$$

$$\xrightarrow{\text{PS}} \frac{e^{i\phi}}{\sqrt{2}} |u\rangle. \tag{5.2}$$

Now, the crucial point is when the photon arrives at BS2. Since the lower pathway is blocked, there is no longer interference here. Hence we have

$$\frac{e^{i\phi}}{\sqrt{2}} |u\rangle \xrightarrow{\text{BS2}} \frac{e^{i\phi}}{\sqrt{2}} (|1\rangle - |2\rangle). \tag{5.3}$$

Therefore, the final state of the photon after it leaves the beam splitter BS2, but before being detected at D1 or D2 is given by

$$|f\rangle = \frac{e^{i\phi}}{2} (|1\rangle - |2\rangle), \tag{5.4}$$

instead of the state in Eq. (2.29).

We note that the prefactor $e^{i\phi}$ in Eq. (5.4) is a global phase factor (i.e., an overall complex factor with unit modulus) since it no longer shifts the relative phase between the two components $|1\rangle$ and $|2\rangle$, which is always π independent of ϕ. Moreover, because $\left|e^{i\phi}\right|^2 = 1$ for arbitrary values of ϕ [see Eq. (2.23)], the global phase factor $e^{i\phi}$ does not change the probabilities (or the square moduli of the probability amplitudes), and therefore state vectors differing by a global phase factor only represent in fact the same physical state. Consequently, Eq. (5.4) can be simplified to

$$|f\rangle = \frac{1}{2}\left(|1\rangle - |2\rangle\right). \tag{5.5}$$

It is easy to see that the final detection probabilities at D1 and D2 will be both $\frac{1}{4}$. This is a very important difference as compared with the state in Eq. (2.29). Indeed, we had found that probabilities in Eqs. (2.33) and (2.34) imply that either D1 or D2 *never* clicks when the relative phase assumes certain limiting values [see Problem 2.8] and that, in general, the probabilities of the two detections are always different except for the case in which $\phi = \frac{\pi}{2}$. Here, on the contrary, we have that both detectors click on average with the same probability. Then, if we let several photons in the same initial state go through the apparatus we can easily apperceive that one of the two path is blocked by simply considering the final detection statistics. By setting different values of the phase shift, we shall be strongly confirmed in this supposition since there is no noticeable change in the detection statistics.

Problem 5.1 Check that for the state $|f\rangle$ in Eqs. (5.4) and (5.5) the final detection probabilities at D1 and D2 are both $\frac{1}{4}$.

Problem 5.2 Perform calculations that are similar to the above ones but assuming that instead of the lower path it is the upper one to be blocked. Compare the detection probabilities with those we have derived above. Are they the same or not? What do these cases tell you about the possibility to infer which is the blocked path?

5.3 Classical and Quantum Probability

This thought experiment shows two very important aspects of quantum theory. The first and most evident one is that quantum mechanical probabilities are inherently different from classical probabilities. Quantum probability can be reduced to a classical probability for particles in the case we have considered in the previous section, but it also can be a weird probability showing a classically inconceivable interference phenomena like it happens with the probabilities in Eqs. (2.33) and (2.34). The difference between the two is determined by the absence or presence of what we have called feature [see Section 4.8], that is, non-local interdependence between components of a quantum system, whose expression, when both paths are unblocked, is indeed represented by the interfering components at BS2. This is evident when considering that a theorem of classical probability theory is violated by quantum mechanical probability.[a]

Indeed, classically, given any two events A and B, the probability that they occur jointly is less than or equal to the sum of the probabilities that they occur separately. That is, we have [see Eqs. (2.2c) and (2.2d)]

$$\wp(A, B) \leq \wp(A) + \wp(B), \qquad (5.6)$$

where $\wp(A)$ is the probability for the event A alone to occur, $\wp(B)$ is the probability for the event B alone to occur, and $\wp(A, B)$ is the joint probability for both events A and B to occur. This is evident, since it is less probable that things occur jointly than they do not (if we throw two dice, it is more probable that we get one six than two of them). In the context of our quantum mechanical examination, we can consider the following two events:

A: The photon takes the *upper* path and D1 clicks.
B: The photon takes the *lower* path and D1 clicks.

From our discussion in the previous section, we see immediately that $\wp(A) = \frac{1}{4}$. To find $\wp(B)$, we consider the alternative case in which instead of the lower path it is the upper one to be blocked [see Problem 5.2]. It can be shown either by symmetry arguments or by explicit calculations that in this case the final detection probabilities at D1 and D2 are both $\frac{1}{4}$ as well. Hence we have $\wp(B) = \frac{1}{4}$. What is the joint probability $\wp(A, B)$, that is, the probability that the photon takes *both* the upper *and* lower paths and D1 clicks? After a little thought, we find that this is precisely the situation of the original interferometer experiment with *unblocked* paths that is considered

---

[a](Auletta *et al.*, 2009, Section 1.4).

in Sections 2.3 and 2.6–2.8. The reason for this is because the state of the photon *before* entering BS2 is given by Eq. (2.25), which is clearly a superposition of the states $|d\rangle$ and $|u\rangle$. Obviously, this cannot happen classically but it is a good quantum analogue of a joint probability for the model that we are considering here. Thus, for the joint probability we have [see Eq. (2.33) and Problem 5.3]

$$\wp(A, B) = \frac{1}{2}(1 + \cos\phi). \qquad (5.7)$$

Collecting the above results, we conclude that the inequality (5.6) is violated whenever $\wp(A, B)$ is greater than $\frac{1}{2}$ (since we always have $\wp(A) + \wp(B) = \frac{1}{2}$). It is noted that maximum violation occurs when $\phi = 0$, for which $\wp(A, B) = 1$. The reason that quantum mechanical probabilities differ inherently from the classical ones is that the former are derived from the square moduli of the corresponding probability amplitudes, which being complex numbers thus carrying phase information allow for interference, while the latter are not.

A model using a screen that obviously reduces the final detection probabilities by absorbing a part of the photons could be considered a bias. Such an objection, although true does not catch the essence of the above argument, since we have learned from the previous chapter that, in a quantum mechanical context, to add an obstacle does not necessarily reduce the probability to get some final detection events. Indeed, as shown in Section 4.7, inserting a polarization filter in the direction **a'** between a vertical and a horizontal filter amounts to having prepared the photons in the state $|a'\rangle$ [see Eq. (4.43)], instead of having projected the photons into the state of vertical polarization $|v\rangle$. However, as we have seen, the probability of detecting a photon after the horizontal polarization filter is higher in the first case. Therefore, the above objection only takes into account the particular context in which we have argued, but it is not a threat to its general conclusion.

This brings us to another important aspect of quantum theory. The fact that the quantum detection statistics is altered by the sole presence of an obstacle tells us that we can know in many cases that there is an obstacle in one of the two paths without having a single photon had interacted with it. Let us suppose that we set the phase shifter PS so that D2 should never click if there is interference at BS2 (for instance, when $\phi = 0$). Now, if there is an obstacle preventing such an interference, we have learned that D1 will click with a probability of $\frac{1}{4}$ (or an occurrence fraction of $\frac{1}{4}$ if we send several photons). This is amazing, since the photons that have been detected are those that have taken the *other* path, that is the upper one, avoiding the obstacle. In other words, we have here a measurement *without interaction*, what is called an interaction-free measurement. This was considered something impossible from a classical point of view until the 1980s. Before that time it was assumed that in order to detect something one needed to interact with it and so at least to acquire a photon! The assumption that there could be no measurement of an object without some kind of interaction is a consequence of the idea that everything could be explained through mechanical causes only [see Section 1.1].

The kind of thought experiment described in the previous section has been also experimentally performed and represents a new important development of quantum mechanics. The fact that quantum mechanics allows for measurements without interaction leads to new technological possibilities. We shall mention here one of particular relevance. This is the possibility to *see in the dark*,[a] that is, to take photographs of objects in full darkness. We may indeed think of many parallel interferometer "dead" paths each one leading to a point of the object of which we like to take a picture and associated with a pixel of the final image. This kind of technology can be particular useful for exploration in space and also for medical purposes, when we need to take a picture of a tissue or an organ that is particularly photosensitive.

Problem 5.3 Perform calculations by using the probabilities that detector D2 clicks and show that the results are essentially the same as those developed in this section.

5.4 Double Slit Experiment

There is a famous thought (also experimentally realized) experiment, the so-called double slit experiment, which is a full analogue of the Mach–Zehnder apparatus presented in Section 5.2. Let us first

---

[a](Kwiat *et al.*, 1996).

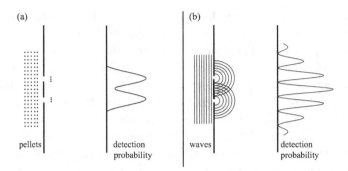

Figure 5.2 A schematic top view of the double slit experiment. (a) The configuration corresponding to classical particles. (b) The configuration corresponding to classical waves. The curve shows the mean of the probability distribution.

consider the classical particle case in which we send macroscopic pellets through an apparatus consisting of two slits and a screen detecting their final positions [see Fig. 5.2(a)]. It is clear that in this case when pellets are fired at random from the left, they will enter the apparatus through *one* of the two slits and reach the detection screen. On average they will have a maximum probability to end in the two stripes that are right behind the slits. Indeed, the probability curves are here represented by the bell-like shapes (or, more accurately, the Poisson distributions) that feature a diminishing probability when moving away from the center. The two probabilities here superpose slightly in the center of the screen, which is the situation that we shall expect when the two slits are sufficiently near. However, when we increase their distance, the two distributions no longer superpose.

Then we consider the classical wave case, for instance sending light through the slits [see Fig. 5.2(b)], which corresponds to the quantum mechanical case in which the two paths of an interferometer are unblocked, we will have a completely different situation. The two component waves going out of the two slits interfere, so now the detection probability shows characteristic bumps even in the places that no classical particles could reasonably reach. Quantum mechanically, this setup corresponds to the detection of quantum wave-like behavior, as displayed in Fig. 2.7. If we like to detect the complementary behavior, that is, the particle-like one, we need to close one slit (which corresponds to the case in which one path of the interferometer is blocked). In this case, there is no longer interference and the photons are detected at the point in which we would expect them also in the classical particle situation.

From the previous examination we draw two very important consequences that we have already stated but without justifying them until now. In classical physics, probabilities only express our *subjective* ignorance about a system. Let us consider the flip of a coin. The outcome of coin flipping is a random event since we cannot predict whether it is a head or a tail. However, this is only because it is very difficult to ascertain all the factors involved in the occurrence of the event. However, from a classical-mechanical point of view we can assume that we can build a very powerful machine able to monitor the whole dynamics of the coin, included a perfect knowledge of the wind, and we can further assume that we are able to perfectly know the weight and even the smallest imperfections of the coin; then, at least in principle, we can completely predict the outcome. This is precisely what is inherently impossible in quantum mechanics. The superposition state $|f\rangle$ in Eq. (5.1) implies that *both* components $|d\rangle$ and $|u\rangle$ are present, although in terms of the probability amplitudes, and it is therefore meaningless to ask whether the photon has taken the lower or the upper path. Therefore, quantum probabilities are irreducible and objective. They are *irreducible* because there is no deterministic descriptions to which quantum probabilities can be reduced. They are *objective* since quantum probabilities express states of affairs and not states of subjective ignorance or technological inability.

The crucial point is to understand which kind of objectivity we are speaking about, a problem that will occupy us across the book. Ultimately, we have to do with probability amplitudes, from which the probabilities are derived. Is there any sense in which they could be said to be real? In Section 4.8 we have introduced the concept of features as those quantum factors responsible for the typical quantum effects. As a matter of fact, if their presence or absence determines different detection statistics, they cannot be sort of ghosts lost in the borderline between being and not being. Or are they?

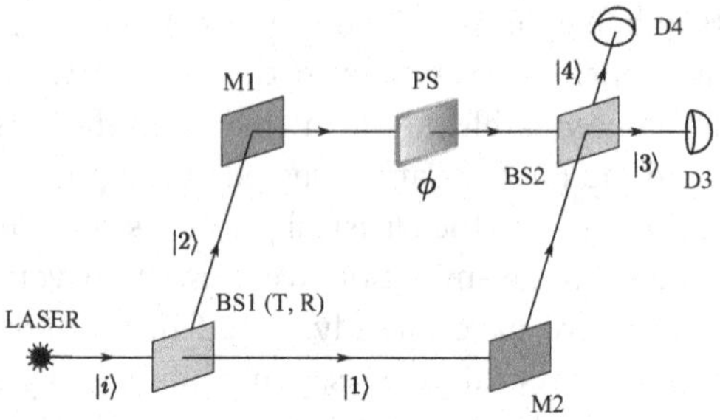

Figure 5.3 Mach–Zehnder interferometer with a variable beam splitter BS1, showing a smooth complementarity between wave-like and particle-like behavior.

Problem 5.4 Verify that the probabilities introduced in the previous section can be applied to the three cases considered here: (a) the classical double slit particle experiment, (b) the classical (and quantum) double slit wave experiment, and (c) the quantum (but also classical) single slit particle experiment.

5.5 Path Predictability and Interference Visibility

To ascertain the problem raised at the end of the previous section, let us consider the arrangement of the Mach–Zehnder interferometer shown in Fig. 5.3. Compared with the previous experiments, the novelty is that now we have the first beam splitter that is no longer a 50–50 one. Instead, suppose that the beam splitter transmits and reflects the incoming photon with probabilities given by T^2 and R^2, respectively (we shall assume that the respective transmission and reflection coefficients T and R are non-negative real numbers for the sake of simplicity). We have also relabeled the states associated with the lower and upper paths as $|1\rangle$ and $|2\rangle$, respectively (and the states associated with the paths leading to the detectors as $|3\rangle$ and $|4\rangle$). In this case, for a photon initially in the state $|i\rangle$ ($= |1\rangle$) the action of BS1 and PS can be written as

$$|i\rangle \xrightarrow{\text{BS1}} T|1\rangle + R|2\rangle$$

$$\xrightarrow{\text{PS}} T|1\rangle + e^{i\phi}R|2\rangle. \tag{5.8}$$

Making use of the following expressions for the action of BS2

$$|1\rangle \xrightarrow{\text{BS2}} \frac{1}{\sqrt{2}}(|3\rangle + |4\rangle) \quad \text{and} \quad |2\rangle \xrightarrow{\text{BS2}} \frac{1}{\sqrt{2}}(|3\rangle - |4\rangle), \quad (5.9)$$

we have

$$T|1\rangle + e^{i\phi}R|2\rangle \xrightarrow{\text{BS2}} \frac{1}{\sqrt{2}}\left[T(|3\rangle + |4\rangle) + e^{i\phi}R(|3\rangle - |4\rangle)\right]. \quad (5.10)$$

Upon collecting the terms, we find the final state of the photon after leaving BS2 is given by

$$|f\rangle = \frac{1}{\sqrt{2}}\left[\left(T + e^{i\phi}R\right)|3\rangle + \left(T - e^{i\phi}R\right)|4\rangle\right]. \quad (5.11)$$

Hence, the final detection probabilities at D3 and D4 are respectively given by [see Problem 5.7]

$$\wp_3 = \frac{1}{2}(1 + 2TR\cos\phi) \quad \text{and} \quad \wp_4 = \frac{1}{2}(1 - 2TR\cos\phi). \quad (5.12)$$

In obtaining the above expressions, we have used the condition $T^2 + R^2 = 1$.

We may choose various values for the coefficients T and R of BS1 as well as for the phase shift ϕ of PS, provided that the condition $T^2 + R^2 = 1$ is satisfied. Let us now introduce a simplification and consider the case in which the phase shift $\phi = 0$ (in radians). In the limiting case in which $T = 0$ (and $R = 1$), the photon is totally reflected and we know for sure that it has taken the upper path. In this case, the first beam splitter behaves as an ordinary mirror. In the opposite limiting case in which $R = 0$ (and $T = 1$), the photon is fully transmitted and therefore it takes the lower path with certainty. In this case it is as if we had taken the first beam splitter away. It is evident that in these two limiting cases we have situations with full predictability of the path followed by the photon, which are analogous to the presence of an obstacle discussed in the previous sections (since after passing BS1 the photon still has a single component). We can mathematically quantify this path predictability by the absolute value of the difference between the probabilities for the photon to be reflected and transmitted, that is,

$$\mathcal{P} = |T^2 - R^2|, \quad (5.13)$$

where the absolute value is justified by the fact that we do not know *a priori* which one of the probabilities (T^2 or R^2) is larger.

The reason of the above formula is that $\mathcal{P}$ has the maximum value 1 when one of the two probabilities is 1 (and the other is 0), and the minimum value 0 when the two probabilities are equal, that is, $T^2 = R^2 = \frac{1}{2}$. Then, a pure particle-like (or local) behavior has $\mathcal{P}$ maximized whereas a pure wave-like (or non-local) behavior has $\mathcal{P}$ minimized. In other words, because of the presence of non-local quantum feature [see Section 4.8], in order to localize the photon we need to be able to *discriminate* between the two paths.

The previous equation can be easily derived as follows. Considering always the case where the phase shift $\phi = 0$, the probability amplitudes (up to a global phase factor) that the photon is detected at D3 and D4 are respectively given by [see Eq. (5.11)]

$$\vartheta_3 = \sqrt{\wp_3} = \frac{1}{\sqrt{2}}|T + R|, \qquad (5.14a)$$

and

$$\vartheta_4 = \sqrt{\wp_4} = \frac{1}{\sqrt{2}}|T - R|. \qquad (5.14b)$$

The product of the two probability amplitudes gives

$$\vartheta_3\vartheta_4 = \frac{1}{2}|T + R||T - R| = \frac{1}{2}|T^2 - R^2|, \qquad (5.15)$$

which, apart from the factor $\frac{1}{2}$, is equal to the path predictability $\mathcal{P}$ in Eq. (5.13). Since the latter is maximum when one of the two probabilities (either T^2 or R^2) is zero, so is the product of the two probability amplitudes $\vartheta_3\vartheta_4$. This is again a peculiarity of quantum probability since classically the product of two mutually exclusive probabilities is maximum when the two probabilities are equal.

On the other hand, we can also introduce a complementary parameter to quantify the visibility of interference fringes. Since the path predictability is maximum when one of the two coefficients (either T or R) is zero, the interference visibility should be proportional to the product of the two coefficients. In other words, interference fringes are visible only in the case in which the path predictability is not maximum. Conventionally, the interference visibility is defined by

$$\mathcal{V} = 2TR, \qquad (5.16)$$

which, as we will see below, expresses precisely the presence of quantum feature. Indeed, in the probabilities $\wp_3$ and $\wp_4$ in Eq. (5.12), the quantity V represents precisely the quantum feature that is added to the purely classical part represented by the number $\frac{1}{2}$ (in the simplest case where $\phi = 0$). It is interesting to see that the interference visibility can be extracted from the difference between the probabilities $\wp_3$ and $\wp_4$, that is,

$$\wp_3 - \wp_4 = \frac{1}{2}\left[(T + R)^2 - (T - R)^2\right]$$
$$= 2TR, \tag{5.17}$$

where we have assumed again that $\phi = 0$. This is understandable since the purely classical part cancels out in the difference of the two quantum probabilities.

From Eq. (5.16) we see that V has the maximum value 1 when the two coefficients T and R are equal, that is, $T = R = \frac{1}{\sqrt{2}}$, and the minimum value 0 when one of the two coefficients is 1 (and the other is 0). This, together with the corresponding behavior of $\mathcal{P}$, allows us to conclude that the path predictability and the interference visibility are complementary quantities in the sense that the path predictability is gained at the expense of the interference visibility, and vice versa. To put it another way, any distinguishability between the paths of an interferometer destroys the visibility quality of the interference fringes. Indeed, it is also easy to verify that

$$\mathcal{P}^2 + V^2 = T^4 + R^4 - 2T^2R^2 + 4T^2R^2$$
$$= \left(T^2 + R^2\right)^2$$
$$= 1. \tag{5.18}$$

This important equation is called the *Greenberger–Yasin equality*,[a] which means that the visibility of interference and the predictability of path are complementary although strictly connected, and even both present in most cases, apart from the limiting ones (in which one of the quantities is zero). In other words, for a photon in an interferometer between the two purely classical behaviors (when T, R $= 0, 1$) there is continuous range of non-local quantum mechanical behaviors (when $0 < T, R < 1$).

The novel results that we have found so far can be summarized as follows:

---

[a] (Greenberger/Yasin, 1988).

(i) Quantum mechanically, there is continuous range of behaviors for a photon in an interferometer that lies between a sort of pure particle-like behavior (which has the maximum path predictability) and a pure wave-like behavior (which has the maximum interference visibility).

(ii) The wave-like and particle-like behaviors of a photon are complementary.

Most people think that a particle going along a certain path is undeniable real. If, however, this behavior is entrenched with non-local feature that is responsible for the interference effects, there is no particular reason not to assign also some status of reality to the latter. Therefore, this examination suggests a smooth complementarity between the wave-like and particle-like behaviors, which we would like to posit with a complementarity principle.

Principle 5.1 (Complementarity Principle) *Local events and non-local features are complementary.*

The exact meaning of the words *local event* and *non-local feature* will be the object of a further thought experiment that we shall discuss below. We also recall that the complementarity principle, although with a different formulation, was communicated to the physics community by Bohr during the Como Conference in 1927 and later on in a paper published in the journal *Nature*.[a]

Problem 5.5 Explain why we have imposed the requirement that $T^2 + R^2 = 1$.

Problem 5.6 Make all the derivation above by making use of the unitary transformation (4.35) for BS2, the result of Problem 4.10, and the following transformation for BS1

$$\hat{U}_{\mathrm{BS1}} = \begin{bmatrix} T & R \\ R & -T \end{bmatrix}, \tag{5.19}$$

where $T^2 + R^2 = 1$. As usual, we have in component form $|i\rangle = |1\rangle = \binom{1}{0}$ and $|2\rangle = \binom{0}{1}$.

Problem 5.7 Derive the probabilities $\wp_3$ and $\wp_4$ in Eq. (5.12) and compare them with the probabilities $\wp_1$ and $\wp_2$ in Eqs. (2.33) and (2.34).

---

[a](Bohr, 1928).

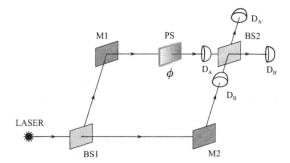

Figure 5.4 In a delayed choice experiment we can choose at the last moment whether to place the detectors before the beam splitter BS2 (as D_A and D_B) or after the beam splitter BS2 (as $D_{A'}$ and $D_{B'}$).

Problem 5.8 Show that the operator $\hat{U}_{BS1}$ in Eq. (5.19) is unitary.

5.6 Delayed Choice Experiment

Let us consider another variant of the interferometer experiments, in particular the setup shown in Fig. 5.4. It is related to the experiment we have considered in Section 5.5, so we only need to grasp the general idea here. Suppose that we perform a usual interferometer experiment like those reported several times in the book *until* the photon's components have been reflected by the two mirrors and the upper one has passed the phase shifter. The novelty comes in when we decide *just before* those components arrive at the beam splitter BS2 whether to place the detectors in the usual position after BS2 or to put them before it. We now consider each case in turn.

- In the first case, we should detect the known interference profile, provided that there is no obstacle in one of the paths [see Fig. 2.7]. Here, we cannot predict which path the photon has taken.
- In the latter position, we prevent the interference at BS2 and are able to tell with certainty which path has been taken by the photon (if it is detected at D_A, it has taken the upper path; if it is detected at D_B, it has taken the lower path).

To a certain extent this ideal experiment is a setup for testing complementarity. However, there is another subtler aspect: there is a certain time, i.e., between the photon passing through BS1 and just before the photon reaching BS2 (or at least before we place the detectors in one of the two alternative positions), in which no (detection) event takes place. We have sent a photon (one at a time) in the interferometer and the photon will be finally detected, no matter how we place the detectors. The obvious conclusion is that there must be some reality in the apparatus, since a photon cannot be lost in the land of nothingness and come out again from this obscure region. However, the problem is that here no event takes place. So, what is the minimum kind of reality that we are allowed to attribute to this situation? Certainly not the single components of the photon (nor the possible paths it could take) since we have seen that in the interference-like setup we can tell nothing about the path (and so *a fortiori* nothing about the components themselves). The only minimum reality is therefore the interdependence between the two components themselves, what we have called *feature*, which we can assume to be present in both alternative experimental setups until the photon reaches a detector in one of the two setups. Since features precisely consist in the long-range non-separability of the superposition states, we are conclusively authorized to say that they are deeply *non-local*, in accordance with our examination in Sections 2.5 and 4.8. Therefore, we are forced to acknowledge that the world is not only made up of local events but also of non-local correlations, and the specific quantum mechanical characteristics (the quantumness) of the latter are represented by features.

On the other hand, when we place the detectors before BS2 and (apparently) destroy the quantum features, it is evident that this gives rise to events that are *local*. Therefore, such local events are incompatible with the presence of non-local features. It is true that when we place the detectors after BS2, we are able to reconstruct an interference profile. However, the single detection event always reduces a superposition state of the photon to one of its component states, so that also in this case it remains true that a (detection) event is incompatible with the presence of quantum feature and their manifestation (interference). Therefore, we have confirmed that local events can be taken to be (smoothly) complementary to non-local features as stated by the complementarity principle.

Problem 5.9 Describe the delayed choice experiment in mathematical terms by making use of the formalism developed in Sections 2.3–2.6. Distinguish between the case in which the detectors are located after BS2 and the case in which they are located before BS2. In the first case, take advantage of the discussion in Section 2.7. In the second case, make use of the examination developed in Section 5.2.

5.7 Summary

In this chapter we have

- Considered the Mach–Zehnder interferometer experiment with a blocked path.
- Learned that quantum probabilities are inherently different from classical probabilities.
- Interpreted the Mach–Zehnder interferometer experiment with a blocked path as a case of measurement without interaction.
- Established the complementarity between path predictability and interference visibility.
- Found out the complementarity between local events and non-local features.
- Demonstrated the reality of non-local features through a delayed choice experiment.

Part II

Formal Issues: Observables

Chapter 6

Position and Momentum

With this second part we shall start more formal developments that allow us to improve our understanding of quantum theory. In particular, we shall deal with the two most basic quantum mechanical observables, position and momentum, which are crucial for understanding the state and the dynamics of a quantum system. Different from classical mechanics, quantum mechanics tells us that it is not possible to measure both of the position and momentum simultaneously. This is a consequence of the non-commutativity of these two observables. Moreover, in this chapter we shall learn the basic formalism of integration and differentiation, a fundamental mathematical tool if one is really interested in understanding quantum mechanics.

6.1 Position Operator: Discrete Case

Until now we have considered observables in general and mostly treated examples dealing with photon polarization or with interferometer experiments. In the last chapter we also introduced the path predictability $\mathcal{P}$ for a photon in an interferometer. However, we have not considered the position operator (or position observable) in all

Quantum Mechanics for Thinkers
Gennaro Auletta and Shang-Yung Wang
Copyright © 2014 Pan Stanford Publishing Pte. Ltd.
ISBN 978-981-4411-71-4 (Hardcover), 978-981-4411-72-1 (eBook)
www.panstanford.com

its generality. Let us introduce its one-dimensional expression as $\hat{x}$ (describing the position of a one-dimensional system along a certain x axis). We first analyze the very elementary case represented precisely by the path predictability, i.e., the case in which the position observable can only occupy two positions: the upper or lower path in an interferometer. In this case, we can associate the upper path with a numerical value, say 1, and the lower path with another numerical value, say -1.[a] For instance, we can conceive the vertical parts of the upper and lower paths as the vertical lines described by the equations $x = 1$ and $x = -1$, respectively (obviously, the lengths of the paths have no meaning here). Then, according to the discussion presented in Section 4.4, we can write the position operator $\hat{x}$ as

$$\hat{x} = |u\rangle\langle u| - |d\rangle\langle d|. \tag{6.1}$$

It is easy to see that the states $|u\rangle$ and $|d\rangle$ are eigenstates of the position operator $\hat{x}$ with corresponding eigenvalues given by 1 and -1, respectively. In other words, we have [see Eq. (4.4)]

$$\hat{x}|u\rangle = |u\rangle \quad \text{and} \quad \hat{x}|d\rangle = -|d\rangle, \tag{6.2}$$

where use has been made of the orthogonality condition $\langle u|d\rangle = 0$. Moreover, theaction of the position operator $\hat{x}$ on a generic state vector $|\psi\rangle$ is given by [see also Eqs. (3.33) and (3.34)]

$$\hat{x}|\psi\rangle = |u\rangle\langle u|\psi\rangle - |d\rangle\langle d|\psi\rangle, \tag{6.3}$$

where $\langle u|\psi\rangle$ and $\langle d|\psi\rangle$ are the probability amplitudes associated with the probability of finding the system in positions $|u\rangle$ and $|d\rangle$, respectively, when we perform a position measurement on the state $|\psi\rangle$.

In general cases, however, we expect that a certain quantum system may occupy more than two positions, perhaps many. We can generalize Eq. (6.1) to the case with n possible positions $x_0, x_1, \ldots, x_{n-1}$ and write the position operator as [see Box 3.2]

$$\hat{x} = x_0|x_0\rangle\langle x_0| + x_1|x_1\rangle\langle x_1| + \cdots + x_{n-1}|x_{n-1}\rangle\langle x_{n-1}|$$

$$= \sum_{j=0}^{n-1} x_j|x_j\rangle\langle x_j|. \tag{6.4}$$

---

[a]We note that this kind of association is completely arbitrary and that this particular association can be realized by connecting the detectors D_A and D_B in Fig. 5.4 to a computer, which is programed in such a way that its display shows 1 when D_A clicks and -1 when D_B clicks.

The real number x_j is the eigenvalue of the position operator $\hat{x}$ and the state $|x_j\rangle$ is the corresponding eigenstate since we have

$$\hat{x}|x_j\rangle = x_j|x_j\rangle, \tag{6.5}$$

where use has been made of the orthonormal condition [see Eq. (3.29)]

$$\langle x_j|x_k\rangle = \delta_{jk}. \tag{6.6}$$

It is evident that Eq. (6.5) is the eigenvalue equation of the position operator in its generality [see again Eq. (4.4)]. If, through a measuring device, the system is found to have a position x_j, then the position state of the system is given by $|x_j\rangle$. Similarly, for an arbitrary state of the system $|\psi\rangle$ we have

$$\hat{x}|\psi\rangle = x_0|x_0\rangle\langle x_0|\psi\rangle + x_1|x_1\rangle\langle x_1|\psi\rangle + \cdots + x_{n-1}|x_{n-1}\rangle\langle x_{n-1}|\psi\rangle$$

$$= \sum_{j=0}^{n-1} x_j|x_j\rangle\langle x_j|\psi\rangle, \tag{6.7}$$

where again the scalar product $\langle x_j|\psi\rangle$ represents the probability amplitude of finding the system in position x_j. These probability amplitudes are the coefficients in the expansion of the state vector $|\psi\rangle$ in the eigenbasis $\{|x_j\rangle\}$ (where $0 \leq j \leq n-1$) of the position operator $\hat{x}$ [see Section 4.5]. Indeed, we have

$$|\psi\rangle = \sum_{j=0}^{n-1} \langle x_j|\psi\rangle|x_j\rangle, \tag{6.8}$$

where $|\psi\rangle$ is again an arbitrary state vector of the system. Upon interchanging the order of the bracket $\langle x_j|\psi\rangle$ and the ket $|x_j\rangle$ in each term on the right-hand side of Eq. (6.8), we can rewrite the above expression as

$$|\psi\rangle = \sum_{j=0}^{n-1} |x_j\rangle\langle x_j|\psi\rangle, \tag{6.9}$$

which implies that

$$\sum_{j=0}^{n-1} |x_j\rangle\langle x_j| = \hat{I}. \tag{6.10}$$

This is precisely the completeness relation for the (discrete) position eigenstates.

Finally, we note that the above results of Eqs. (6.4), (6.6), and (6.10) are very general in nature and can be generalized to all observables. For instance, we can consider an arbitrary observable $\hat{O}$ with discrete eigenvalues o_j and the corresponding eigenstates $|o_j\rangle$ (where j is again some index labeling the eigenvalues). Obviously, the eigenvalues o_j need to be real numbers and the eigenstates $|o_j\rangle$ satisfy the orthonormal conditions [see Eq. (3.29)]

$$\langle o_j|o_k\rangle = \delta_{jk}. \tag{6.11}$$

Then, it can be shown that the following completeness relation holds:

$$\sum_j |o_j\rangle\langle o_j| = \hat{I}. \tag{6.12}$$

The orthonormal conditions (6.11) and completeness relation (6.12) imply that the eigenstates $|o_j\rangle$ constitute a complete orthonormal basis. Moreover, the observable $\hat{O}$ can be expressed as

$$\hat{O} = \sum_j o_j |o_j\rangle\langle o_j|. \tag{6.13}$$

Since the set of all eigenvalues of an observable is called its spectrum, the above expression is referred to as the *spectral decomposition* of the observable $\hat{O}$ as well as Eq. (6.4) represents the spectral decomposition of the one–dimensional position observable in particular.

Problem 6.1 Consider a quantum system occupying four possible positions, write down in component form the eigenbasis of the position operator $\hat{x}$. Write down in component form an arbitrary ket $|\psi\rangle$ in terms of this eigenbasis.

6.2 From Summation to Integration

It is quite common that quantum systems may potentially occupy many (or in fact infinite) positions in a continuous way. A typical example is the so-called free particle, i.e., a particle that does not interact with other systems or is not subject to external forces [see Section 7.6]. This is a kind of confirmation in negative way of Principle 4.1. In these cases, a summation will not work and we need another mathematical tool, namely, integration.

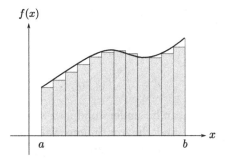

Figure 6.1 The integral of a function $f(x)$ over x from a to b is the net signed area bounded by the graph of $f(x)$, the x axis, and the vertical lines $x = a$ and $x = b$. This area can be approximated by small rectangles.

Let us consider the function $f(x)$ whose graph is shown in Fig. 6.1. We would like to calculate the *net signed* area A that is bounded by the graph of $f(x)$, the x axis, and the vertical lines $x = a$ and $x = b$ (with $a < b$). A region above the x axis is considered positive area, while a region below the x axis is considered negative area. A very intuitive way to do this is to partition this area in very small rectangles. To this end, we first subdivide the interval $[a, b]$ into n equal subintervals by means of the intermediate points $x_1, x_2, \ldots, x_{n-1}$ and set $a = x_0$ and $b = x_n$. The length in each subinterval is given by the quantity $\Delta x = x_{j+1} - x_j$ (where $j = 0, 1, \ldots, n - 1$). If we consider the n rectangles of width Δx and height $f(x_j)$, we shall approximate the area A by summing over the area of the n rectangles, that is,

$$A \approx \sum_{j=0}^{n-1} f(x_j)\Delta x. \tag{6.14}$$

The symbol $\approx$ means that the quantity on its left is approximately equal to the quantity on its right. Evidently, the approximation becomes better as we increase the number of subintervals, n. In the limiting case in which n approaches infinity, the rectangles become so tiny that their upper or lower sides coincide with the graph of $f(x)$. Hence we have precisely made a summation on the continuum and obtained the area A as an infinite series [see Box 3.2], that is,

$$A = \lim_{n \to \infty} \sum_{j=0}^{n-1} f(x_j)\Delta x, \tag{6.15}$$

where the notation $\lim_{n\to\infty}$ denotes the limit as n approaches infinity. It is a notational as well as a conceptual convenience to define the infinite series on the right-hand side of Eq. (6.15) by an integral

$$A = \int_a^b f(x)dx, \qquad (6.16)$$

where the symbol $\int$ can be understood as a stretched sigma (Σ), hinting at continuity and the continuous variable x replace the discrete parameter x_j. The above expression precisely means that we have performed the integral of the function $f(x)$ over x from a to b, where a and b are respectively the lower and upper limits of the integration. In mathematical terms, the function $f(x)$ that is being integrated is called the integrand, the variable x is the variable of integration, the closed interval $[a, b]$ is the interval of integration (which includes also the endpoints a and b of the interval), and dx is the differential of x, which for the moment can just be thought of as the continuum counterpart of Δx and whose significance will be clear below. It will be important to note that the variable of integration x is a "dummy" variable in that the integral itself is not a function of this dummy variable, but of the lower and upper limits it takes. We could have written the integral in Eq. (6.16) as

$$\int_a^b f(t)dt, \quad \int_a^b f(x')dx', \quad \text{or} \quad \int_a^b f(y)dy. \qquad (6.17)$$

All of the above represent the same integral since they have the same lower and upper limits.

There are many properties of integrals, here we summarize the most basic ones. The first property is linearity, which means that

$$\int_a^b [\alpha f(x) + \beta g(x)]dx = \alpha \int_a^b f(x)dx + \beta \int_a^b g(x)dx, \qquad (6.18)$$

where α, β are (real) constants. Another one is that reversing the limits of integration changes the sign of the integral, that is,

$$\int_a^b f(x)dx = -\int_b^a f(x)dx. \qquad (6.19)$$

This, together with setting $a = b$, implies

$$\int_a^a f(x)dx = 0. \qquad (6.20)$$

The third property states that if $a < c < b$, then we can decompose the interval of integration $[a, b]$ into two subintervals:

$$\int_a^b f(x)dx = \int_a^c f(x)dx + \int_c^b f(x)dx. \qquad (6.21)$$

Moreover, integrals are invariant under translations, i.e.,

$$\int_a^b f(x)dx = \int_{a+c}^{b+c} f(x-c)dx, \qquad (6.22)$$

where $f(x - c)$ is the function obtained from $f(x)$ under a translation to the right by a distance c. Finally, integrals satisfy the inequality

$$\left| \int_a^b f(x)dx \right| \le \int_a^b |f(x)|dx, \qquad (6.23)$$

which is called the triangle inequality for integrals. We note that all of the above properties can be proved by using the definition of integral (6.15).

If the function $f(x)$ is a continuous real-valued function, then the integral in Eq. (6.16) may be readily calculated provided that the primitive function (or indefinite integral [see Box 6.1]) of the function $f(x)$ is known. Let the latter be denoted by $F(x)$, then we have

$$\int_a^b f(x)dx = F(x) \Big|_{x=a}^{x=b}$$

$$= F(b) - F(a), \qquad (6.24)$$

where on the right-hand side of the first equality a long vertical bar with a superscript and a subscript is a shorthand notation denoting that the difference is taken between the primitive function evaluated at the upper and lower limits of integration. In other words, the integral of the function $f(x)$ over x from a to b is calculated as the *difference* of the primitive function $F(x)$ evaluated at the upper limit b (i.e. $F(b)$) and at the lower limit a (i.e. $F(a)$). The primitive functions of some elementary functions are listed in Table 6.1. In particular, from the primitive function for the power function we have

$$\int_a^b x^\alpha dx = \frac{x^{\alpha+1}}{\alpha+1} \Big|_{x=a}^{x=b} = \frac{b^{\alpha+1} - a^{\alpha+1}}{\alpha+1} \quad (\alpha \ne -1) \qquad (6.25)$$

Table 6.1 Primitive functions of some elementary functions

Type	Function	Primitive function
Constant	1	x
Power	x^α	$\dfrac{x^{\alpha+1}}{\alpha+1} \quad (\alpha \neq -1)$
	x^{-1}	$\ln|x|$
Exponential	e^x	e^x
Trigonometric	$\sin x$	$-\cos x$
	$\cos x$	$\sin x$

Box 6.1 Definite and indefinite integrals

The integral in Eq. (6.16) with definite lower and upper limits is referred to as the *definite integral*. Since the definite integral of a function $f(x)$ is related to its primitive function $F(x)$, we can think of this relationship as an association between $f(x)$ and $F(x)$. This leads to another kind of integral called the *indefinite integral*. An indefinite integral of a function $f(x)$ is written as

$$\int f(x)dx, \tag{6.26}$$

i.e., using an integral symbol with no limits. This association however is not unique because a constant added to a primitive function will still correspond to the same definite integral. Hence, the indefinite integral of a function $f(x)$ is often written in the form

$$\int f(x)dx = F(x) + c, \tag{6.27}$$

where $F(x)$ is the primitive function of $f(x)$ and c is an arbitrary constant known as the constant of integration. For this reason, the primitive function $F(x)$ is also called the indefinite integral of the function $f(x)$. We note that for the sake of convenience both the definite and indefinite integrals are usually referred to simply as the integral whenever no confusion may arise or the specific meaning can be determined from the context.

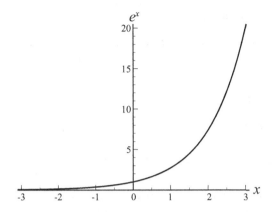

Figure 6.2 Plot of the exponential function e^x.

and

$$\int_a^b \frac{1}{x}dx = \ln|x|\Big|_{x=a}^{x=b} = \ln|b| - \ln|a| = \ln\left|\frac{b}{a}\right| \quad (ab > 0), \quad (6.28)$$

where $\ln x$ is the logarithmic function with base e.[a] The above result is important because power functions appear throughout mathematics and physics. The exponential function e^x is important in part because it is the only function which is its own primitive. Hence, we have

$$\int_a^b e^x dx = e^b - e^a. \quad (6.29)$$

A plot of the exponential function is shown in Fig. 6.2 [see also Box 2.4]. Moreover, the integrals of the sine and cosine functions can be calculated in a similar manner, the results are given by

$$\int_a^b \cos x dx = \sin b - \sin a \quad \text{and} \quad \int_a^b \sin x dx = -(\cos b - \cos a),$$
$$(6.30)$$

respectively.

The concept of integral can be generalized straightforwardly to functions of several variables, for instance, $f(x, y)$ or $f(x, y, z)$.

---

[a]We will discuss the logarithmic function in more detail in Section 9.6. By now, let us take it as a particular kind of function having the properties $\ln x + \ln y = \ln(xy)$ and $\ln x - \ln y = \ln\left(\frac{x}{y}\right)$.

Box 6.2 Examples of integration

Equation (6.25) allows us to present some easy examples with numbers. For instance, consider the following integral

$$I_1 = \int_{-2}^{2} x \, dx = \frac{x^2}{2} \Big|_{x=-2}^{x=2} = \frac{2^2}{2} - \frac{(-2)^2}{2} = 0. \qquad (6.31)$$

This is a particular simple case since the graph of the function $f(x) = x$ is a straight line at 45 degrees, which allows us to compute the area through traditional geometric means and to make a comparison. Indeed, the area defined by the above integral actually consists of two identical triangles, one lies above the x axis and the other lies below. Since for each triangle the base is 2 and the height is 2, its area is equal to $2 \times 2/2 = 2$. However, for the triangle that lies below the x axis, its area picks up a minus sign and hence is equal to -2. Therefore, the net signed area is zero, which agrees with the result obtained by direct integration. Another example is the following integral:

$$I_2 = \int_{1}^{3} \sqrt{x} \, dx = \int_{1}^{3} x^{\frac{1}{2}} dx = \frac{2x^{\frac{3}{2}}}{3} \Big|_{1}^{3} = \frac{2}{3}\left(3^{\frac{3}{2}} - 1^{\frac{3}{2}}\right) = 2\sqrt{3} - \frac{2}{3},$$
$$(6.32)$$

where for the sake of notational simplicity the superscript and subscript of the long vertical bar have been simplified and use has been made of the properties

$$a^{\frac{1}{2}} = \sqrt{a} \quad \text{and} \quad a^{\frac{3}{2}} = \sqrt{a^3} = a\sqrt{a}, \qquad (6.33)$$

which are valid for arbitrary $a \geq 0$.

Integrals of a function of two variables are called double integrals, and those of a function of three variables are called triple integrals.

Problem 6.2 Compute the following integrals:

$$\frac{1}{4}\int_{-2}^{3} x^3 dx, \quad \int_{0}^{\pi} \sin x \, dx, \quad \text{and} \quad \int_{0}^{t} e^{kx} dx \quad (k \neq 0).$$

Problem 6.3 Evaluate the following integral:

$$\int_{0}^{1} \frac{3x}{x^2 + x - 2} dx.$$

(*Hint*: Use partial fraction expansion and the property that integrals are invariant under translations (6.22).)

6.3 Position Operator: Continuous Case

We can now go over from the discrete case of the position operator discussed in Section 6.1 to the continuous case. Let us consider a quantum system that may occupy infinite possible positions in the interval $-L \leq x \leq L$ in a continuous manner. Again, subdividing the interval $[-L, L]$ into n equal subintervals by means of the intermediate points $x_1, x_2, \ldots, x_{n-1}$ and setting $x_0 = -L$ and $x_n = L$, we can approximately expand the state vector $|\psi\rangle$ in a superposition of the position eigenstates $|x_j\rangle$ (with $j = 0, 1, \ldots, n - 1$) as [see Eqs. (6.8) and (6.14)]

$$|\psi\rangle \approx \sum_{j=0}^{n-1} |x_j\rangle\langle x_j|\psi\rangle \Delta x, \qquad (6.34)$$

where $\Delta x = x_{j+1} - x_j$. In the limit that n approaches infinity, we recover the continuum limit and obtain that [see Eq. (6.15)]

$$|\psi\rangle = \lim_{n\to\infty} \sum_{j=0}^{n-1} |x_j\rangle\langle x_j|\psi\rangle \Delta x, \qquad (6.35)$$

or in terms of integral [see Eq. (6.16)],

$$|\psi\rangle = \int_{-L}^{L} |x\rangle\langle x|\psi\rangle dx, \qquad (6.36)$$

where we recall that dx for the moment can be thought of as the continuum counterpart of Δx. In the above expression the state $|x\rangle$ is the eigenstate of the position operator (or position eigenstate for short) and its corresponding eigenvalue is given by x (where $-L \leq x \leq L$). That is, we have the eigenvalue equation

$$\hat{x}|x\rangle = x|x\rangle, \qquad (6.37)$$

which is a generalization of Eq. (6.5) to the continuous case. The reader could now get confused by the fact that we have also used kets of similar type, like $|h\rangle$ and $|v\rangle$, but for the discrete (two-dimensional) case. How can we distinguish the latter kets from the kets like $|x\rangle$? Actually, there is no problem here, since the kind of (whether continuous or discrete) expansion of a state vector that is allowed or even necessary depends on the problem at hand.

In the case of a free particle the interval of integration is in fact the whole interval of real numbers from minus infinity (the limit of the descending series of negative real numbers) to plus infinity (the limit of the ascending series of positive real numbers). Upon taking the limit that L approaches infinity, we can rewrite Eq. (6.36) as

$$|\psi\rangle = \int_{-\infty}^{+\infty} |x\rangle\langle x|\psi\rangle dx, \qquad (6.38)$$

which in turn implies the following completeness relation for the (continuous) position eigenstates [see Eq. (6.10)]

$$\int_{-\infty}^{+\infty} |x\rangle\langle x|dx = \hat{1}. \qquad (6.39)$$

Likewise, the spectral decomposition [see Section 6.1] of the position operator $\hat{x}$ is given by

$$\hat{x} = \int_{-\infty}^{+\infty} x|x\rangle\langle x|dx, \qquad (6.40)$$

whose validity will be justified shortly. It is noted that the scalar product $\langle x|\psi\rangle$ in Eq. (6.38) is the continuous expansion coefficient of the state vector $|\psi\rangle$ in the position operator eigenbasis $\{|x\rangle\}$ (where $-\infty < x < \infty$) and hence, for a given state vector $|\psi\rangle$, can be understood as a continuous function of the position operator eigenvalue x. Indeed, the scalar product $\langle x|\psi\rangle$ is referred to as the *wave function in position space*. Here by position space we mean the real space as we know it. In other words, $\langle x|\psi\rangle$ describes the state of a quantum system as a function of the position and therefore can be written in the usual function form as

$$\psi(x) = \langle x|\psi\rangle. \qquad (6.41)$$

In terms of the wave function $\psi(x)$, Eq. (6.38) can be rewritten as

$$|\psi\rangle = \int_{-\infty}^{+\infty} \psi(x)|x\rangle dx. \qquad (6.42)$$

We note that, unlike the discrete case, the square modulus of the probability amplitude $|\langle x|\psi\rangle|^2$ does *not* give the probability of finding a quantum system *at* the position x. As a matter of fact, in the continuum limit such a probability is identically zero because it requires infinite precision to exactly identify a real number (we encounter here a first limitation to the assumption of error-free measurement in classical mechanics [see Section 1.2]). Instead, we speak of the probability of finding a quantum system *in proximity of* the position x, or more precisely between x and $x + dx$, where dx (which we have already met in an integral like that in Eq. (6.42)) is an arbitrarily small number called the *differential* of x. Such a probability is instead given by the square modulus of the probability amplitude multiplied by this arbitrarily small number, i.e., $|\psi(x)|^2 dx$. Note that here we should take $dx > 0$ since the probability is always non-negative. Therefore, in the continuous case the square modulus of the wave function $|\psi(x)|^2$ gives probability density instead of probability. It is also important to note that the case in which the position eigenvalue x lies between minus and plus infinity is actually the most general situation in one dimension that goes far beyond the free particle case. The special case in which a particle is confined in a certain region can always be recovered by requiring that the corresponding wave function vanishes identically outside that region.

The orthonormal conditions of the position eigenstates $|x\rangle$ are given by

$$\langle x | x' \rangle = \delta(x - x'), \tag{6.43}$$

where $\delta(x - x')$ is the Dirac delta function (or delta function for short), which is the continuous analogue of the Kronecker delta [see Eq. (3.29)]. The definition and some of the most important properties of the delta function can be found in Box 6.3. With the help of the orthonormal conditions (6.43), we can now justify the validity of Eq. (6.40) by using it to derive Eq. (6.37):

$$\hat{x} |x\rangle = \left(\int_{-\infty}^{+\infty} x' |x'\rangle \langle x'| dx' \right) |x\rangle$$

$$= \int_{-\infty}^{+\infty} x' |x'\rangle \langle x'|x\rangle dx'$$

$$= \int_{-\infty}^{+\infty} x' |x'\rangle \delta(x' - x) dx'$$

$$= x |x\rangle, \tag{6.44}$$

Box 6.3 Delta function

The delta function $\delta(x)$ can be thought of as an infinitely high, infinitely thin spike at the origin, with unit area under the spike. Therefore, it has two defining properties:

$$\delta(x) = 0 \quad \text{for all} \quad x \neq 0, \quad \text{and} \quad \int_{-\infty}^{+\infty} \delta(x)dx = 1. \quad (6.45)$$

There are many properties of the delta function, here we summarize the most important ones. The fundamental property of the delta function is that

$$\int_{-\infty}^{+\infty} f(x)\delta(x - a)dx = f(a), \quad (6.46)$$

where $f(x)$ is a continuous function. It is noted that this property is tantamount to the replacement

$$f(x)\delta(x - a) = f(a)\delta(x - a). \quad (6.47)$$

The delta function also satisfies the following scaling property

$$\delta(ax) = \frac{1}{|a|}\delta(x), \quad (6.48)$$

where a is a non-zero real number. In particular, the delta function is an even function, in the sense that

$$\delta(-x) = \delta(x). \quad (6.49)$$

A very useful integral representation of the delta function is given by

$$\delta(x - a) = \frac{1}{2\pi} \int_{-\infty}^{+\infty} e^{ik(x-a)}dk. \quad (6.50)$$

This expression is of great use when we discuss the eigenfunctions of the momentum observable in Section 6.6.

where the dummy variable of integration has been changed to x' so as to avoid confusion with the position eigenvalue x, and use has been made of Eq. (6.46).

Finally, let us consider the following expression

$$\langle x|\hat{x}|\psi\rangle = \int_{-\infty}^{+\infty} x'\langle x|x'\rangle\langle x'|\psi\rangle dx'$$

$$= \int_{-\infty}^{+\infty} x'\delta(x'-x)\langle x'|\psi\rangle dx'$$

$$= x\langle x|\psi\rangle, \tag{6.51}$$

where again we have used Eqs. (6.39), (6.43), and (6.46). We recall that $\langle x|\hat{x}|\psi\rangle$ is a shorthand notation for the scalar product of the kets $|x\rangle$ and $\hat{x}|\psi\rangle$, i.e., $\langle x|\hat{x}|\psi\rangle = \langle x|\hat{x}\psi\rangle$, where $|\hat{x}\psi\rangle = \hat{x}|\psi\rangle$ [see Eq. (3.54)]. Since Eq. (6.51) is valid for arbitrary $|\psi\rangle$, it follows that

$$\langle x|\hat{x} = x\langle x|, \tag{6.52}$$

which we note can also be obtained by taking the conjugate transpose of Eq. (6.37).

The meaning of Eqs. (6.37) and (6.52) is that when the position eigenstates $|x\rangle$ are used as the basis states, the action of the position operator $\hat{x}$ in this basis is just like that of a real number x. Here and henceforth, the use of the position eigenstates as basis states to represent operators and states is referred to as the *position representation*. Similarly, the wave function $\psi(x) = \langle x|\psi\rangle$ is the state $|\psi\rangle$ expressed in the position representation. While there are other representations in quantum mechanics, the position representation is the most intuitive and hence the most often used one. Therefore, whenever no confusion may arise, use of the latter in many cases is implicitly assumed and in this way we can conveniently write Eq. (6.51) as

$$\hat{x}\psi(x) = x\psi(x). \tag{6.53}$$

It is noted that the scalar product $\langle x|x_0\rangle$ for fixed x_0 is called the *position eigenfunction*, that is, the particular position eigenstate $|x_0\rangle$ expressed in the position representation. Writing $\varphi_{x_0}(x) = \langle x|x_0\rangle$, from the orthonormal conditions (6.43) we have

$$\varphi_{x_0}(x) = \delta(x - x_0). \tag{6.54}$$

The corresponding eigenvalue equation for $\varphi_{x_0}(x)$ is given by

$$\hat{x}\varphi_{x_0}(x) = x_0\varphi_{x_0}(x), \tag{6.55}$$

which can be obtained by right multiplying Eq. (6.52) by $|x_0\rangle$ and then using Eq. (6.47).

The generalization to the three-dimensional case is straightforward. In Cartesian coordinates, the position operators $\hat{\mathbf{r}}$ is a three-component vector operator

$$\hat{\mathbf{r}} = (\hat{x}, \hat{y}, \hat{z}), \tag{6.56}$$

which has been written here in the canonical row–vector formulation that is also usual in classical mechanics. The position eigenstate is now denoted by

$$|\mathbf{r}\rangle = |x, y, z\rangle, \tag{6.57}$$

which satisfies the eigenvalue equation

$$\hat{\mathbf{r}}|\mathbf{r}\rangle = \mathbf{r}|\mathbf{r}\rangle, \tag{6.58}$$

where the eigenvalues $\mathbf{r} = (x, y, z)$ (with $-\infty < x, y, z < \infty$) represent the possible position (Euclidean) vectors that a quantum system may potentially occupy. The use of the position eigenstates $|\mathbf{r}\rangle$ as basis states to represent operators and states is the position representation in three dimensions. In particular, we have

$$\langle\mathbf{r}|\hat{\mathbf{r}} = \mathbf{r}\langle\mathbf{r}|, \tag{6.59}$$

namely, the action of the position operator $\hat{\mathbf{r}}$ in the position representation is just like that of a spatial vector $\mathbf{r}$. The scalar product $\psi(\mathbf{r}) = \langle\mathbf{r}|\psi\rangle$ is the three-dimensional wave function, and the position eigenfunction is given by

$$\varphi_{\mathbf{r}_0}(\mathbf{r}) = \delta^{(3)}(\mathbf{r} - \mathbf{r}_0), \tag{6.60}$$

where

$$\delta^{(3)}(\mathbf{r} - \mathbf{r}') = \delta(x - x')\delta(y - y')\delta(z - z') \tag{6.61}$$

is the three-dimensional delta function. Moreover, the completeness relation and orthonormal conditions are respectively given by

$$\int d^3r |\mathbf{r}\rangle\langle\mathbf{r}| = \hat{1} \quad \text{and} \quad \langle\mathbf{r}|\mathbf{r}'\rangle = \delta^{(3)}(\mathbf{r} - \mathbf{r}'), \tag{6.62}$$

where

$$\int d^3r = \int_{-\infty}^{+\infty} dx \int_{-\infty}^{+\infty} dy \int_{-\infty}^{+\infty} dz \qquad (6.63)$$

means the integral is taken over the whole three-dimensional space.

Problem 6.4 Show that in the position representation the normalization condition $\langle \psi | \psi \rangle = 1$ can be written in one dimension and three dimensions as

$$\int_{-\infty}^{+\infty} |\psi(x)|^2 dx = 1 \quad \text{and} \quad \int |\psi(\mathbf{r})|^2 d^3r = 1,$$

respectively. Explain how the above equations justify that $|\psi(x)|^2$ and $|\psi(\mathbf{r})|^2$ give the probability density.

Problem 6.5 Suppose that a quantum system of a one-dimensional particle is described by the wave function

$$\psi(x) = \mathcal{N} e^{-\lambda |x|/\hbar} \quad (-\infty < x < \infty),$$

where $\lambda > 0$ is a constant. Find the normalization constant $\mathcal{N}$, which can be chosen to be real and positive. (*Hint*: Use the normalization condition and take advantage of the formula $\int_0^\infty e^{-\alpha x} dx = \frac{1}{\alpha}$, where $\alpha > 0$.)

6.4 Derivatives: From Finite to Infinitesimal Quantities

We would like now to study the other fundamental basic observable of quantum (and classical) mechanics called momentum.[a] In order to deal with the concept of momentum, we need first to introduce the mathematical notion of derivative. Loosely speaking, a derivative can be thought of as a measure of how much a function is changing in response to changes in its argument. For instance, as we will see below, the derivative of the position of a moving object with respect to time is the object's instantaneous velocity and the derivative of the latter with respect to time is the acceleration of the object.[b]

---

[a] It would be appropriate for the reader to take a breath before proceeding further.
[b] (Cullerne/Machacek, 2008, pp. 6–9).

Let us consider a function $f(x)$ and we would like to know how $f(x)$ changes with respect to x. The simplest case is when $f(x)$ is a linear function of x, meaning that the graph of $f(x)$ is a straight line. In this case, we have $f(x) = ax$, where the constant a is the slope of the line. Let Δx and Δf denote the change in x and the corresponding change in $f(x)$, respectively. Then the ratio of the differences (or difference quotient) $\frac{\Delta f}{\Delta x}$ is given by

$$
\begin{aligned}
\frac{\Delta f}{\Delta x} &= \frac{f(x + \Delta x) - f(x)}{\Delta x} \\
&= \frac{a(x + \Delta x) - ax}{\Delta x} \\
&= \frac{a\Delta x}{\Delta x} \\
&= a,
\end{aligned}
\tag{6.64}
$$

which is precisely the slope of the line. It is noted that for the above equation to be valid we must have $\Delta x \neq 0$. Moreover, in this simplest case $\frac{\Delta f}{\Delta x}$ is independent of both x and Δx, meaning that the line has a constant slope a so that Δf is proportional to Δx with the proportional constant given by a.

Now comes the question of how to go beyond the simplest linear case? Again, we consider the simplest non-linear case in which $f(x)$ is a quadratic function of x, namely, $f(x) = ax^2$. The graph of the function $f(x)$ is a parabola as shown in Fig. 6.3. The difference quotient $\frac{\Delta f}{\Delta x}$ is given by

$$
\begin{aligned}
\frac{\Delta f}{\Delta x} &= \frac{f(x + \Delta x) - f(x)}{\Delta x} \\
&= \frac{a(x + \Delta x)^2 - (ax)^2}{\Delta x} \\
&= \frac{2ax\Delta x + (\Delta x)^2}{\Delta x} \\
&= 2ax + \Delta x.
\end{aligned}
\tag{6.65}
$$

It is noted that $\frac{\Delta f}{\Delta x}$ now depends on both x and Δx. The dependence on x is not unexpected, as a non-linear function does not possess a constant slope but there are many tangents to the curve, as it is evident by an inspection of Fig. 6.3. However, the dependence on Δx is troublesome since it means that in determining $\frac{\Delta f}{\Delta x}$ there is

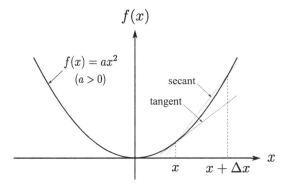

$$f(x) = ax^2$$
$$(a > 0)$$

secant

tangent

$$x \qquad x + \Delta x \qquad x$$

Figure 6.3 Geometrically, the derivative of a function at a point equals the slope of the tangent line to the graph of the function at that point.

arbitrariness in the choice of Δx (in other words, we could choose an interval Δx of any size). As a result, the quantity $\frac{\Delta f}{\Delta x}$ cannot be an intrinsic property of the function $f(x)$. In order to get rid of this unwanted dependence of Δx, we take the limit in which Δx becomes smaller and smaller (but still remains nonzero). In this limit we obtain an intrinsic property of the function $f(x)$, called the *derivative* of $f(x)$ with respect to x. In mathematical terms, the latter is defined by

$$\frac{df}{dx} = \lim_{\Delta x \to 0} \frac{\Delta f}{\Delta x}, \tag{6.66}$$

where the notation $\lim_{\Delta x \to 0}$ denotes the limit Δx is so exceedingly small that it is virtually zero (but is not exactly equal to zero). We note that a quantity that is explicitly nonzero and yet smaller in absolute value than any real positive quantity is called an *infinitesimal*. While the use of the notation $\frac{df}{dx}$ to denote the derivative of $f(x)$ with respect to x seems at first obscure, its significance will be clear below. By now, we remark that we have already met the quantity dx as the limiting quantity of a finite interval Δx [see comments to Eq. (6.16)]. We have also called dx the *differential* of x, i.e., the infinitesimal limit of Δx, when discussing the probability density [see p. 119]. Something similar also happens for the infinitesimal limit of Δf, and the resulting infinitesimal quantity df is called the differential of the function f. Therefore, as the notation suggests, the derivative $\frac{df}{dx}$ of a

Table 6.2 Derivatives of some elementary functions. It is noted that as can be seen clearly by comparing with Table 6.1, differentiation is the inverse operation of integration

Type	Function	Derivative
Power	x^α	$\alpha x^{\alpha-1}$
Exponential	e^x	e^x
Logarithmic	$\ln x$	x^{-1}
Trigonometric	$\sin x$	$\cos x$
	$\cos x$	$-\sin x$

function $f(x)$ with respect to x can be thought of as the ratio of two infinitesimal quantities df and dx called the differentials.

If the limit on the right-hand side of Eq. (6.66) exists, then $f(x)$ is said to be differentiable at x. The process of finding a derivative is called *differentiation*. From Eq. (6.65) and the above definition of derivative, we find that the derivative of $f(x) = ax^2$ with respect to x is given by

$$\frac{df}{dx} = \frac{d(ax^2)}{dx} = \lim_{\Delta x \to 0} (2ax + \Delta x) = 2ax. \qquad (6.67)$$

Geometrically, the derivative of a function at a point equals the slope of the tangent line to the graph of the function at that point [see Fig. 6.3]. Since a point has dimension zero, we begin to grasp the concept of infinitesimal. Note also that the derivative of a straight line is its slope, in accordance with Eq. (6.64):

$$\frac{d(ax)}{dx} = a. \qquad (6.68)$$

In general, the derivative of a power function is given by

$$\frac{d(ax^n)}{dx} = nax^{n-1}. \qquad (6.69)$$

The derivative of a function can, in principle, be computed from the definition (6.66) by considering the difference quotient and computing its limit. In practice, once the derivatives of a few simple functions are known, the derivatives of other functions are more easily computed by using rules for obtaining derivatives of more complicated functions from simpler ones. The derivatives of some elementary functions are listed in Table 6.2. Special attention needs to be paid to the exponential function e^x [see Fig. 6.2], which is the only function whose derivative is itself. That is, we have [see again Table 6.2]

$$\frac{de^x}{dx} = e^x. \tag{6.70}$$

The most basic derivative rules are summarized below.

- Constant rule: the derivative of a constant is zero, i.e.,

$$\frac{df}{dx} = 0 \quad \text{if } f(x) = \text{constant.} \tag{6.71}$$

- Sum rule: if $f(x)$ and $g(x)$ are differentiable functions, then

$$\frac{d}{dx}(af + bg) = a\frac{df}{dx} + b\frac{dg}{dx}, \tag{6.72}$$

where a and b are constants.

- Product rule: if $f(x)$ and $g(x)$ are differentiable functions, then

$$\frac{d(fg)}{dx} = \frac{df}{dx}g + f\frac{dg}{dx}. \tag{6.73}$$

- Quotient rule: if $f(x)$ and $g(x)$ are differentiable functions with $g(x) \neq 0$, then

$$\frac{d}{dx}\left(\frac{f}{g}\right) = \frac{1}{g^2}\left(\frac{df}{dx}g - f\frac{dg}{dx}\right). \tag{6.74}$$

- Chain rule: if $y = f(u)$ and $u = g(x)$ are differentiable functions, then the derivative of the composite function $y(x) = f(g(x))$ with respect to x is given by

$$\frac{dy}{dx} = \frac{dy}{du}\frac{du}{dx}. \tag{6.75}$$

Two examples of the application of derivative rules can be found in Box. 6.4.

What precedes shows that the procedure of differentiation can be reiterated. The derivative of $\frac{df}{dx}$, if it exists, is denoted by $\frac{d^2f}{dx^2}$ and is called the second derivative of $f(x)$. For instance, if $2ax$ is the derivative of the function ax^2, then $2a$ is its second derivative. Similarly, the derivative of the second derivative of $f(x)$, if it exists, is denoted by $\frac{d^3f}{dx^3}$ and is called the third derivative of $f(x)$. These repeated derivatives are called higher-order derivatives. For obvious reason the derivative $\frac{df}{dx}$ is also referred to as the first derivative. For

---

Box 6.4 Examples of differentiation

As specific examples, let us calculate the derivative of the functions $f(x) = x \sin x^2$ and $g(x) = \frac{\sin x^2}{x}$. Using the product rule (6.73), we have

$$\frac{d}{dx}(x \sin x^2) = \frac{dx}{dx} \sin x^2 + x \frac{d}{dx} \sin x^2$$

$$= \sin x^2 + x \frac{d}{dx} \sin x^2. \qquad (6.76)$$

To compute $\frac{d}{dx} \sin x^2$, we use the chain rule (6.75) by rewriting $\sin x^2$ as a composite function of $y = \sin u$ and $u = x^2$, and obtain

$$\frac{d}{dx} \sin x^2 = \frac{du}{dx} \frac{d \sin u}{du}$$

$$= 2x \cos u$$

$$= 2x \cos x^2, \qquad (6.77)$$

where in obtaining the last equality we have replaced u by x^2. Collecting the results, we find

$$\frac{d}{dx}(x \sin x^2) = \sin x^2 + 2x^2 \cos x^2. \qquad (6.78)$$

Similarly, we use the quotient rule (6.74) to obtain

$$\frac{d}{dx}\left(\frac{\sin x^2}{x}\right) = \frac{1}{x^2}\left(x \frac{d}{dx} \sin x^2 - \frac{dx}{dx} \sin x^2\right)$$

$$= \frac{1}{x^2}(2x^2 \cos x^2 - \sin x^2), \qquad (6.79)$$

where use has been made of Eq. (6.77). A few more examples can be found at the end of this section as a problem and are left as an exercise for the reader [see Problem 6.6].

---

the sake of notational simplicity, the derivative of a function $f(x)$ is often denoted by $f'(x)$. Similarly, the second and third derivatives are denoted by $f''(x)$ and $f'''(x)$, respectively.

We are now in a position to reveal the significance of the notation $\frac{df}{dx}$ as well as to make a direct connection between differentiation and integration. As mentioned, the derivative $\frac{df}{dx}$ can be thought of as the ratio of two infinitesimals df and dx called the differentials.

Let us consider an infinitesimal change dx in the variable x. The corresponding infinitesimal change df in the function $f(x)$ is given by

$$df = \frac{df}{dx}dx. \tag{6.80}$$

Integrating the above equation over x from a to b on the right-hand side, and over f from the corresponding $f(a)$ to $f(b)$ on the left-hand side, we obtain (by inverting the two sides)

$$\int_a^b \frac{df}{dx}dx = \int_{f(a)}^{f(b)} 1\, df = f(b) - f(a), \tag{6.81}$$

where in the second equality we have used the fact that the primitive function of unity is f when the latter is the variable of integration (as it is evident from the first line of Table 6.1). Hence, we conclude that the primitive function of $\frac{df}{dx}$ is $f(x)$, or in terms of the indefinite integral [see Box 6.1]

$$\int \frac{df}{dx}dx = f(x). \tag{6.82}$$

Since the integral of the derivative of a function is equal to the function, we are authorized to conclude that the indefinite integral of a function $f(x)$ is a function $F(x)$ [see Eqs. (6.24) and (6.27)] whose derivative is equal to $f(x)$, i.e.,

$$\frac{dF}{dx} = f(x). \tag{6.83}$$

We remark that since the derivative of a constant is zero (according to the constant rule of derivatives), Eq. (6.27) easily follows. Therefore, we have definitely established that differentiation is the inverse operation of integration.

We recall that in classical mechanics, a linear function is used to describe the distance traveled by an object moving at a *constant velocity* and a quadratic function is used to describe the distance traveled by an object moving at a *constant acceleration*. Specifically, let us consider the one-dimensional case in which an object moves along the x axis and has position $x(t)$ at time t. The velocity $v(t)$ of the object is the derivative of $x(t)$ with respect to t (or the time derivative of position)

$$v(t) = \frac{dx}{dt}. \tag{6.84}$$

In other words, the time derivative of position expresses how the position of a certain object varies in time, which is precisely the intuitive definition of velocity. If an object moves at a velocity $v(t)$, the displacement of the object between $t = 0$ and t is given by

$$x(t) - x(0) = \int_0^t v(t')dt'. \tag{6.85}$$

which for a constant velocity $v(t) = v$ reduces to

$$x(t) = vt + x_0, \tag{6.86}$$

where $x_0 = x(0)$ is the initial position of the object at $t = 0$. We note that in order to avoid possible confusion, the dummy variable of integration in the time integral on the right-hand side of Eq. (6.85) has been changed to t'. Moreover, the acceleration $a(t)$ of the object is the derivative of its velocity $v(t)$ with respect to t (or the time derivative of velocity)

$$a(t) = \frac{dv}{dt}, \tag{6.87}$$

which, together with Eq. (6.84), means that

$$a(t) = \frac{d}{dt}\left(\frac{dv}{dt}\right) = \frac{d^2x}{dt^2}. \tag{6.88}$$

Therefore, acceleration is the second time derivative of position. If an object moves at an acceleration $a(t)$, the change in the object's velocity between $t = 0$ and t is given by

$$v(t) - v(0) = \int_0^t a(t')dt', \tag{6.89}$$

which for a constant acceleration $a(t) = a$ reduces to

$$v(t) = at + v_0, \tag{6.90}$$

where $v_0 = v(0)$ is the initial velocity of the object at $t = 0$. Upon substituting $v(t)$ into Eq. (6.86), we obtain the familiar result

$$x(t) = \frac{1}{2}at^2 + v_0t + x_0, \tag{6.91}$$

which describes the position of an object moving at a constant acceleration as a function of time (the factor $\frac{1}{2}$ in the quadratic term is due to mathematical reasons, see Box 6.5).

Problem 6.6 Compute the derivatives of the following functions:

$$\sqrt{x}, \quad x\cos x, \quad \frac{\sin x}{x}, \quad \text{and} \quad \frac{1}{1 + \sqrt{x}}.$$

Box 6.5 Taylor series

The considerations at the end of Section 6.4 allow for a generalization. In specific, a function can be represented by an infinite series that are calculated from the values of the function's derivatives at a given point [see also Box 3.2]. The resulting series is called the Taylor series (or Taylor expansion) of the function about that point. In particular, the Taylor series of the function $f(x)$ about the point $x = a$ is given by

$$f(x) = f(a) + f'(a)(x - a) + \frac{f''(a)}{2!}(x - a)^2$$

$$+ \frac{f'''(a)}{3!}(x - a)^3 + \ldots = \sum_{n=0}^{\infty} \frac{f^{(n)}(a)}{n!}(x - a)^n, \quad (6.92)$$

where $n! = n \times (n-1) \times \cdots 1$ (with $0! = 1$) denotes the factorial of n and $f^{(n)}(a)$ denotes the nth derivative of $f(x)$ evaluated at the point $x = a$. For $x - a \ll 1$ (i.e., for $x - a$ much smaller than 1) we may keep only the first few terms in the expansion and use the resulting polynomial as an *approximation* for $f(x)$. An important example is the Taylor series for the exponential function e^x about $x = 0$:

$$e^x = 1 + x + \frac{x^2}{2!} + \frac{x^3}{3!} + \cdots = \sum_{n=0}^{\infty} \frac{x^n}{n!}. \quad (6.93)$$

The above expansion holds because, as can be seen in Table 6.2, the derivative of e^x with respect to x is also e^x and $e^0 = 1$. This leaves x^n in the numerator and $n!$ in the denominator for each term in the infinite sum. In the limit x approaches zero, we can simply keep the lowest order term in the above expansion and obtain

$$e^x \approx 1 + x \quad \text{for} \quad x \to 0, \quad (6.94)$$

which will be of great use in the discussion below for the momentum operator [see Section 6.6]. Similarly, the trigonometric functions $\sin x$ and $\cos x$ in the limit x approaches zero (measured in radians) can be approximated by

$$\sin x \approx x, \quad \cos x \approx 1 \quad \text{for} \quad x \to 0, \quad (6.95)$$

which again will be of great use in our discussion below for the angular momentum operator [see Section 8.1].

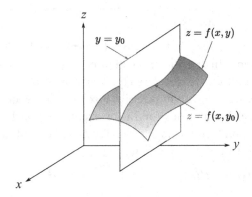

Figure 6.4 Partial derivative with respect to x of the function $z = f(x, y)$, whose graph is represented here by the surface in grey scale. The partial derivative with respect to x with $y = y_0$ held constant is the slope of the tangent to the curve $z = f(x, y_0)$, which is the intersection of the plane $y = y_0$ and the surface $z = f(x, y)$.

Problem 6.7 Use the Taylor series of the exponential function (6.93) to verify that $\frac{d}{dx} e^x = e^x$.

6.5 Partial and Total Derivatives

The concept of derivative can be generalized to functions of several variables, for instance, $f(x, y)$ and $f(x, y, z)$. This is a necessary step for dealing with a three-dimensional treatment of quantum systems (and therefore also with observables like the angular momentum, as we shall see). In these cases, we distinguish between *total* and *partial* derivatives. We shall first deal with the latter kind and come back to total derivatives at the end of this section. In specific, a partial derivative of a function of several variables is its derivative with respect to one of those variables while the others are held constant. Like ordinary derivatives, the partial derivative also has a geometric representation as the slope of the tangent at a given point. Suppose that $f(x, y)$ is a function of two variables. The graph of $f(x, y)$ defines a surface $z = f(x, y)$ in a three-dimensional space as shown in Fig. 6.4. To every point on this surface, there is an infinite number of tangent lines. Partial differentiation corresponds to selecting a particular one of these lines and finding its slope. For the partial derivative of $f(x, y)$ with respect x, we are considering y as constant and x as variable, and for partial derivative with respective to y, we are considering x as constant and y as variable. In the former case, we are able to trace a curve $z = f(x, y_0)$ (in other words, we are selecting a single curve out of a surface) and the partial derivative $\partial f(x, y)/\partial x$ taken at a certain point gives the slope of the tangent to the curve at that point along the x direction [see Fig. 6.4]. Similarly, in the latter case we are able to trace a curve $z = f(x_0, y)$, and the partial derivative $\partial f(x, y)/\partial y$ taken at a certain point gives the slope of the tangent to the curve at that point along the y direction. The rules for partial derivatives are the same as those for ordinary derivatives, except that the variables that are held constant are treated as ordinary constants. Higher-order partial derivatives like $\partial^2 f/\partial x \partial y$, $\partial^2 f/\partial y \partial x$, $\partial^2 f/\partial^2 x$, $\partial^2 f/\partial^2 y$, etc., can be considered in a similar manner as for ordinary derivatives. A very useful property of the mixed second partial derivatives is that

$$\frac{\partial^2 f}{\partial x \partial y} = \frac{\partial^2 f}{\partial y \partial x}, \tag{6.96}$$

provided that the mixed derivatives are continuous. An example of partial differentiation can be found in Box. 6.6.

Gradient is a very important differential operator in mathematics and physics. The gradient of a function at a point is the vector that points in the direction of maximum rate of increase of the function at that point, and whose magnitude is precisely that rate of increase [see Fig. 6.5]. Examples are represented by the gravitational potential (in this case the negative of the gradient points towards the gravitational field) or the concentration of a chemical (here the gradient points towards the maximum concentration). The gradient of a function $f(x, y, z)$ is denoted by ∇f, where the symbol ∇ is called the gradient operator (or *del*, or *nabla*). The form of the gradient ∇f depends on the coordinate system used. In Cartesian coordinates, it is defined in terms of partial derivative operators by

$$\nabla = \mathbf{e}_x \frac{\partial}{\partial x} + \mathbf{e}_y \frac{\partial}{\partial y} + \mathbf{e}_z \frac{\partial}{\partial z}, \tag{6.97}$$

where we recall that $\mathbf{e}_x$, $\mathbf{e}_y$, and $\mathbf{e}_z$ are unit vectors in the x, y, and z directions, respectively. In component form the gradient operator is

Box 6.6 Example of partial differentiation

As an example of partial derivatives, let us consider the function

$$f(x, y) = x \sin y + y^2 \cos xy.$$

In this case, we can compute the two partial derivatives

$$\frac{\partial f(x, y)}{\partial x} = \sin y - y^3 \sin xy, \tag{6.98a}$$

$$\frac{\partial f(x, y)}{\partial y} = x \cos y + 2y \cos xy - xy^2 \sin xy, \tag{6.98b}$$

where the term $-y^3 \sin xy$ in Eq. (6.98a) results from applying the chain rule (6.75), while the last two terms in Eq. (6.98b) are the result of the application of the product rule (6.73) and the chain rule (6.75). Similarly, the mixed second partial derivatives are given by

$$\frac{\partial^2 f(x, y)}{\partial y \partial x} = \cos y - 3y^2 \sin xy - xy^3 \cos xy, \tag{6.99a}$$

$$\frac{\partial^2 f(x, y)}{\partial x \partial y} = \cos y - 2y^2 \sin xy - y^2 \sin xy - xy^3 \cos xy$$

$$= \cos y - 3y^2 \sin xy - xy^3 \cos xy, \tag{6.99b}$$

which clearly verify the property (6.96).

given by

$$\nabla = \left(\frac{\partial}{\partial x}, \frac{\partial}{\partial y}, \frac{\partial}{\partial z} \right), \tag{6.100}$$

which is a row-vector formulation used also in classical mechanics. Hence, the gradient ∇f is a vector whose Cartesian components are the partial derivatives of $f(x, y, z)$, that is (again in the row-vector formulation)

$$\nabla f = \left(\frac{\partial f}{\partial x}, \frac{\partial f}{\partial y}, \frac{\partial f}{\partial z} \right). \tag{6.101}$$

The gradient, being itself a partial derivative operator, satisfies the usual derivative rules discussed in Section 6.4. As a quick example, the gradient of the function $f(x, y, z) = x^3 + 2xy + yz^2$ is given by

$$\nabla f = (3x^2 + 2y)\mathbf{e}_x + (2x + z^2)\mathbf{e}_y + 2yz\mathbf{e}_z.$$

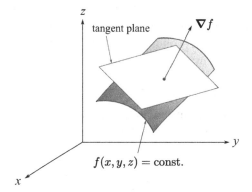

Figure 6.5 The gradient of a function $f(x, y, z)$ at a certain point can be visualized as the vector normal to the plane tangent at that point to the surface $f(x, y, z) = $ const. (represented here in grey scale). The gradient points in the direction of the maximum rate of increase of $f(x, y, z)$ at that point, and whose magnitude is that rate of increase.

Another important differential operator closely related to the gradient operator is the Laplacian, usually denoted by $\mathbf{\nabla}^2$ (del squared). In Cartesian coordinates, the Laplacian is defined in terms of partial derivative operators by

$$\mathbf{\nabla}^2 = \frac{\partial^2}{\partial x^2} + \frac{\partial^2}{\partial y^2} + \frac{\partial^2}{\partial z^2}. \tag{6.102}$$

The Laplacian of a function $f(x, y, z)$ is given by

$$\mathbf{\nabla}^2 f = \frac{\partial^2 f}{\partial x^2} + \frac{\partial^2 f}{\partial y^2} + \frac{\partial^2 f}{\partial z^2}. \tag{6.103}$$

The value of $\mathbf{\nabla}^2 f$ at a point can be thought of as the rate at which a certain average value of $f(x, y, z)$ over a sphere centered at that point, deviates from the actual value of $f(x, y, z)$ as the radius of the sphere approaches zero.

Sometimes a function of several variables may depend on one of its variables not only directly but also indirectly. For instance, the function $f(t, x(t))$ depends on t directly as well as indirectly via the variable $x(t)$ [see Eq. (6.75)]. In this case, in addition to the partial derivative of $f(t, x(t))$ with respect to t, for which the variable $x(t)$ is taken as constant, we can also consider the *total derivative* of $f(t, x(t))$ with respect to t, for which the variable $x(t)$ that depends on t is taken as variable as well. In other words, the total derivative takes the indirect dependence of t into account and gives the overall dependence of $f(t, x(t))$ on t. The total derivative of $f(t, x(t))$ with respect to t is given by

$$\frac{df}{dt} = \frac{\partial f}{\partial t} + \frac{\partial f}{\partial x}\frac{dx}{dt}. \tag{6.104}$$

Multiplying both sides of the above equation by the differential dt and using the fact that $dx = (dx/dt)dt$, we obtain

$$df = \frac{df}{dt}dt = \frac{\partial f}{\partial t}dt + \frac{\partial f}{\partial x}\frac{dx}{dt}dt = \frac{\partial f}{\partial t}dt + \frac{\partial f}{\partial x}dx, \tag{6.105}$$

where df is called the *total differential* of f.

Problem 6.8 Compute the first and the mixed second partial derivatives of the function

$$f(x, y) = \frac{x - y}{x + y}.$$

Problem 6.9 Find the partial derivative $\partial f/\partial t$ and the total derivative df/dt for the function

$$f(t, x) = t^2 - 8tx - x^2, \quad \text{where} \quad x(t) = \sin t.$$

6.6 Momentum as Generator of Space Translations

After this mathematical parenthesis we can come back to physics. In Section 6.4, we have discussed the concept of velocity in classical mechanics [see Eq. (6.84)]. It is certainly an important physical quantity. However, both in classical and in quantum mechanics physicists prefer dealing with another physical quantity called *momentum*. As we have already mentioned [see Section 1.1], in classical mechanics the momentum of a particle is defined by the product of the mass and velocity of the particle:

$$\mathbf{p} = m\mathbf{v} = m\frac{d\mathbf{r}}{dt}, \tag{6.106}$$

where m, $\mathbf{r}$, $\mathbf{v}$, and $\mathbf{p}$ are respectively the mass, position, velocity, and momentum of the particle. Here we have considered the general three-dimensional case. In other words, the position, velocity, and momentum are expressed in Cartesian coordinates as three-component spatial vectors $\mathbf{r} = (x, y, z)$, $\mathbf{v} = (v_x, v_y, v_z)$, and

$\mathbf{p} = (p_x, p_y, p_z)$, respectively (again the row-vector formulation has been used). The momentum of a system of particles is the sum of the momenta of the individual particles

$$\mathbf{p} = \sum_{i=1}^{n} m_i \mathbf{v}_i, \qquad (6.107)$$

where m_i and $\mathbf{v}_i$ are the respective mass and velocity of the ith particle, and n is the number of particles in the system. Momentum turns out to be related to the capability of a moving system to have a certain dynamical impact on other systems. For instance, we have already mentioned the circumstance that the heavier a car is or the faster it is moving, the more damage it will cause during an accident. Indeed, the net external force applied on an object is equal to the time derivative of its momentum

$$\mathbf{F} = \frac{d}{dt}(m\mathbf{v}), \qquad (6.108)$$

which, for an object of constant mass, is simply a reformulation of Newton's second law (1.2) [see also Eq. (6.87)].

It is very important that we understand what momentum means from a fundamental point of view: momentum is a conserved quantity in a closed system, i.e., a system that is not subject to external forces [see Section 1.1]. Moreover, since the momentum of a system is what describes its spatial translations (changes of position), it can be understood as (or thought of as being strictly related to) the generator of space translations, that is, as the dynamical parameter or operator that is associated with changes in the kinematic parameter represented by the position. To understand what this means, let us first consider an *infinitesimal* space translation in the x direction given by

$$x \xrightarrow{\varepsilon} x' = x + \varepsilon, \qquad (6.109)$$

where ε is an infinitesimal distance. The translated position x' can be expressed in terms of the initial position x as

$$x' = \left(1 + \varepsilon \hat{D}_x\right) x, \qquad (6.110)$$

where $\hat{D}_x = \frac{d}{dx}$ is the generator of space translations in the x direction, which simply is the derivative operator with respect to x.

To go over from an infinitesimal to a *finite* space translation in the x direction

$$x \xrightarrow{a} x' = x + a, \tag{6.111}$$

where a is an finite distance, we may think of the latter as a result of an infinite number of repeated applications of the infinitesimal translation (6.110). This is achieved by first subdividing the finite distance a into n equal segments, each of length a/n, and then taking the limit that n approaches infinity, namely, with a reversal of the methodology that we have already applied when introducing the formalism of differentiation [see Section 6.4]. Since the effect of each infinitesimal translation is given by Eq. (6.110) and all of them together constitute the finite displacement a, we can reiterate each of these infinitesimal translations, that is, we can multiply these infinitesimal displacement operations and express the translated position x' in terms of the initial position x as

$$
\begin{aligned}
x' &= \lim_{n \to \infty} \overbrace{\left(1 + \frac{a}{n}\hat{D}_x\right) \cdots \left(1 + \frac{a}{n}\hat{D}_x\right)}^{\text{total of } n \text{ factors}} x \\
&= \lim_{n \to \infty} \left(1 + \frac{a}{n}\hat{D}_x\right)^n x \\
&= e^{a\hat{D}_x} x, \tag{6.112}
\end{aligned}
$$

where in the last equality use has been made of the mathematical formula

$$\lim_{n \to \infty} \left(1 + \frac{x}{n}\right)^n = e^x. \tag{6.113}$$

This formula is a consequence of the Taylor expansion (6.93). Indeed, for fixed x, the ratio $\frac{x}{n}$ becomes exceedingly small as n approaches infinity, hence from Eq. (6.94) we can approximate $1 + \frac{x}{n}$ by $e^{\frac{x}{n}}$ and formula (6.113) follows since $\left(e^{\frac{x}{n}}\right)^n = e^x$. The expression $e^{a\hat{D}_x}$ denotes the exponential of the operator $a\hat{D}_x$, which is itself an operator, called the space translation operator by a distance a in the x direction. Having established a connection with the Taylor expansion, we note that in general the exponential of an operator $\hat{A}$ is defined by [see Eq. (6.93)]

$$e^{\hat{A}} = \hat{I} + \hat{A} + \frac{\hat{A}^2}{2!} + \frac{\hat{A}^3}{3!} + \cdots = \sum_{n=0}^{\infty} \frac{\hat{A}^n}{n!}, \tag{6.114}$$

which allows us to finally write

$$e^{a\hat{D}_x} = \hat{I} + a\hat{D}_x + \frac{a^2}{2}\hat{D}_x^2 + \cdots,$$

$$= 1 + a\frac{d}{dx} + \frac{a^2}{2}\frac{d^2}{dx^2} + \cdots. \tag{6.115}$$

It is easy to check that when the operator $e^{a\hat{D}_x}$ is applied to x, i.e., to the left-hand side of Eq. (6.111), we get its right-hand side because only the constant term and the first derivative term in the expansion (6.115) are different from zero. It is precisely in this sense that $\hat{D}_x$ is referred to as the *generator* of space translations in the x direction. As a consequence, the momentum operator in quantum mechanics is closely related to the derivative operator $\hat{D}_x$. We shall prove this result below in an explicit way.

We confine by now the exposition to the one-dimensional case. In the position representation, the one-dimensional momentum operator takes the form[a]

$$\hat{p}_x = -i\hbar\frac{\partial}{\partial x}, \tag{6.116}$$

or more formally [see Eqs. (6.37) and (6.52)],

$$\hat{p}_x|x\rangle = -i\hbar\frac{\partial}{\partial x}|x\rangle \quad \text{and} \quad \langle x|\hat{p}_x = -i\hbar\frac{\partial}{\partial x}\langle x|, \tag{6.117}$$

which precisely express the required variation of the position x. In the above expressions, $\hbar$ is called the reduced Planck constant, i.e., the Planck constant h divided by 2π [see Eq. (4.2)] and the factor $-i$ is there to ensure that as an observable the momentum operator $\hat{p}_x$ is a Hermitian operator. Analogous to what happens for the position operator, there will also be eigenvalues p_x (where $-\infty < p_x < \infty$) and the corresponding eigenstates $|p_x\rangle$ of the one-dimensional momentum operator $\hat{p}_x$ determined by the eigenvalue equation [see Eq. (4.4)]:

$$\hat{p}_x|p_x\rangle = p_x|p_x\rangle. \tag{6.118}$$

Likewise, the orthonormal conditions of the momentum eigenstates $|p_x\rangle$ are given by [see Eq. (6.43)]

$$\langle p_x|p_x'\rangle = \delta(p_x - p_x'), \tag{6.119}$$

---

[a]We use here the partial derivative $\frac{\partial}{\partial x}$ instead of the ordinary derivative $\frac{d}{dx}$ in order to remind the reader that the one-dimensional case is indeed a special case of the three-dimensional one.

and the completeness relation for the momentum eigenstates is given by [see Eq. (6.39)]

$$\int_{-\infty}^{+\infty} |p_x\rangle\langle p_x| dp_x = \hat{1}. \tag{6.120}$$

To find the *momentum eigenfunction* $\varphi_p(x) = \langle x|p_x\rangle$, i.e., the counterpart for the momentum operator of what the position eigenfunction is for the position operator [see Eqs. (6.54)–(6.55)], we shall solve the eigenvalue equation (6.118) in the position representation. Left multiplying both sides of Eq. (6.118) by $\langle x|$ and using Eqs. (6.117) to rewrite the left-hand side in terms of the momentum operator in the position representation, we obtain

$$\hat{p}_x\varphi_p(x) = p_x\varphi_p(x), \tag{6.121}$$

or equivalently,

$$-i\hbar\frac{\partial}{\partial x}\varphi_p(x) = p_x\varphi_p(x). \tag{6.122}$$

The above equation is our first example of the so-called *differential equation*, that is, an equation that involves an unknown function as well as its derivatives [see Box 6.7]. Differential equations play a very important and useful role in mathematics, physics, and other disciplines. Various techniques have been developed for finding the solutions of differential equations. While this is a topic far beyond the scope of this book, the basic methods of solving the simplest differential equations can be found in Boxes 6.8 and 7.1. The momentum eigenfunction, i.e., the solution to the above differential equation, is given by [see Box 6.8]

$$\varphi_p(x) = \langle x|p_x\rangle = \frac{1}{\sqrt{2\pi\hbar}} e^{\frac{i}{\hbar}p_x x} \quad (-\infty < p_x < \infty), \tag{6.123}$$

where the normalization constant $\frac{1}{\sqrt{2\pi\hbar}}$ is there to fulfill the orthonormal conditions (6.119) [see Eq. (6.50)]. The validity of the solution can be verified by direct substitution into the eigenvalue equation (6.122).

We note that the space translation operator $e^{a\hat{D}_x}$ in Eq. (6.112) can be written in terms of the momentum operator $\hat{p}_x$ as[a]

$$\hat{U}_x(a) = e^{-\frac{i}{\hbar}a\hat{p}_x}, \tag{6.124}$$

---

[a]The careful reader may have noticed that there is a sign difference in the exponent between the space translation operator in Eq. (6.124) and that in p. 138. This seeming sign difference is due to the fact that the former is written as an active transformation while the latter is written as a passive one. We will return to this point in Section 7.4.

Box 6.7 Differential equations

Differential equations can be classified into two types: an *ordinary differential equation* (ODE) is an equation containing a function of one independent variable and its derivatives, and a *partial differential equation* (PDE) is an equation that contains a multivariable function and its partial derivatives. We shall discuss here only the ordinary differential equations, which are further classified according to the *order* of the highest derivative of the dependent variable with respect to the independent variable appearing in the equation. The most important cases for applications are first-order and second-order ordinary differential equations.

An ODE is said to be *linear* if the unknown function and its derivatives appear at most only once in each term of the equation. For instance, a linear second-order ordinary differential equation is of the form

$$a_2(x)\, y''(x) + a_1(x)\, y'(x) + a_0(x)\, y(x) = b(x). \qquad (6.125)$$

The function $b(x)$ is called the source term, leading to two further important classifications: If $b(x) = 0$ then the equation is called homogeneous (i.e. it is a linear ODE which does not have terms independent of the unknown function and its derivatives); otherwise it is called nonhomogeneous. In general, an nth order homogeneous linear ODE has n linearly independent solutions. Furthermore, any linear combination of linearly independent solutions is also a solution.

---

which is a unitary operator [see Section 4.6 and Problem 6.11]. As a result, the momentum eigenstates $|p_x\rangle$ are also eigenstates of $\hat{U}_x(a)$, namely, we have

$$\hat{U}_x(a)|p_x\rangle = e^{-\frac{i}{\hbar} p_x a}|p_x\rangle. \qquad (6.126)$$

Let us consider the action of $\hat{U}_x(a)$ on the position eigenstate $|x\rangle$

$$|x\rangle \xrightarrow{\ a\ } \hat{U}_x(a)|x\rangle. \qquad (6.127)$$

Box 6.8 Linear differential equations with constant coefficients I

A very important class of linear differential equations are the homogeneous ones with constant coefficients. Let us first consider the basic equation with widespread application in physical sciences (what is said in this box is also true for partial differential equations):

$$\frac{dy}{dx} + ky = 0, \tag{6.128}$$

where k is a constant. It is a first order homogeneous linear differential equation. Indeed, the expression (6.70) allows us to write

$$y(x) = e^{-kx}. \tag{6.129}$$

Since the differential equation (6.128) is homogeneous, it follows that if y is a solution so is a constant multiple of y. Therefore, the general solution is given by

$$y(x) = ce^{-kx}, \tag{6.130}$$

where c is some constant.

By inserting the identity operator (6.120) between $\hat{U}_x(a)$ and $|x\rangle$, we obtain

$$
\begin{aligned}
\hat{U}_x(a)|x\rangle &= \int_{-\infty}^{+\infty} \hat{U}_x(a)|p_x\rangle\langle p_x|x\rangle dp_x \\
&= \int_{-\infty}^{+\infty} \langle p_x|x\rangle \hat{U}_x(a)|p_x\rangle \\
&= \frac{1}{\sqrt{2\pi\hbar}} \int_{-\infty}^{+\infty} e^{-\frac{i}{\hbar}p_x x} e^{-\frac{i}{\hbar}p_x a} |p_x\rangle dp_x \\
&= \frac{1}{\sqrt{2\pi\hbar}} \int_{-\infty}^{+\infty} e^{-\frac{i}{\hbar}p_x(x+a)} |p_x\rangle dp_x \\
&= \int_{-\infty}^{+\infty} |p_x\rangle\langle p_x|x+a\rangle dp_x \\
&= |x+a\rangle, \tag{6.131}
\end{aligned}
$$

where use has been made of Eqs. (6.126), the first property of the scalar product (3.21), which implies that $\langle p_x|x\rangle = \langle x|p_x\rangle^*$, and the fact that $\langle x|p_x\rangle$ is the momentum eigenfunction $\varphi_p(x)$ given by Eq. (6.123). The result means that under a space translation by a distance a, the state $|x\rangle$ transforms into the state $|x + a\rangle$. This definitively proves that momentum is indeed the generator of space translations as advertised earlier.

Problem 6.10 Check that the Eq. (6.123) is a solution of Eq. (6.122).

Problem 6.11 Show that the space translation operator $\hat{U}_x(a) = e^{-\frac{i}{\hbar}a\hat{p}_x}$ is a unitary operator.

Problem 6.12 Show that $\hat{U}_x^\dagger(a)\hat{x}\hat{U}_x(a) = \hat{x} + a$. (*Hint*: Use Eq. (6.131) to express the matrix elements $\langle x|\hat{U}_x^\dagger(a)\hat{x}\hat{U}_x(a)|x'\rangle$ in term of those of $\hat{x}$.)

6.7 Momentum Representation

Like the position eigenstates, the momentum eigenstates also constitute a complete orthonormal basis. The use of the momentum eigenstates as basis states to represent operators and states is referred to as the *momentum representation*. In other words, in the momentum representation the momentum operator $\hat{p}_x$ is simply represented by the real number p_x, as it is clear by Eq. (6.118) or its conjugate transpose [see Eq. (6.52) for position]

$$\langle p_x|\hat{p}_x = p_x\langle p_x|. \tag{6.132}$$

Then, any state vector $|\psi\rangle$ can be written as [see Eq. (6.38)]

$$|\psi\rangle = \int_{-\infty}^{+\infty} |p_x\rangle\langle p_x|\psi\rangle dp_x, \tag{6.133}$$

where use has been made of the completeness relation (6.120). The scalar product $\langle p_x|\psi\rangle$ in the above equation is the state vector $|\psi\rangle$ expressed in the momentum representation, and can be understood as a continuous function of the momentum eigenvalue p_x, that is,

$$\widetilde{\psi}(p_x) = \langle p_x|\psi\rangle \tag{6.134}$$

which is called the *wave function in momentum space*. Here we use the notation $\widetilde{\psi}(p_x)$ rather than the somewhat confusing one

$\psi(p_x)$ in order to highlight the fact that for a given state vector $|\psi\rangle$, the wave function in position space $\psi(x) = \langle x|\psi\rangle$ and the corresponding wave function in momentum space $\tilde{\psi}(p_x) = \langle p_x|\psi\rangle$ in general have different functional forms. Analogous to the wave function in position space $\psi(x)$ [see Eq. (6.41)], the wave function in momentum space $\tilde{\psi}(p_x)$ is the probability amplitude whose square modulus $|\tilde{\psi}(p_x)|^2$ gives probability density in momentum space. In other words, the probability of finding a quantum system with momentum between p_x and $p_x + dp_x$ is given by $|\tilde{\psi}(p_x)|^2 dp_x$.

We now have two representations that we may use to express operators and states. In other words, for any state vector $|\psi\rangle$ we can either use the position or the momentum representation to expand it. In the former case, we have the position space wave function $\psi(x)$, while in the latter case we have the momentum space wave function $\tilde{\psi}(p_x)$. It will be important to note that both representations are equally good for expressing the state vector $|\psi\rangle$. In fact, the choice between these two representations is only a matter of convenience, and the wave functions $\psi(x)$ and $\tilde{\psi}(p_x)$ are simple related by a change of basis [see Section 4.6], but this time using integrals due to the continuous nature of the observables involved. To find the connection between $\psi(x)$ and $\tilde{\psi}(p_x)$, and hence the connection between these two representations, we left multiply Eq. (6.133) by $\langle x|$, and Eq. (6.38) by $\langle p_x|$, to obtain

$$\psi(x) = \langle x|\psi\rangle = \int_{-\infty}^{+\infty} \langle x|p_x\rangle\langle p_x|\psi\rangle dp_x = \int_{-\infty}^{+\infty} \langle x|p_x\rangle \tilde{\psi}(p_x) dp_x$$

(6.135a)

and

$$\psi(p_x) = \langle p_x|\psi\rangle = \int_{-\infty}^{+\infty} \langle p_x|x\rangle\langle x|\psi\rangle dx = \int_{-\infty}^{+\infty} \langle p_x|x\rangle \psi(x) dx,$$

(6.135b)

respectively. Using the fact that $\langle x|p_x\rangle$ is the momentum eigenfunction $\varphi_p(x)$ given by Eq. (6.123) and the property that $\langle p_x|x\rangle = \langle x|p_x\rangle^*$, we can rewrite the above equations as

$$\psi(x) = \frac{1}{\sqrt{2\pi\hbar}} \int_{-\infty}^{+\infty} \tilde{\psi}(p_x) e^{\frac{i}{\hbar}p_x x} dp_x$$

(6.136a)

and

$$\tilde{\psi}(p_x) = \frac{1}{\sqrt{2\pi\hbar}} \int_{-\infty}^{+\infty} \psi(x) e^{-\frac{i}{\hbar}p_x x} dx.$$

(6.136b)

These equations are the sought connection between the position and momentum representations and are called the *Fourier transforms*. In general, quantities like position and momentum which are related to each other by the Fourier transforms are called canonical conjugate quantities. Moreover, as we will see in Section 6.8, such a connection shows that the position and momentum observables of a quantum system are to a certain extent complementary [see Chapter 5].

It is straightforward to generalize the above exposition to the three-dimensional case, in which we need to consider each Cartesian component separately and so to make use of partial derivatives. The momentum operator in the position representation is given by (here we use, as usual, the row-vector formulation when we have expressions that have specific classical counterparts)

$$\hat{\mathbf{p}} = -i\hbar\nabla = -i\hbar\left(\frac{\partial}{\partial x}, \frac{\partial}{\partial y}, \frac{\partial}{\partial z}\right), \qquad (6.137)$$

or more formally,

$$\hat{\mathbf{p}}|\mathbf{r}\rangle = -i\hbar\nabla|\mathbf{r}\rangle, \qquad (6.138)$$

where ∇ is the gradient operator [see Eq. (6.100)]. The momentum eigenstate is now denoted by

$$|\mathbf{p}\rangle = |p_x, p_y, p_z\rangle, \qquad (6.139)$$

which satisfies the eigenvalue equations

$$\hat{\mathbf{p}}|\mathbf{p}\rangle = \mathbf{p}|\mathbf{p}\rangle, \qquad (6.140)$$

where the eigenvalues $\mathbf{p} = (p_x, p_y, p_z)$ (with $-\infty < p_x, p_y, p_z < \infty$) represent the possible momentum vectors that a quantum system may potentially have. The completeness relation and orthonormal conditions are respectively given by

$$\int d^3p|\mathbf{p}\rangle\langle\mathbf{p}| = \hat{I} \quad \text{and} \quad \langle\mathbf{p}|\mathbf{p}'\rangle = \delta^{(3)}(\mathbf{p} - \mathbf{p}'), \qquad (6.141)$$

where

$$\int d^3p = \int_{-\infty}^{+\infty} dp_x \int_{-\infty}^{+\infty} dp_y \int_{-\infty}^{+\infty} dp_z \qquad (6.142)$$

means the integral is taken over the whole three-dimensional momentum space. Moreover, the scalar product

$$\widetilde{\psi}(\mathbf{p}) = \langle\mathbf{p}|\psi\rangle \qquad (6.143)$$

is the three-dimensional momentum space wave function. In particular, the three-dimensional momentum eigenfunction [see Eq. (6.123)] is given by

$$\varphi_\mathbf{p}(\mathbf{r}) = \langle \mathbf{p}|\mathbf{r}\rangle = \frac{1}{(2\pi\hbar)^{3/2}}e^{\frac{i}{\hbar}\mathbf{p}\cdot\mathbf{r}}, \qquad (6.144)$$

where $\mathbf{p}\cdot\mathbf{r} = xp_x+yp_y+zp_z$ is the dot product of the spatial vectors $\mathbf{p}$ and $\mathbf{r}$ [see Box 3.1]. The momentum eigenfunction $\varphi_\mathbf{p}(\mathbf{r})$ allows us to write the following Fourier transforms between the wave functions $\psi(\mathbf{r})$ and $\tilde\psi(\mathbf{p})$:

$$\psi(\mathbf{r}) = \frac{1}{(2\pi\hbar)^{3/2}}\int d^3p\,\tilde\psi(\mathbf{p})\,e^{\frac{i}{\hbar}\mathbf{p}\cdot\mathbf{r}} \qquad (6.145a)$$

and

$$\tilde\psi(\mathbf{p}) = \frac{1}{(2\pi\hbar)^{3/2}}\int d^3r\,\psi(\mathbf{r})\,e^{-\frac{i}{\hbar}\mathbf{p}\cdot\mathbf{r}}. \qquad (6.145b)$$

Last but not the least, the space translation operator by a displacement $\mathbf{a}$ in three dimensions is given by

$$\hat{U}_\mathbf{r}(\mathbf{a}) = e^{-\frac{i}{\hbar}\mathbf{a}\cdot\hat{\mathbf{p}}}, \qquad (6.146)$$

where $\mathbf{a}\cdot\hat{\mathbf{p}} = a_x\hat{p}_x + a_y\hat{p}_y + a_z\hat{p}_z$.

Problem 6.13 Show that the normalization condition $\langle\psi|\psi\rangle = 1$ can be written in the momentum representation as [see also Problem 6.4]

$$\int_{-\infty}^{+\infty} |\tilde\psi(p_x)|^2 dp_x = 1.$$

Explain how the above equation justifies that $|\tilde\psi(p_x)|^2$ gives the probability density in momentum space.

Problem 6.14 For the rectangle wave function $\psi(x)$ given by

$$\psi(x) = \begin{cases} 1 & 0 \le x \le 1, \\ 0 & \text{otherwise}, \end{cases}$$

find the corresponding momentum space wave function $\tilde\psi(p_x)$.

6.8 Commutation and Uncertainty Relations

During 1925 and 1926, with the help of the new mathematical tools of matrices and operators developed by Heisenberg, an important relation between the position and momentum was discovered.[a] It was found that the quantum operators for the position and momentum do not commute, i.e., $\hat{x}\hat{p}_x \neq \hat{p}_x\hat{x}$ [see Section 4.7].[b] Later in 1927, Heisenberg realized that the non-commutativity of position and momentum implies that these is a fundamental limit on the accuracy with which the position and momentum of a particle can be simultaneously known. Heisenberg's uncertainty principle, for which he is well known, is stated in his own words as follows.[c]

Principle 6.1 (Uncertainty Principle) *The more precisely the position is determined, the less precisely the momentum is known in this instant, and vice versa.*

We shall proceeds in three steps. First, we shall derive the so-called commutation relations for position and momentum. Then, we shall derive the uncertainly relations for arbitrary observables. Finally, we shall formulate the uncertainty relation for position and momentum. To see that the position and momentum observables do not commute, let us consider the following expression (in the position representation)

$$
\begin{aligned}
(\hat{x}\hat{p}_x - \hat{p}_x\hat{x})\,\psi(x) &= x\left(-\mathrm{i}\hbar\frac{\partial}{\partial x}\right)\psi(x) + \mathrm{i}\frac{\partial}{\partial x}[x\psi(x)] \\
&= -\mathrm{i}\hbar x\frac{\partial}{\partial x}\psi(x) + \mathrm{i}\hbar x\frac{\partial}{\partial x}\psi(x) + \mathrm{i}\hbar\psi(x) \\
&= \mathrm{i}\hbar\psi(x),
\end{aligned}
\tag{6.147}
$$

where we have made use of the product rule for derivatives (6.73) and the fact that the wave function $\psi(x)$ is a function of the position eigenvalue x while the momentum acts on it as the operator given by

---

[a](Heisenberg, 1925), (Born/Jordan, 1925), (Born *et al.*, 1926).

[b]The patient reader who is not a physicist and has followed us thus far has shown a remarkable determination. Arriving at this point already represents a significant step. We would again suggest the reader to take a breath before moving on. Moreover, when he or she shall arrive at the end of this chapter, we suggest to read it again as many times as necessary to get a full understanding of this matter.

[c](Heisenberg, 1927).

Eq. (6.116). Since we have made no specific assumption on the wave function $\psi(x)$, we are allowed to say that for an arbitrary quantum state we always have the relation

$$[\hat{x}, \hat{p}_x] = i\hbar\hat{I}, \tag{6.148}$$

where the expression

$$[\hat{x}, \hat{p}_x] = \hat{x}\hat{p}_x - \hat{p}_x\hat{x} \tag{6.149}$$

is called the *commutator* of $\hat{x}$ and $\hat{p}_x$ [see Box 6.9]. A relation between operators like that given by Eq. (6.148) is called the *commutation relation*. We note that while the commutation relation (6.148) is derived here in the position representation, it holds true also in the momentum representation and in fact it is representation independent. It is straightforward to check that similar commutation relations hold true for the other Cartesian components of the two observables while different Cartesian components always commute. Indeed, we have

$$[\hat{x}, \hat{p}_x] = [\hat{y}, \hat{p}_y] = [\hat{z}, \hat{p}_z] = i\hbar\hat{I} \tag{6.150a}$$

and

$$[\hat{x}, \hat{p}_y] = [\hat{x}, \hat{p}_z] = [\hat{y}, \hat{p}_x] = [\hat{y}, \hat{p}_z] = [\hat{z}, \hat{p}_x] = [\hat{z}, \hat{p}_y] = 0. \tag{6.150b}$$

Box 6.9 Properties of commutators

It is easy to check that commutator satisfies the following properties:

$$[\hat{A}, \hat{B}] = -[\hat{B}, \hat{A}], \tag{6.151a}$$

$$[\alpha\hat{A} + \beta, \hat{B}] = \alpha[\hat{A}, \hat{B}], \tag{6.151b}$$

$$[\alpha\hat{A} + \beta\hat{B}, \hat{C}] = \alpha[\hat{A}, \hat{C}] + \beta[\hat{B}, \hat{C}], \tag{6.151c}$$

$$[\hat{A}\hat{B}, \hat{C}] = \hat{A}[\hat{B}, \hat{C}] + [\hat{B}, \hat{C}]\hat{A}, \tag{6.151d}$$

$$[\hat{A}, [\hat{B}, \hat{C}]] + [\hat{C}, [\hat{A}, \hat{B}]] + [\hat{B}, [\hat{C}, \hat{A}]] = 0, \tag{6.151e}$$

where α and β are complex numbers and the last equality (6.151e) is known as the Jacobi identity.

Because position and momentum are canonical conjugate quantities, the commutation relations of position and momentum (6.150) are also referred to as the canonical commutation relations.

Various commutation relations hold true for many quantum mechanical observables (all those pairs that are not jointly measurable or that do not have a common eigenbasis) and imply very interesting consequences: the uncertainty relations. Since the consequences are very general, it is convenient to generalize our treatment to generic quantum observables [see Section 4.7]. To this purpose, we introduce the useful concepts of expectation value, variance, and uncertainty of a quantum observable. We recall that the possible outcomes of a measurement are the eigenvalues of the observable being measured, and that because of the inherent probabilistic nature of quantum phenomena, the measurement outcome of an experiment will generally not be the same if the experiment is repeated several times [see Sections 3.2, 5.3, and 5.4]. Quantum mechanics does not, in fact, predict the results of individual measurements, but only their statistical mean, i.e., the weighted average of all possible measurement outcomes. Therefore, this predicted mean value is called the *expectation value* of the observable that is being measured. Let $\hat{O}$ be an observable with eigenvalues o_j's (which we take here as discrete for the sake of simplicity) and corresponding eigenstates $|o_j\rangle$'s (where j is some index labeling the eigenvalues). If the observable $\hat{O}$ is measured on a system in a state $|\psi\rangle$, then the measured result will be one of its eigenvalue o_j and the probability of obtaining a given eigenvalue o_j is given by $|\langle o_j|\psi\rangle|^2$. Hence, the expectation value of the observable $\hat{O}$ for a quantum system in the state $|\psi\rangle$, which is written as $\langle\hat{O}\rangle_\psi$, is given by the statistical mean

$$\begin{aligned}
\langle\hat{O}\rangle_\psi &= \langle\psi|\hat{O}|\psi\rangle \\
&= \sum_j o_j\langle\psi|o_j\rangle\langle o_j|\psi\rangle \\
&= \sum_j o_j|\langle o_j|\psi\rangle|^2,
\end{aligned} \tag{6.152}$$

where use has been made of the spectral representation of the observable $\hat{O}$ [see Eq. (6.13)]. For instance, consider the observable

$\hat{O}$ and the state $|\psi\rangle$ given by

$$\hat{O} = |h\rangle\langle h| - |v\rangle\langle v| \quad \text{and} \quad |\psi\rangle = \frac{1}{\sqrt{2}}(|h\rangle + |v\rangle), \tag{6.153}$$

respectively, where $|h\rangle$ and $|v\rangle$ are respectively the horizontal and vertical photon polarization states. The expectation value of $\hat{O}$ in the state $|\psi\rangle$ will be

$$\begin{aligned}
\langle \hat{O}\rangle_\psi &= \frac{1}{2}\big((\langle h| + \langle v|)(|h\rangle\langle h| - |v\rangle\langle v|)(|h\rangle + |v\rangle)\big) \\
&= \frac{1}{2}\big(\langle h|h\rangle\langle h|h\rangle - \langle v|v\rangle\langle v|v\rangle\big) \\
&= 0.
\end{aligned} \tag{6.154}$$

The generalization to observables with continuous eigenvalues is straightforward, and Eq. (6.152) remains valid in the continuous case. For the expectation value of the position operator $\hat{x}$, we have

$$\langle \hat{x}\rangle_\psi = \langle \psi|\hat{x}|\psi\rangle = \int_{-\infty}^{+\infty} x\,|\psi(x)|^2 dx. \tag{6.155}$$

A similar expression holds for the momentum operator $\hat{p}_x$, namely,

$$\langle \hat{p}_x\rangle_\psi = \langle \psi|\hat{p}_x|\psi\rangle = \int_{-\infty}^{+\infty} p_x\,|\widetilde{\psi}(p_x)|^2 dp_x. \tag{6.156}$$

It is noted that since the eigenvalues of an observable (represented by a Hermitian operator) are real, the expectation value of an observable must also be real. The expectation value also satisfies the following properties:

$$\big\langle (a\hat{O} + b)\big\rangle_\psi = a\langle \hat{O}\rangle_\psi + b, \tag{6.157a}$$

$$\big\langle (a\hat{O}_1 + b\hat{O}_2)\big\rangle_\psi = a\langle \hat{O}_1\rangle_\psi + b\langle \hat{O}_2\rangle_\psi, \tag{6.157b}$$

where a and b are real numbers.

The *variance* of an observable is a measure of the spread of the measurement results about its expectation value, taking account of all possible results and their probabilities. The variance of the observable $\hat{O}$ in the state $|\psi\rangle$ is defined by

$$\begin{aligned}
\Delta_\psi^2 \hat{O} &= \Big\langle \big(\hat{O} - \langle \hat{O}\rangle_\psi\big)^2\Big\rangle_\psi \\
&= \Big\langle \psi\Big|\big(\hat{O} - \langle \psi|\hat{O}|\psi\rangle\big)^2\Big|\psi\Big\rangle.
\end{aligned} \tag{6.158}$$

In other words, it is the expectation value of the squared deviation of an observable from its expectation value with both taken on the same state. Since $\hat{O}$ is a Hermitian operator (i.e., $\hat{O} = \hat{O}^\dagger$) and $\langle \hat{O} \rangle_\psi$ is real, the variance $\Delta^2_\psi \hat{O}$ can be thought of as the norm squared of the following state [see Eq. (3.23)]

$$|\psi'\rangle = (\hat{O} - \langle \hat{O} \rangle_\psi)|\psi\rangle, \tag{6.159}$$

that is,

$$\Delta^2_\psi \hat{O} = \||\psi'\rangle\|^2 = \langle \psi'|\psi'\rangle \geq 0, \tag{6.160}$$

where the equality holds if and only if $|\psi'\rangle = 0$. This means that the variance of an observable must be real and non-negative. The *uncertainty* of the observable $\hat{O}$ in the state $|\psi\rangle$ is defined by the square root of its variance in that state (or as the norm of $|\psi'\rangle$), i.e.,

$$\Delta_\psi \hat{O} = \sqrt{\left\langle \left(\hat{O} - \langle \hat{O} \rangle_\psi\right)^2\right\rangle_\psi}. \tag{6.161}$$

Indeed, this explains the use of the notation $\Delta^2_\psi \hat{O}$ for the variance of the observable $\hat{O}$ in the state $|\psi\rangle$.

We now establish the *uncertainty relation* for non-commuting observables. Let us consider two non-commuting observables $\hat{O}$ and $\hat{O}'$ and, without loss of generality, assume that the expectation values $\langle \hat{O} \rangle_\psi$ and $\langle \hat{O}' \rangle_\psi$ of the two observables in the state $|\psi\rangle$ are both equal to zero (this, however, does not imply that the expectation values in all states are zero). With these assumptions, the uncertainties of the observable $\hat{O}$ and $\hat{O}'$ in the state $|\psi\rangle$ reduce to

$$\Delta_\psi \hat{O} = \sqrt{\langle \psi|\hat{O}^2|\psi\rangle} \quad \text{and} \quad \Delta_\psi \hat{O}' = \sqrt{\langle \psi|\hat{O}'^2|\psi\rangle}. \tag{6.162}$$

To proceed further, we consider the states

$$|\varphi\rangle = \hat{O}|\psi\rangle \quad \text{and} \quad |\varphi'\rangle = \hat{O}'|\psi\rangle, \tag{6.163}$$

which allow us to write $\Delta_\psi \hat{O}$ and $\Delta_\psi \hat{O}'$ in terms of their norms:

$$\Delta_\psi \hat{O} = \sqrt{\langle \varphi|\varphi\rangle} \quad \text{and} \quad \Delta_\psi \hat{O}' = \sqrt{\langle \varphi'|\varphi'\rangle}. \tag{6.164}$$

From a pure mathematical point of view, we always have the inequality (known as the Cauchy–Schwarz inequality)

$$|\langle \varphi|\varphi'\rangle| \leq \sqrt{\langle \varphi|\varphi\rangle}\sqrt{\langle \varphi'|\varphi'\rangle}, \tag{6.165}$$

since the scalar product of two arbitrary (normalized) states $|a\rangle$ and $|b\rangle$ satisfies $0 \le |\langle a|b\rangle| \le 1$ and we always have $\langle a|a\rangle = \langle b|b\rangle = 1$ [see Section 3.4]. By substituting Eqs. (6.163) into the inequality (6.165), we obtain

$$\left|\langle\psi|\hat{O}\hat{O}'|\psi\rangle\right| \le \sqrt{\langle\psi|\hat{O}^2|\psi\rangle}\sqrt{\langle\psi|\hat{O}'^2|\psi\rangle} = \Delta_\psi\hat{O}\,\Delta_\psi\hat{O}', \quad (6.166)$$

where we have used Eq. (6.162) and the fact that $\hat{O}$ and $\hat{O}'$ are Hermitian operators. Obviously, a similar result can be obtained if we interchange the positions of the two observables, that is,

$$\left|\langle\psi|\hat{O}'\hat{O}|\psi\rangle\right| \le \sqrt{\langle\psi|\hat{O}^2|\psi\rangle}\sqrt{\langle\psi|\hat{O}'^2|\psi\rangle} = \Delta_\psi\hat{O}\,\Delta_\psi\hat{O}'. \quad (6.167)$$

Let us consider the last two equations. From the triangle inequality (2.7c), it follows that if z_1 and z_2 are two complex numbers and a is a real non-negative number such that $|z_1|, |z_2| \le a$, then we also have

$$|z_1 - z_2| \le 2a. \quad (6.168)$$

This, together with the fact that the uncertainty of an observable is real and non-negative, implies that

$$\left|\langle\psi|\hat{O}\hat{O}'|\psi\rangle - \langle\psi|\hat{O}'\hat{O}|\psi\rangle\right| \le 2\,\Delta_\psi\hat{O}\,\Delta_\psi\hat{O}', \quad (6.169)$$

or equivalently,

$$\Delta_\psi\hat{O}\,\Delta_\psi\hat{O}' \ge \frac{1}{2}\left|\langle[\hat{O}, \hat{O}']\rangle_\psi\right|, \quad (6.170)$$

where is usually referred to as the generalized uncertainty relation. Since this inequality is true for an arbitrary pair of observables (if they commute the right-hand side vanishes and the inequality is obviously satisfied), it must also be true for the position and momentum observables. By taking into account again the commutation relations (6.148), we finally obtain the uncertainty relation for position and momentum[a]

$$\Delta x \,\Delta p_x \ge \frac{\hbar}{2}. \quad (6.171)$$

Here, we have dropped any reference to the state since this conclusion is of general validity and does not depend on the specific state we have chosen. This result means that we cannot get a perfect determination of *both* position and momentum at the same instant.

---

[a](Heisenberg, 1927).

Every time we try to reduce the uncertainty in position, this is done at the expenses of the increase of the uncertainty in momentum, and vice versa.

Problem 6.15 Compute the commutators $[\hat{x}^2, \hat{p}_x]$, $[\hat{x}, \hat{p}_x^2]$, and $[\hat{x}\hat{p}_x, \hat{p}_x\hat{x}]$.

Problem 6.16 Check the correctness of Eqs. (6.162).

Problem 6.17 Find the variance $\Delta_\psi^2 \hat{O}$ for the observable $\hat{O}$ and the state $|\psi\rangle$ given by Eq. (6.153).

6.9 Conceptual Aspects of the Uncertainty Relations

The uncertainty relations have extraordinary conceptual conse-quences. As we have stressed, all classical quantities commute. This means that for a classical system we can in principle measure perfectly *both* momentum and position simultaneously, or at least be able to put together these two pieces of information in order to get a full knowledge of the state of a classical system at a certain time. This is very important since in classical mechanics the state of a system is described by these two variables. Given that the two variables are perfectly determined, they can be represented by a unique point in the phase space (the Cartesian space whose axes are position and momentum and in which all possible states of a system are represented) as shown in Fig. 6.6(a). In other words, in accordance with what we have explained in Chapter 1, classical systems are assumed to be both perfectly determined and knowable. Another way to say this is that it is assumed that the whole information that is contained in the state of a classical system is perfectly accessible and therefore acquirable by an observer, at least in principle.

Things are totally different when considering quantum mechan-ics. Here, due to the uncertainty relation (6.171) the quantum mechanical phase space can at best be thought of as being divided into cells, each of an approximate area $\Delta x\,\Delta p_x \approx \hbar/2$, and the state of a quantum system is roughly represented by an elliptical spot that is bounded by the cell [see Fig. 6.6(b)]. Any attempt at reducing the uncertainty Δx will immediately induce a corresponding increase

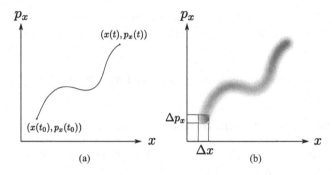

Figure 6.6 (a) In classical mechanics, given the initial state of a system $(x(t_0), p_x(t_0))$ at time t_0, the state of the system $(x(t), p_x(t))$ at later time t is uniquely determined [see Section 7.1]. The state of a classical system at any given time is represented by a point in the phase space, and its time evolution is represented by a trajectory. (b) According to the uncertainty relation $\Delta x \, \Delta p_x \geq \hbar/2$, the quantum mechanical phase space can be thought of as being divided into cells, each of an approximate area $\Delta x \, \Delta p_x \approx \hbar/2$, and the state of a quantum system is roughly represented by an elliptical spot that is bounded by the cell. Because of the uncertainty principle it is not possible to ascribe a definite phase space trajectory to a quantum system. The most probable states of a quantum system can at best be thought of as a fuzzy contour.

of the uncertainty Δp_x, and vice versa, in accordance with the uncertainty relation. In other words, although we are free in the choice of the observable to measure, the consequences of measurement are not controllable by us. If we decide to measure the observable $\hat{x}$, we shall obtain a certain kind of information that is contained in the state of the system. On the other hand, if we decide to measure the observable $\hat{p}_x$, we shall obtain another kind of information. The crucial problem is that we cannot put together these two pieces of information, which shows that we cannot choose to measure both the observables $\hat{x}$ and $\hat{p}_x$, and therefore that we are not completely free even in the single measurement act, at least regarding its consequences that we cannot control, although we can choose alternative experimental setups. How is it possible? To a certain extent, both classically and quantum mechanically, the experimental contexts for measuring position and momentum are not compatible (classically, to measure momentum we need a mobile device while for measuring position we need a fixed grid).

Hence we cannot measure these two quantities jointly, instead we need a succession of at least two measurements. In classical mechanics, the assumption of continuity and the deterministic character of classical laws [see Section 1.2] ensures us that we can combine the information obtained through these two consecutive measurements in order to fully reconstruct the state of the system. However, this is precisely what cannot happen in quantum mechanics. In Section 4.7, we have indeed learned that the order in which we perform two or more measurements does matter and therefore the decision to measure first, say, the position will influence the final result. Moreover, since each measurement will inevitably change the state of the system, we cannot cumulate the information extracted from the subsequent measurements.

This implies that the observer of a quantum system is not a detached spectator who can observe systems without determining certain consequences on those systems. Knowledge is not a contemplative activity but is deeply and dynamically involved in what we call reality. This statement, which can sound quite surprising and whose full meaning can be understood only later on, seems to imply that also the uncertainty relations themselves depend on certain measurements that we perform on quantum systems. This operationist epistemology was also supported by Heisenberg himself.[a] However, already in 1960 it was shown that this is not necessarily the case.[b] This conclusion has been confirmed by interaction-free measurements [see Sections 5.2 and 5.3] and by the so-called non-demolition measurements.[c] The reason is that uncertainty relations are a consequence of quantum features [see Section 4.8] and not of the operations that we may perform on a quantum system. On the contrary, the results we can obtain through these operations are deeply conditioned and framed by the constraints imposed by quantum features through uncertainty relations (as we shall see, these constraints are not of traditional physical kind), so that the results and their possible influences on a subsequent measurement cannot be controlled by the observer.

---

[a] (Heisenberg, 1927).
[b] (Renninger, 1960).
[c] (Braginsky/Khalili, 1992).

This again shows that we cannot avoid assigning a certain reality to quantum features [see Section 5.6].

6.10 Summary

In this chapter we have

- Studied the position and momentum observables.
- Learned the mathematical operations of integration and differentiation.
- Dealt with the continuum formulation of quantum mechanics.
- Shown that momentum is the generator of space translations.
- Learned the wave function in position and momentum spaces.
- Derived the position and momentum eigenfunctions.
- Introduced the position and momentum representations.
- Established the Fourier transform as the transformation from the position representation to the momentum representation (and vice versa with the inverse transformation).
- Derived the commutation and uncertainty relations for position and momentum.
- Discussed some conceptual consequences of the uncertainty relations.

Chapter 7

Energy and Quantum Dynamics

In this chapter we shall deal with quantum dynamics and in particular with the basic equation that governs quantum dynamics, the Schrödinger equation. We shall consider two of the most important model systems in quantum mechanics: the free particle and the harmonic oscillator. Moreover, we shall deal with the density matrix formalism and composite systems. Another crucial concept that will be introduced here is that of entanglement, a particular manifestation of quantum features.

7.1 Hamiltonian and Classical Dynamics

Dynamics is the study of how a system changes in time. Where dynamics is involved, also force and energy are involved. A kind of forces or fields called the conservative force is of particular importance in classical mechanics. When a system is subject to a conservative force and moving from one position to another, the work[a] done on the system by the conservative force is independent

---

[a]The work W done by a force $\mathbf{F}$ along a path C is the integral of its scalar tangential component over the distance through which it acts, i.e.,

$$W = \int_C \mathbf{F} \cdot d\mathbf{r},$$

where $\mathbf{r}$ is the position vector that specifies the path.

Quantum Mechanics for Thinkers
Gennaro Auletta and Shang-Yung Wang
Copyright © 2014 Pan Stanford Publishing Pte. Ltd.
ISBN 978-981-4411-71-4 (Hardcover), 978-981-4411-72-1 (eBook)
www.panstanford.com

of the path taken by the system. As a consequence, if a system travels in a closed loop and returns to the starting position, the net work done by a conservative force is always zero. Therefore, we can identify the negative of the work done by a conservative force on a system from position a to position b as the difference in the potential energy of the system at positions a and b. The gravitational force, spring force, and electrostatic force (in a time-independent magnetic field) are the most familiar examples of conservative forces, while friction and air drag are classical examples of non-conservative forces. In classical mechanics the total mechanical energy E of a system is the sum of the kinetic energy T, which the system possesses due to its motion and is therefore related to its momentum, and the potential energy V, which is determined by the force applied to it. For a conservative force, the potential energy of a system depends only on its position and the force is given by the negative gradient of the potential energy, namely, we have

$$\mathbf{F}(\mathbf{r}) = -\nabla V(\mathbf{r}), \tag{7.1}$$

where $\mathbf{F}(\mathbf{r})$ is the conservative force and $V(\mathbf{r})$ is the corresponding potential energy. Hence, the total mechanical energy of a system can be written as

$$E = T(\mathbf{p}) + V(\mathbf{r}) = \frac{\mathbf{p}^2}{2m} + V(\mathbf{r}), \tag{7.2}$$

where m is the mass of the system and the form of the potential term $V(\mathbf{r})$ depends on the specific conservative forces involved. Moreover, the law of conservation of mechanical energy states that if a system is subjected only to conservative forces, the total mechanical energy of the system remains constant in time.

It is very important that we understand what energy means from a fundamental point of view. To this end, we introduce the so-called Hamiltonian formulation of classical mechanics, in which momentum and position are treated dynamically on an equal footing. In the Hamiltonian formulation, Newton's second law (1.2) is reformulated as the *Hamilton equations*:

$$\frac{dr_j}{dt} = \frac{\partial H(\mathbf{r}, \mathbf{p})}{\partial p_j}, \qquad \frac{dp_j}{dt} = -\frac{\partial H(\mathbf{r}, \mathbf{p})}{\partial r_j}, \tag{7.3}$$

where t is the time variable and $j = x, y, z$ with r_x corresponding to x, r_y to y, and r_z to z. The quantity $H(\mathbf{r}, \mathbf{p})$ is called the *Hamiltonian function* (or *Hamiltonian* for short), whose value is the total energy E of the system being described, given by Eq. (7.2). Given the initial state of a classical system $(\mathbf{r}(t_0), \mathbf{p}(t_0))$ at time t_0, the state of the system $(\mathbf{r}(t), \mathbf{p}(t))$ at later time t is uniquely determined through the Hamilton equations (7.3). As a result, the state of a classical system at any given time is represented by a point in the phase space, and its time evolution is represented by a trajectory [see Fig. 6.6(a) for a one-dimensional case]. The equivalence of the Hamilton equations and Newton's second law can be established rigorously. Here we shall illustrate the equivalence by considering a simple example that a one-dimensional object of mass m moves in a constant gravitational field. The Hamiltonian of the object is given by [see Eq. (7.2)]

$$H(z, p_z) = \frac{p_z^2}{2m} + mgz, \qquad (7.4)$$

where p_z is the momentum of the object, g is the acceleration of gravity, and we have chosen the gravitational field pointing in the negative z direction. The Hamilton equations can be found by direct differentiation, yielding

$$\frac{dz}{dt} = \frac{p_z}{m}, \qquad \frac{dp_z}{dt} = -mg. \qquad (7.5)$$

Taking the time derivative of the first equation in the above expression and using the second to eliminate dp_z/dt, we find

$$\frac{d^2z}{dt^2} = -g, \qquad (7.6)$$

which is precisely what would be obtained by using Newton's second law. Indeed, the Hamiltonian formulation provides not only a new and equivalent way of looking at Newtonian mechanics, but also a deeper insight that the dynamics of a system is governed by its Hamiltonian (or total energy).

Problem 7.1 Find the Hamiltonian and the Hamilton equations for a one-dimensional object of mass m attached to the end of a spring with spring constant k.

7.2 Schrödinger Equation

In quantum mechanics the total energy of a system is certainly a physical observable, as is evident from the fact that the energy levels of the hydrogen atom are responsible for the observed emission spectrum of atomic hydrogen [see Section 4.2]. Thus, here comes the question of how to construct the energy observable in quantum mechanics. This is the fundamental content of the correspondence principle,[a] which was stated by Niels Bohr in 1920, though he had previously made use of it as early as 1913 in developing his model of the atom.

Principle 7.1 (Correspondence Principle) *For the limiting cases of large energies and of orbits of large dimensions, quantum mechanics passes over into classical mechanics.*

The energy observable in quantum mechanics is referred to as the *Hamiltonian operator* (or *Hamiltonian* for short). According to the correspondence principle, we need only to replace position and momentum in classical mechanics by the position operator $\hat{\mathbf{r}}$ and the momentum operator $\hat{\mathbf{p}}$, respectively. As a result, the Hamiltonian operator takes the form [see Eq. (7.2)]

$$\hat{H} = \frac{\hat{\mathbf{p}}^2}{2m} + V(\hat{\mathbf{r}}), \qquad (7.7)$$

where $\hat{\mathbf{p}}^2 = \hat{p}_x^2 + \hat{p}_y^2 + \hat{p}_z^2$ is the magnitude squared of the momentum operator [see Eq. (3.19)]. Since both the momentum and position operators are Hermitian operators, the Hamiltonian is a Hermitian operator as well. In general, the sum of Hermitian operators is itself a Hermitian operator, this however does not necessarily holds true for the product of Hermitian operators. In the three-dimensional case the Hamiltonian in the position representation can be written as [see Eqs. (6.59) and (6.138)]

$$\hat{H} = -\frac{\hbar^2}{2m}\nabla^2 + V(\hat{\mathbf{r}}), \qquad (7.8)$$

---

[a](Bohr, 1920).

where ∇^2 is the Laplacian in the Cartesian coordinates [see Eq. (6.102)]. Moreover, in the one-dimensional case the Hamiltonian reduces to

$$\hat{H} = \frac{\hat{p}_x^2}{2m} + V(\hat{x}), \tag{7.9}$$

which in the position representation takes the form [see Eqs. (6.52) and (6.117)]

$$\hat{H} = -\frac{\hbar^2}{2m}\frac{\partial^2}{\partial x^2} + V(x). \tag{7.10}$$

We expect that, by analogy with the position and momentum operators, for the Hamiltonian $\hat{H}$ there will be real eigenvalues E and corresponding eigenstates $|E\rangle$, such that the following eigenvalue equation holds:

$$\hat{H}|E\rangle = E|E\rangle. \tag{7.11}$$

It is noted that the energy eigenvalues E represent the possible values of the total energy of a quantum system and, like the position and momentum eigenstates (as shown in the previous chapter), the energy eigenstates constitute a complete orthonormal basis. The use of the energy eigenstates as basis states to represent operators and states is referred to as the *energy representation*.

Since the state of a quantum system is described by a state vector, quantum dynamics is the study of how state vectors change in time. To highlight this time dependence of state vectors, we will write a generic state vector of a quantum system as $|\psi(t)\rangle$ with t being the time variable. In quantum mechanics, the time derivative of the state vector $|\psi(t)\rangle$ is governed by the Hamiltonian of the system through the equation

$$i\hbar\frac{d}{dt}|\psi(t)\rangle = \hat{H}|\psi(t)\rangle, \tag{7.12}$$

where the time derivative d/dt is understood as the total time derivative [see Section 6.5]. The above equation is known as the *Schrödinger equation* and has a specificity distinct from all the other equations written down so far precisely because it describes the time evolution of a quantum system.[a] In other words, given

---

[a](Schrödinger, 1926a)–(Schrödinger, 1926d).

the quantum state at some initial time (say $t_0 = 0$), we can solve the Schrödinger equation to obtain the quantum state at any subsequent time. Using the one-dimensional form for the Hamiltonian in the position representation (7.10), we can write the Schrödinger equation (7.12) in terms of the (now time-dependent) wave function $\psi(x, t) = \langle x | \psi(t) \rangle$ as [see Section 6.3]

$$i\hbar \frac{\partial}{\partial t} \psi(x, t) = -\frac{\hbar^2}{2m} \frac{\partial^2}{\partial x^2} \psi(x, t) + V(x)\psi(x, t). \qquad (7.13)$$

It is noted that in the above equation the total time derivative has reduced to the partial time derivative because the position eigenstates $|x\rangle$, and consequently the position variable x, are taken to be time independent. Similarly, the three-dimensional counterpart can be easily written down as

$$i\hbar \frac{\partial}{\partial t} \psi(\mathbf{r}, t) = -\frac{\hbar^2}{2m} \nabla^2 \psi(\mathbf{r}, t) + V(\mathbf{r})\psi(\mathbf{r}, t), \qquad (7.14)$$

where $\psi(\mathbf{r}, t) = \langle \mathbf{r} | \psi(t) \rangle$ is the three-dimensional time-dependent wave function. Using the above equation, Schrödinger calculated in late 1925 the energy levels of the hydrogen atom by treating the electron as a de Broglie matter wave [see Section 2.1] moving in a Coulomb potential created by the positively charged nucleus. The result of Schrödinger's calculation agreed accurately with the observed emission spectrum of atomic hydrogen [see Section 8.5] and subsequently created a revolution in quantum theory.

Problem 7.2 Write down the Schrödinger equation for a one-dimensional particle of mass m in a constant gravitational field.

7.3 Time Evolution as a Unitary Transformation

We have seen that momentum is the generator of space translations and therefore it is expressed as the partial derivative with respect to space [see Section 6.6]. Similarly, since the Hamiltonian governs the dynamics of a quantum system, it is the generator of time translations in the sense that it transforms the state vector at a given time to an infinitesimally later time, thus generating the time evolution of quantum states. To elaborate on this point, we first note that the solution to the time-dependent Schrödinger equation (7.12) is formally given by[a]

$$|\psi(t)\rangle = e^{-\frac{i}{\hbar}\hat{H}(t-t_0)}|\psi(t_0)\rangle, \tag{7.15}$$

where $|\psi(t_0)\rangle$ is the state vector at some initial time t_0 and t is any later time. The validity of the solution (7.15) can be verified straightforwardly by its direct substitution into Eq. (7.12) and by using the derivative of the exponential function (6.70) and the chain rule (6.75). It proves convenient to rewrite the above equation as

$$|\psi(t)\rangle = \hat{U}_t(t-t_0)|\psi(t_0)\rangle, \tag{7.16}$$

where $\hat{U}_t(t-t_0)$ is the time evolution operator

$$\hat{U}_t(t-t_0) = e^{-\frac{i}{\hbar}\hat{H}(t-t_0)}. \tag{7.17}$$

Note that since $\hat{H}$ is a Hermitian operator, $\hat{U}_t(t-t_0)$ is a unitary operator [see Section 4.6]. In other words, making use of a change of variables $\tau = t - t_0$, we have

$$\hat{U}_t^{\dagger}(\tau)\hat{U}_t(\tau) = \hat{U}_t(\tau)\hat{U}_t^{\dagger}(\tau) = \hat{I}, \tag{7.18}$$

where $\hat{U}_t^{\dagger}(\tau)$ is the conjugate transpose of $\hat{U}_t(\tau)$, namely,

$$\hat{U}_t^{\dagger}(\tau) = \left(e^{-\frac{i}{\hbar}\hat{H}\tau}\right)^{\dagger} = e^{+\frac{i}{\hbar}\hat{H}\tau}. \tag{7.19}$$

As a consequence, time evolution in quantum mechanics is a unitary transformation and hence preserving the scalar products of state vectors. In particular, we have

$$\langle\psi(t)|\psi(t)\rangle = \langle\psi(t_0)|\hat{U}_t^{\dagger}(\tau)\hat{U}_t(\tau)|\psi(t_0)\rangle = \langle\psi(t_0)|\psi(t_0)\rangle = 1, \tag{7.20}$$

where the unitarity property (7.18) has been used. Moreover, multiplying both sides of Eq. (7.16) by

$$\hat{U}_t^{\dagger}(\tau) = \hat{U}_t(-\tau), \tag{7.21}$$

and using the unitarity property of $\hat{U}_t(\tau)$, we obtain

$$|\psi(t_0)\rangle = \hat{U}_t(-\tau)|\psi(t)\rangle, \tag{7.22}$$

---

[a] For the sake of simplicity, we consider only the case in which the Hamiltonian does not explicit depend on time. The case of an explicitly time-dependent Hamiltonian can be treated accordingly by employing more involved mathematics and therefore is beyond the scope of this book.

namely, time evolution in quantum mechanics is reversible. This is precisely what we expect from the unitarity of the transformation [see Section 4.6]. In other words, while $\hat{U}_t(\tau)$ is the "forward" time evolution operator that brings the system from an initial state $|\psi(t_0)\rangle$ to a final state $|\psi(t)\rangle$ (with $t > t_0$), $\hat{U}_t^\dagger(\tau)$ is the "backward" time evolution operator that brings the system from a final state $|\psi(t)\rangle$ to an initial state $|\psi(t_0)\rangle$.

Now, in order to deal with the general problem of the time evolution of an arbitrary state vector, let us first consider the simplest situation, namely, when the initial state $|\psi(0)\rangle = |E\rangle$ (where for the sake of simplicity we have set $t_0 = 0$) is an eigenstate of the Hamiltonian $\hat{H}$ with the corresponding eigenvalue given by E [see Eq. (7.11)]. In this case, we can rewrite Eq. (7.15) as

$$|\psi(t)\rangle = e^{-\frac{i}{\hbar}Et}|\psi(0)\rangle, \qquad (7.23)$$

where use has been made of the fact that the energy eigenstate $|E\rangle$ is also an eigenstate of the time evolution operator $\hat{U}_t(t)$ with $e^{-\frac{i}{\hbar}Et}$ being the eigenvalue. Note that the factor $e^{-\frac{i}{\hbar}Et}$ on the right-hand side of the above equation is a global phase factor (i.e., an overall complex factor with unit modulus) and is therefore physically irrelevant [see Section 5.2]. This shows that energy eigenstates do not evolve with time, which allows us to write them without showing time dependence. For this reason, energy eigenstates are usually referred to as *stationary states*. Considering here the discrete case, let us make use of some labeling index n to denote the nth energy eigenstate by $|\psi_n\rangle$ and the corresponding eigenvalue by E_n. Then, we can reformulate the eigenvalue equation as [see Eq. (7.11)]

$$\hat{H}|\psi_n\rangle = E_n|\psi_n\rangle. \qquad (7.24)$$

This way of expressing the energy eigenvalues and energy eigenstates is not only conventional but universally used. We recall that since the energy eigenstates constitute a complete orthonormal basis, an arbitrary initial state vector $|\psi(0)\rangle$ can be expanded as a superposition of the energy eigenstates $|\psi_n\rangle$ as

$$|\psi(0)\rangle = \sum_n c_n(0)|\psi_n\rangle, \qquad (7.25)$$

where $c_n(0) = \langle\psi_n|\psi(0)\rangle$ are the coefficients of the expansion at the initial time. The state vector $|\psi(t)\rangle$ at a later time $t > 0$ can be found by utilizing the time evolution operator $\hat{U}_t(t)$ (since, having set $t_0 = 0$, we have $\tau = t$). From Eq. (7.16) we obtain

$$|\psi(t)\rangle = \hat{U}_t(t)|\psi(0)\rangle$$
$$= \sum_n c_n(0)\hat{U}_t(t)|\psi_n\rangle$$
$$= \sum_n c_n(0)e^{-\frac{i}{\hbar}E_n t}|\psi_n\rangle, \tag{7.26}$$

where used has been made of the fact that $|\psi_n\rangle$ are stationary states with energy E_n. It is noted that the factor $e^{-\frac{i}{\hbar}E_n t}$ in each term of the summation is a *relative* phase factor and is therefore physically relevant. Since $|\psi(t)\rangle$ can also be expanded as a superposition of the energy eigenstates $|\psi_n\rangle$, the above expression implies that the resultant expansion coefficients $c_n(t) = \langle\psi_n|\psi(t)\rangle$ are given by

$$c_n(t) = e^{-\frac{i}{\hbar}E_n t}c_n(0). \tag{7.27}$$

Similar expressions can be found for the continuous case, where we need to use integration instead of summation.

In this way, if the energy eigenvalues and energy eigenstates of a quantum system are known, then given the initial state of the system and its probability amplitudes in the energy eigenstates, we are able to follow the time evolution of the system and to obtain the corresponding probability amplitudes at arbitrary later times. In other words, the problem of studying the dynamics of a quantum system is reduced to that of finding the energy eigenvalues and energy eigenstates of the system. The latter are determined by the eigenvalue equation (7.11), which in the position representation is given by

$$-\frac{\hbar^2}{2m}\frac{\partial^2}{\partial x^2}\psi_E(x) + V(x)\psi_E(x) = E\psi_E(x), \tag{7.28}$$

where E is the energy eigenvalue and $\psi_E(x) = \langle x|E\rangle$ is the corresponding energy eigenfunction. Due to the mentioned stationarity of energy eigenstates (or eigenfunctions), the above equation is called the *time-independent* Schrödinger equation while its time-dependent counterpart Eq. (7.13) is referred to as the *time-dependent* Schrödinger equation. In three dimensions, the time-independent Schrödinger equation takes the form

$$-\frac{\hbar^2}{2m}\nabla^2\psi_E(\mathbf{r}) + V(\mathbf{r})\psi_E(\mathbf{r}) = E\psi_E(\mathbf{r}), \tag{7.29}$$

where E is the energy eigenvalue and $\psi_E(\mathbf{r}) = \langle \mathbf{r}|E \rangle$ is the corresponding energy eigenfunction. Moreover, given the knowledge of the initial state, since the same state can be expanded in different bases corresponding to the different observables being measured [see Section 4.5], thanks to Eq. (7.26) and the existence of quantum features [see Section 4.8], we are able to compute the probability amplitudes and therefore also the probabilities at any later time for the measurement outcomes associated with *any* observable. This clearly shows that quantum mechanics is full deterministic when considering the laws that govern the dynamics, precisely as it is the case for classical mechanics. The difference, however, is that in quantum mechanics measurement outcomes as detection events are random and therefore are not ruled by those laws [see Sections 5.4–5.5 and 6.9]. What quantum laws govern are only the probabilities for certain outcomes and not the outcomes themselves. It is a sort of determinism of probabilities.[a]

Problem 7.3 Check the validity of the solution given by Eq. (7.15).

Problem 7.4 In terms of the orthonormal basis $\{|1\rangle, |2\rangle\}$, the Hamiltonian $\hat{H}$ of a two-state system has eigenstates

$$|E_1\rangle = \frac{1}{\sqrt{2}}(|1\rangle + |2\rangle), \qquad |E_2\rangle = \frac{1}{\sqrt{2}}(|1\rangle - |2\rangle).$$

The corresponding energy eigenvalues are E_1 and E_2, respectively, where $E_1 \neq E_2$ are constants of energy dimension. Given that the initial state at time $t = 0$ is $|\psi(0)\rangle = |1\rangle$, compute the probability of finding the system in the state $|1\rangle$ as a function of t.

7.4 Active and Passive Transformations

In quantum mechanics we can consider the dynamics of a quantum system from two alternative (but equivalent) points of view: either by keeping the observables constant and considering the time evolution of the state, or by keeping the state constant and

---

[a]We suggest the reader to work out the problems at the end of this section and to take a stop here for a while, reading again the previous pages. It is also suggested that the reader takes again a pause after reading the next two sections and before going to concrete examples of dynamics.

considering the time evolution of the observables. In the first case it is the state vector that undergoes the transformation (this is called an *active transformation*); in the second we have the eigenstates of the observable under consideration undergoing the inverse transformation (this is called a *passive transformation*), in such a way that the two transformations are equivalent. The active transformation is also called the Schrödinger picture and was actually considered in the previous two sections. The passive transformation is called the Heisenberg picture, and it has been considered previously [see Sections 4.5 and 4.6] although without a specific treatment.

To illustrate the active and passive transformations with a simple example, let us introduce a two-state system and a generic observable $\hat{X}$ (which could be thought of as a kind of two-dimensional position observable like the one considered in Section 6.1). The eigenstates of the observable $\hat{X}$ are given by $|x_1\rangle$ and $|x_2\rangle$ (with x_1 and x_2 denoting the corresponding eigenvalues). The observable $\hat{X}$ that is under consideration is given by

$$\hat{X} = x_1|x_1\rangle\langle x_1| + x_2|x_2\rangle\langle x_2|. \qquad (7.30)$$

A certain initial state vector $|\psi\rangle$ that makes an angle $75°$ with respect to the basis state $|x_1\rangle$ can be written as

$$|\psi\rangle = \frac{\sqrt{3}-1}{2\sqrt{2}}|x_1\rangle + \frac{\sqrt{3}+1}{2\sqrt{2}}|x_2\rangle. \qquad (7.31)$$

Indeed, the two coefficients above correspond to the cosine of $75°$ (≈ 0.2588) and the sine of $75°$ (≈ 0.9659), respectively, as shown in Fig. 7.1(a). Let us write the basis states $|x_1\rangle$ and $|x_2\rangle$ as

$$|x_1\rangle = \begin{pmatrix} 0 \\ 1 \end{pmatrix} \quad \text{and} \quad |x_2\rangle = \begin{pmatrix} 1 \\ 0 \end{pmatrix}. \qquad (7.32)$$

We may now consider the following transformation in the *active* sense

$$\hat{U} = \frac{1}{\sqrt{2}} \begin{bmatrix} 1 & 1 \\ 1 & -1 \end{bmatrix}, \qquad (7.33)$$

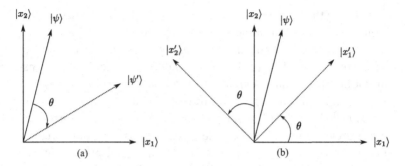

Figure 7.1 Active and passive transformations. A transformation may be considered from two equivalent viewpoints: (a) From the active point of view, the state vector $|\psi\rangle$ is transformed (here represented by a clockwise rotation by an angle θ) into the state vector $|\psi'\rangle$ while the basis vectors $|x_1\rangle$ and $|x_2\rangle$ are kept fixed. (b) From the passive point of view, the basis vectors $|x_1\rangle$ and $|x_2\rangle$ are transformed in a reverse manner (here represented by a counterclockwise rotation by the same angle θ) to the basis vectors $|x_1'\rangle$ and $|x_2'\rangle$, respectively. Note that the state $|\psi\rangle$ in the new basis $\{|x_1'\rangle, |x_2'\rangle\}$ is equivalent to the transformed state $|\psi'\rangle$ in the old basis $\{|x_1\rangle, |x_2\rangle\}$.

which induces a clockwise rotation of $45°$ on the state vector $|\psi\rangle$ (and is structurally similar to beam splitting [see Eq. (4.35)]). Then, the state vector $|\psi\rangle$ is transformed into

$$|\psi'\rangle = \hat{U}\,|\psi\rangle$$
$$= \frac{1}{\sqrt{2}}\,\frac{1}{2\sqrt{2}}\begin{bmatrix} 1 & 1 \\ 1 & -1 \end{bmatrix}\begin{pmatrix} \sqrt{3}-1 \\ \sqrt{3}+1 \end{pmatrix}$$
$$= \frac{1}{2}\begin{pmatrix} \sqrt{3} \\ 1 \end{pmatrix}. \tag{7.34}$$

In other words, we have

$$|\psi'\rangle = \hat{U}\,|\psi\rangle = \frac{\sqrt{3}}{2}|x_1\rangle + \frac{1}{2}|x_2\rangle. \tag{7.35}$$

Indeed, the coefficients $\frac{\sqrt{3}}{2}$ and $\frac{1}{2}$ in the above expansion are the cosine and sine of $30°$, respectively. This is precisely the transformed state where the initial state vector $|\psi\rangle$ ends after a clockwise rotation of $45°$, as shown again in Fig. 7.1(a). Moreover, it is easy to see that the sum of the respective probabilities is equal to 1, as expected for a unitary transformation. We then consider the inverse of $\hat{U}$, i.e., a counterclockwise rotation of $45°$:

$$\hat{U}^{-1} = \frac{1}{\sqrt{2}} \begin{bmatrix} 1 & 1 \\ 1 & -1 \end{bmatrix}. \tag{7.36}$$

Note that in this particular case $\hat{U}$ is indeed the inverse of itself. This transformation brings the original basis $\{|x_1\rangle, |x_2\rangle\}$ into the new basis $\{|x_1'\rangle, |x_2'\rangle\}$ [see Fig. 7.1(b)]:

$$\hat{U}^{-1}|x_1\rangle = \frac{1}{\sqrt{2}} \begin{bmatrix} 1 & 1 \\ 1 & -1 \end{bmatrix} \begin{pmatrix} 1 \\ 0 \end{pmatrix} = \frac{1}{\sqrt{2}} \begin{pmatrix} 1 \\ 1 \end{pmatrix} = |x_1'\rangle, \tag{7.37a}$$

$$\hat{U}^{-1}|x_2\rangle = \frac{1}{\sqrt{2}} \begin{bmatrix} 1 & 1 \\ 1 & -1 \end{bmatrix} \begin{pmatrix} 0 \\ 1 \end{pmatrix} = \frac{1}{\sqrt{2}} \begin{pmatrix} 1 \\ -1 \end{pmatrix} = |x_2'\rangle. \tag{7.37b}$$

We recall that a transformation of a basis means a *passive* transformation of the observable of which the basis is an eigenbasis. Since the eigenvalues of the observable are not affected by the transformation, the transformed observable is given by

$$\hat{X}' = x_1|x_1'\rangle\langle x_1'| + x_2|x_2'\rangle\langle x_2'|. \tag{7.38}$$

From Eqs. (7.37), the above expression can be cast into the following compact form

$$\begin{aligned} \hat{X}' &= x_1\hat{U}^{-1}|x_1\rangle\langle x_1|\hat{U} + x_2\hat{U}^{-1}|x_2\rangle\langle x_2|\hat{U} \\ &= \hat{U}^{-1}\left(x_1|x_1\rangle\langle x_1| + x_2|x_2\rangle\langle x_2|\right)\hat{U} \\ &= \hat{U}^{-1}\hat{X}\hat{U}, \end{aligned} \tag{7.39}$$

or equivalently,

$$\hat{X}' = \hat{U}^{\dagger}\hat{X}\hat{U}, \tag{7.40}$$

where use has been made of the unitarity of $\hat{U}$. An important consequence of the expression (7.40) is that commutation relations are invariant under a unitary transformation. In specific, we have [see Problem 7.5]

$$[\hat{X}, \hat{Y}] = \hat{Z} \xrightarrow{\hat{U}} [\hat{X}', \hat{Y}'] = \hat{Z}', \tag{7.41}$$

where $\hat{X}'$, $\hat{Y}'$, and $\hat{Z}'$ are respectively the transformed observables of $\hat{X}$, $\hat{Y}$, and $\hat{Z}$ under the unitary transformation $\hat{U}$.

Now we claim that the passive inverse transformation of the basis (7.37) is equivalent to the active transformation of the state vector (7.35). While the above example is very elementary, our conclusion is very general and remains valid for arbitrary unitary transformations. First, let us consider the clockwise rotation on the new basis $\{|x_1'\rangle, |x_2'\rangle\}$ that brings it back to the original basis $\{|x_1\rangle, |x_2\rangle\}$. Indeed, we have

$$\hat{U}|x_1'\rangle = \hat{U}\hat{U}^{-1}|x_1\rangle = |x_1\rangle, \tag{7.42}$$

$$\hat{U}|x_2'\rangle = \hat{U}\hat{U}^{-1}|x_2\rangle = |x_2\rangle. \tag{7.43}$$

Makinge use of these transformations, we can rewrite the expansion (7.35) as

$$\begin{aligned}|\psi'\rangle &= \frac{\sqrt{3}}{2}|x_1\rangle + \frac{1}{2}|x_2\rangle \\ &= \frac{\sqrt{3}}{2}\hat{U}|x_1'\rangle + \frac{1}{2}\hat{U}|x_2'\rangle \\ &= \hat{U}\left(\frac{\sqrt{3}}{2}|x_1'\rangle + \frac{1}{2}|x_2'\rangle\right) \\ &= \hat{U}|\psi\rangle, \end{aligned} \tag{7.44}$$

which implies

$$|\psi\rangle = \frac{\sqrt{3}}{2}|x_1'\rangle + \frac{1}{2}|x_2'\rangle, \tag{7.45}$$

since the transformation $\hat{U}^{-1}$ brings the state vector $|\psi'\rangle$ back to $|\psi\rangle$. In this way, a comparison between the previous equation and Eq. (7.35) provides the first observation that the state vector $|\psi\rangle$ in the new basis $\{|x_1'\rangle, |x_2'\rangle\}$ is equivalent to transformed state vector $|\psi'\rangle$ in the old basis $\{|x_1\rangle, |x_2\rangle\}$. The second observation is that the expectation value of the observable $\hat{X}$ in the (actively) transformed state vector $|\psi'\rangle$ is the same as that of the transformed observable $\hat{X}'$ in the state vector $|\psi\rangle$, namely, we have [see Eq. (6.152)]

$$\langle\psi'|\hat{X}|\psi'\rangle = \langle\psi|\hat{X}'|\psi\rangle. \tag{7.46}$$

The proof of the above equality is straightforward and is left as an exercise for the reader [see Problem 7.6]. We note that the left-hand and right-hand sides of the above equation are the *same* expectation value (of the observable that is being measured) calculated respectively from the active and passive viewpoints of the *same* transformation. Therefore, the equality (7.46) provides direct evidence for the equivalence between the active and passive transformations.

Problem 7.5 Prove the relation given by Eq. (7.41), namely, commutation relations are invariant under a unitary transformation.

Problem 7.6 Prove the equality $\langle \psi' | \hat{X} | \psi' \rangle = \langle \psi | \hat{X}' | \psi \rangle$ given by Eq. (7.46).

7.5 Schrödinger and Heisenberg Pictures

Having discussed a concrete example, we can now deal with the issue of active (Schrödinger) and passive (Heisenberg) transformations at a general level in the context of quantum dynamics. In what follows when we indicate an observable $\hat{O}$ and its eigenstates $|o_j\rangle$ without time dependence, they are considered in the Schrödinger picture and therefore correspond to the untransformed ones in the Heisenberg picture. Reciprocally, when the state vector $|\psi(t)\rangle$ carries time dependency, it is always considered in the Schrödinger picture.

We can write down the expectation value of, for instance, the position observable $\hat{x}$ (that could be continuous or discrete) in the state $|\psi(t)\rangle$ at time t in the Schrödinger picture as $\langle \psi(t) | \hat{x} | \psi(t) \rangle$ [see Eq. (6.152)]. By making use of Eq. (7.16) with $t_0 = 0$, we have

$$\langle \psi(t) | \hat{x} | \psi(t) \rangle = \langle \psi(0) | \hat{U}_t^\dagger(t) \, \hat{x} \, \hat{U}_t(t) | \psi(0) \rangle, \qquad (7.47)$$

which allows us to displace the action of the time evolution operator from the state vector to the observable, and therefore to go over to the Heisenberg picture. Define the position observable $\hat{x}(t)$ and the state vector $|\psi\rangle$ in the Heisenberg picture by [see Eq. (7.40)]

$$\hat{x}(t) = \hat{U}_t^\dagger(t) \, \hat{x} \, \hat{U}_t(t), \qquad (7.48a)$$

$$|\psi\rangle = |\psi(0)\rangle = \hat{U}_t^\dagger(t) | \psi(t) \rangle, \qquad (7.48b)$$

respectively. The expectation value of the position observable $\hat{x}(t)$ in the state $|\psi\rangle$ at time t in the Heisenberg picture can be expressed in terms of the observable and state vector in the Schrödinger picture as

$$\langle \psi | \hat{x}(t) | \psi \rangle = \langle \psi(0) | \hat{U}_t^\dagger(t) \, \hat{x} \, \hat{U}_t(t) | \psi(0) \rangle. \qquad (7.49)$$

Comparing Eqs. (7.47) and (7.49), we arrive at the important relation

$$\langle \psi(t) | \hat{x} | \psi(t) \rangle = \langle \psi | \hat{x}(t) | \psi \rangle, \tag{7.50}$$

which establishes the equivalence between the Schrödinger and Heisenberg pictures and can be therefore considered the generalization of Eq. (7.46) that we are looking for. In other words, in accordance with what is discussed in the previous section, the time evolution of an observable in the Heisenberg picture is precisely expressed as the time evolution of the basis vectors that constitute the eigenbasis of the observable. Moreover, this generalization allows us to write the equivalent of the Schrödinger equation in the Heisenberg picture by taking the time derivative of $\hat{x}(t)$ and by making use of the following expression that can be considered a reformulation of the Schrödinger equation (7.12) [see also Eq. (7.17)]

$$\frac{d\hat{U}_t(t)}{dt} = \frac{d}{dt} e^{-\frac{i}{\hbar}\hat{H}t} = -\frac{i}{\hbar}\hat{H}\hat{U}_t(t) = -\frac{i}{\hbar}\hat{U}_t(t)\hat{H}. \tag{7.51}$$

Here, we have made use of the rules (6.70) and (6.75) in the derivative of the exponential function, and the last equality in the above equation is due to the fact that $\hat{H}$ and $\hat{U}_t(t)$ commute. From Eqs. (7.48a) and (7.51), the time derivative of $\hat{x}(t)$ is found to be given by

$$\begin{aligned}
\frac{d\hat{x}(t)}{dt} &= \frac{d}{dt}\left[\hat{U}_t^\dagger(t)\,\hat{x}\,\hat{U}_t(t)\right] \\
&= \frac{d\hat{U}_t^\dagger(t)}{dt}\,\hat{x}\,\hat{U}_t(t) + \hat{U}_t^\dagger(t)\,\hat{x}\,\frac{d\hat{U}_t(t)}{dt} \\
&= \frac{i}{\hbar}\hat{H}\hat{U}_t^\dagger(t)\,\hat{x}\,\hat{U}_t(t) - \frac{i}{\hbar}\hat{U}_t^\dagger(t)\,\hat{x}\,\hat{U}_t(t)\hat{H} \\
&= \frac{i}{\hbar}\left[\hat{H}\hat{x}(t) - \hat{x}(t)\hat{H}\right], \tag{7.52}
\end{aligned}$$

where $\hat{x}$ is treated as a constant (since it has no dependence in time) and in the second line use has been made of the product rule (6.73). By multiplying the above equation by $i\hbar$ and by expressing the last line in terms of the commutator, we finally have

$$i\hbar\frac{d\hat{x}(t)}{dt} = \left[\hat{x}(t), \hat{H}\right], \tag{7.53}$$

where use has been made of the antisymmetric property of the commutator [see Eq. (6.151a)]. The above equation is called the *Heisenberg equation* for the position observable. These considerations are obviously true for a generic quantum observable in the Heisenberg picture, provided that the counterpart of the latter in the Schrödinger picture does *not* depend explicitly on time. Let $\hat{O}$ be a generic Schrödinger-picture quantum operator which does not have explicit time dependence. The corresponding operator in the Heisenberg picture $\hat{O}(t)$ is defined by

$$\hat{O}(t) = \hat{U}_t^\dagger(t)\, \hat{O}\, \hat{U}_t(t),\tag{7.54}$$

whose time evolution is governed by the Heisenberg equation

$$i\hbar\frac{d}{dt}\hat{O}(t) = \left[\hat{O}(t),\, \hat{H}\right].\tag{7.55}$$

In classical mechanics there is no equivalent of the Heisenberg picture. As a matter of fact, since states are treated as collections of properties and therefore are considered observables from any point of view [see Section 6.9], this necessity does not arise at all, with the consequence that the time evolution of observables (or variables) was traditionally not considered as such. It is the richness and powerfulness of quantum mechanics to have thrown more light on this issue from both a formal and a conceptual points of view. We could say that it is a less "flat" theory than classical mechanics.

Before ending this section, we also mention that a third picture is possible, which is called the *Dirac picture* (or interaction picture) and is based on the split of the Hamiltonian into a part that depends on the interaction with other systems and a part that does not. Such a picture is especially useful when (time-dependent) perturbations are involved, a subject that goes far beyond the scope of this book.[a]

Problem 7.7 Show that the Hamiltonian in the Heisenberg picture is the same as that in the Schrödinger picture.

Problem 7.8 Let $\hat{O} = [\hat{O}_1, \hat{O}_2]$, where $\hat{O}$, $\hat{O}_1$, and $\hat{O}_2$ are operators in the Schrödinger picture (a commutator of operators can be itself an operator). Show that in the Heisenberg picture we have $\hat{O}(t) = [\hat{O}_1(t), \hat{O}_2(t)]$, where $\hat{O}(t)$, $\hat{O}_1(t)$, and $\hat{O}_2(t)$ are the corresponding Heisenberg-picture operators.

---

[a]The interested reader may have a look at (Auletta *et al.*, 2009, Chapter 10).

7.6 Free Particle

In this and the following section, we shall consider two specific examples of one-dimensional quantum systems: the free particle and the harmonic oscillator. Let us first determine the possible range of the energy eigenvalue E for a generic one-dimensional quantum system as described by the time-independent Schrödinger equation (7.28),[a] which we rewrite as the eigenvalue equation

$$\hat{H}|\psi\rangle = E|\psi\rangle. \tag{7.56}$$

Here the subscript E in the energy eigenstate $|\psi_E\rangle$ has been suppressed for the sake of notational simplicity (provided that no confusion may arise). By left multiplying this equation by $\langle\psi|$ and using Eq. (7.9), we obtain

$$E = \langle\psi|\hat{H}|\psi\rangle = \frac{1}{2m}\langle\psi|\hat{p}_x^2|\psi\rangle + \langle\psi|V(\hat{x})|\psi\rangle, \tag{7.57}$$

where use has been made of the fact that the energy eigenstate $|\psi\rangle$ is normalized, i.e., $\langle\psi|\psi\rangle = 1$. The expectation value $\langle\psi|\hat{p}_x^2|\psi\rangle$ is non-negative because it can be written as the norm squared of the ket $|\hat{p}_x\psi\rangle = \hat{p}_x|\psi\rangle$, namely,

$$\langle\psi|\hat{p}_x^2|\psi\rangle = \langle\hat{p}_x\psi|\hat{p}_x\psi\rangle \geq 0. \tag{7.58}$$

While the expectation $\langle\psi|V(\hat{x})|\psi\rangle$ does not have a definite sign, it does satisfy the inequality

$$\langle\psi|V(\hat{x})|\psi\rangle \geq V_{\min}, \tag{7.59}$$

provided that the potential energy $V(x)$ has a minimum $V_{\min}$, i.e., $V(x) \geq V_{\min}$ for all values of x. This inequality can be easily derived as follows. In the position representation we have

$$\langle\psi|V(\hat{x})|\psi\rangle = \int_{-\infty}^{+\infty} dx|\psi(x)|^2 V(x) \geq V_{\min} \int_{-\infty}^{+\infty} dx|\psi(x)|^2 = V_{\min}, \tag{7.60}$$

where in the first equality use has been made of the completeness condition (6.39)—an operation that is always allowed regardless of the specific meaning of $|\psi\rangle$—while in the last equality use has been

---

[a]For the sake of simplicity and of later application, we consider here the one-dimensional case. Nevertheless, the analysis can be generalized straightforwardly to three dimensions.

made of the fact that the energy eigenfunction $\psi(x) = \langle x|\psi \rangle$ is normalized, i.e., $\int_{-\infty}^{+\infty} dx |\psi(x)|^2 = 1$. Collecting the results given by Eqs. (7.57)–(7.60), we arrive at the following important property of the energy eigenvalue for a one-dimensional system

$$E \geq V_{\min}. \tag{7.61}$$

Indeed, this property is not totally unexpected. Since the kinetic energy of a system is always greater than or equal to zero, its total energy (i.e., the sum of the kinetic energy and the potential energy [see Section 7.1]) should be greater than or equal the lowest possible potential energy.

A free particle is a particle that is not subject to external forces, or equivalently, a particle moving in a region where its potential energy is constant in space. Since the potential energy is defined up to an additive constant, it is convenient to choose the constant such that the potential energy vanishes. Therefore, we have $V = 0$ for a free particle. Then the time-independent Schrödinger equation (7.28) reduces to

$$-\frac{\hbar^2}{2m} \frac{\partial^2}{\partial x^2} \psi(x) = E \psi(x), \tag{7.62}$$

where E is the energy eigenvalue and $\psi(x)$ is the corresponding energy eigenstate. For the sake of notational simplicity, we have again suppressed the subscript E in the energy eigenfunction. Since we have $V = 0$, we also have $V_{\min} = 0$ and $E \geq 0$. It proves convenient to define a parameter called the *wave number*

$$k = \frac{\sqrt{2mE}}{\hbar}, \tag{7.63}$$

which is real, non-negative, and of inverse length dimension. In fact, we have $k = 2\pi/\lambda$, where λ is the wavelength associated with the free-particle wave function $\psi(x)$. Note that for a free particle, the relation between the energy and momentum can be written as $E = p_x^2/2m$ [see also Eqs. (7.8) and (7.9)]. This allows us to rewrite Eq. (7.63) as $k = p_x/\hbar$, from which the fundamental relation $p_x = h/\lambda$ follows for the one-dimensional case (it can be easily generalized to the three-dimensional case as $p = h/\lambda$, where p is the magnitude of the momentum vector $\mathbf{p}$). This relation is known as the *de Broglie equation*, which relates the momentum of a free particle (a particle-like property) to its wavelength (a wave-like property). In terms of k the eigenvalue equation (7.62) can be rewritten as

$$\psi''(x) + k^2 \psi(x) = 0, \tag{7.64}$$

which is a homogeneous linear second-order ordinary differential equation with constant coefficients [see also Eq. (6.128) although being a first-order one]. The energy eigenfunctions, i.e., the solutions to the above differential equation, are given by [see Box 7.1 and, in particular, Eq. (7.71), and also Eq. (6.130)]

$$\psi(x) = \frac{1}{\sqrt{2\pi\hbar}} e^{\pm ikx} \quad (k \geq 0), \tag{7.65}$$

Box 7.1 Linear differential equations with constant coefficients II

A second order homogeneous linear differential equation with constant coefficients can be written as

$$ay''(x) + by'(x) + cy(x) = 0, \tag{7.66}$$

where a, b, and c are real constants with $a \neq 0$. From our experience with the first order case, we expect at least some of the solutions to be exponentials. So let us find all such solutions by setting $y(x) = e^{-rx}$, where r is a constant to be determined. Substituting the expression into the differential equation (7.66), we obtain after some algebra the *characteristic equation* of this differential equation

$$ar^2 + br + c = 0. \tag{7.67}$$

This is a quadratic equation and has two solutions given by the so-called quadratic formula

$$r = \frac{-b \pm \sqrt{b^2 - 4ac}}{2a}. \tag{7.68}$$

The quantity under the square root, $b^2 - 4ac$, is called the discriminant of the quadratic equation, whose sign determines the nature of the roots. There are three cases to consider.

(i) The discriminant is positive. In this case, the roots r_1 and r_2 are real and distinct. The corresponding solutions $e^{r_1 x}$ and $e^{r_2 x}$ are linearly independent, therefore the general solution is

$$y(x) = c_1 e^{r_1 x} + c_2 e^{r_2 x}, \qquad (7.69)$$

where c_1 and c_2 are constants.

(ii) The discriminant is zero. In this case, the roots are real and identical (called a double root). If this double root is denoted simply by r, then one of the solution is e^{rx}. It can be shown that the other linearly independent solution is xe^{rx}. Hence, the general solution is

$$y(x) = c_1 e^{rx} + c_2 x e^{rx}, \qquad (7.70)$$

where c_1 and c_2 are constants.

(iii) The discriminant is negative. In this case, the roots $r_1 = \alpha + i\beta$ and $r_2 = \alpha - i\beta$ (with α and β being real) are distinct complex conjugate numbers. The corresponding solutions $e^{\alpha x} e^{i\beta x}$ and $e^{\alpha x} e^{-i\beta x}$ are linearly independent, therefore the general solution is

$$y(x) = e^{\alpha x}(c_+ e^{i\beta x} + c_- e^{-i\beta x}), \qquad (7.71)$$

where c_+ and c_- are constants. The solution can be equivalently expressed in terms of the sine and cosine functions as [see Eqs. (2.16) and (2.18)]

$$y(x) = e^{\alpha x}(c_1 \cos \beta x + c_2 \sin \beta x), \qquad (7.72)$$

where c_1 and c_2 are constants.

where the constant $1/\sqrt{2\pi \hbar}$ is there to ensure the usual normalization condition. Recall that the set of all eigenvalues of an observable is called its spectrum. Since for each $k \geq 0$ there is a corresponding $E \geq 0$, the energy spectrum for a free particle is the continuous interval $E \geq 0$ [see also Section 6.2].

Moreover, from Eq. (6.123) it can be easily recognized that the free-particle energy eigenfunctions corresponding to the energy eigenvalue $E > 0$ are momentum eigenfunctions with eigenvalues

$p_x = \hbar k$ and $p_x = -\hbar k$. In other words, the energy eigenvalue $E > 0$ is degenerate in the sense that there are two distinct (i.e., orthogonal) eigenfunctions associated with each free particle energy eigenvalues $E > 0$. Indeed, since a free particle only has kinetic energy, it is clear that a moving free particle (hence with $E > 0$) can move in either one of the two opposite directions parallel to its (straight) trajectory. Let the energy eigenstate of the free particle with energy E and momentum $\hbar k$ be denoted by $|E, k\rangle$ and that with energy E and momentum $-\hbar k$ by $|E, -k\rangle$. Then we have

$$\hat{H}|E, \pm k\rangle = E|E, \pm k\rangle \quad \text{and} \quad \hat{p}_x|E, \pm k\rangle = \pm\hbar k|E, \pm k\rangle, \quad (7.73)$$

where it is noted that the use of the plus–minus sign ($\pm$) is a shorthand notation to present *two* equations in one expression, in which the upper and lower signs are interrupted separately and independently. Hence $|E, \pm k\rangle$ are common eigenstates of the Hamiltonian $\hat{H}$ and the momentum observable $\hat{p}_x$. Indeed, it can be shown that the necessary and sufficient condition that two observables have common eigenstates is that they commute. In such a case, these two observables are also jointly measurable [see Section 6.8]. Therefore, the condition for a degenerate energy spectrum is that there exists at least one nontrivial observable that commutes with the Hamiltonian of the system. For the case of the free particle, the observable is the momentum operator [see Problem 7.9]. We note that this case is our first example of a degenerate energy spectrum, another example will be discussed in Chapter 8.

Problem 7.9 Verify that for a free particle the Hamiltonian and the momentum operator commute.

7.7 Harmonic Oscillator

One of most important examples of quantum dynamics is provided by the one-dimensional harmonic oscillator (about which we can only provide here a very sketchy presentation). A classical oscillator is a very simple system like a pendulum that swings back and forth about the equilibrium point. The oscillation of a classical point particle is represented in Fig. 7.2, which shows that the more it

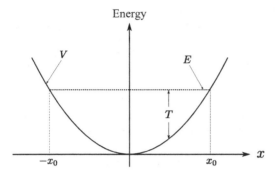

Figure 7.2 The potential energy V, kinetic energy T, and total energy E of a classical harmonic oscillator are plotted as a function of position x. The region $|x| > x_0$, where the total energy is less than the potential energy and in classical mechanics the oscillator cannot appear, is called the classically forbidden region.

swings (farther distance from the equilibrium point located at the origin), the more energy it has (higher level on the vertical energy axis). Something similar applies to quantum oscillators (actually electrons in an atom can be understood as sort of oscillators), with the proviso that the energy levels are here quantized.

The potential energy of a classical harmonic oscillator features a characteristic quadratic (or harmonic) form [again Fig. 7.2; see also Fig. 6.3]

$$V(x) = \frac{1}{2}m\omega^2 x^2, \tag{7.74}$$

where m is the mass, $\omega = 2\pi\nu$ is the angular frequency of the harmonic oscillator, and ν is the frequency, as usual. The Hamiltonian of a quantum harmonic oscillator is given by [see Eq. (7.9)]

$$\hat{H} = \frac{\hat{p}_x^2}{2m} + \frac{1}{2}m\omega^2 \hat{x}^2. \tag{7.75}$$

With the above Hamiltonian, we obtain the time-independent Schrödinger equation for the one-dimensional harmonic oscillator [see Eq. (7.28)]

$$\psi''(x) + \frac{2m}{\hbar^2}\left(E - \frac{1}{2}m\omega^2 x^2\right)\psi(x) = 0, \tag{7.76}$$

where for the sake of notational simplicity we have suppressed the subscript E in the energy eigenfunction. However, following the convention already adopted in Eq. (7.24), we shall make use of the (discrete) index n to label the energy levels and the corresponding eigenfunctions and eigenvalues. Indeed, as already mentioned, explicit calculation shows that only certain solutions $\psi_n(x)$ (i.e., the energy eigenfunctions) and the corresponding discrete energy eigenvalues E_n are allowed, where $n = 0, 1, 2, \ldots$ is called the *quantum number* of the harmonic oscillator.

It is very interesting and instructive to realize that the harmonic oscillator problem can be solved by exploiting only the resource of the canonical commutation relations.[a] To see how this is possible, let us introduce the *annihilation* (lowering) and *creation* (raising) operators

$$\hat{a} = \sqrt{\frac{m}{2\hbar\omega}} \left(\omega\hat{x} + i\frac{\hat{p}_x}{m} \right), \tag{7.77a}$$

$$\hat{a}^\dagger = \sqrt{\frac{m}{2\hbar\omega}} \left(\omega\hat{x} - i\frac{\hat{p}_x}{m} \right), \tag{7.77b}$$

respectively. The specific meaning of these two operators will be clear below. By now, let us say that the action of the annihilation operator represents the emission (loss) of a "photon" (or an energy quantum) while the action of the creation operator represents the absorption (gain) of a "photon" (or an energy quantum).[b] From the canonical commutation relation (6.148), it can be shown [see Problem 7.10] that the annihilation and creation operators satisfy the following important commutation relation

$$[\hat{a}, \hat{a}^\dagger] = \hat{I}. \tag{7.78}$$

We note that the annihilation and creation operators are not Hermitian operators (one is the Hermitian conjugate of the other) and therefore they are not observables (we had already anticipated

---

[a](Auletta *et al.*, 2009, Section 4.4).

[b]We note that actually photons do not have a mass but do possess a momentum. The momentum of a photon is along its propagation direction and the magnitude of the momentum p is given by the de Broglie equation $p = h/\lambda$, where λ is the wavelength of the photon [see p. 175]. As a result, strictly speaking the expression $\hat{p}_x/m$ is not well-defined for photons. However, we can think of m as a small fictitious photon mass that is introduced by hand solely for the purpose of analysis. This is perfectly fine provided that the fictitious photon mass does not appear in the final results of relevant physical quantities (like energy and the number of energy quanta).

this possibility in Section 4.4). Instead, an observable is the operator $\hat{N} = \hat{a}^\dagger \hat{a}$, which is Hermitian and can be expressed in terms of $\hat{x}$ and $\hat{p}_x$ as

$$\hat{a}^\dagger \hat{a} = \frac{m}{2\hbar\omega} \left(\omega\hat{x} - i\frac{\hat{p}_x}{m} \right) \left(\omega\hat{x} + i\frac{\hat{p}_x}{m} \right)$$

$$= \frac{m}{2\hbar\omega} \left(\omega^2\hat{x}^2 + \frac{i\omega}{m}\hat{x}\hat{p}_x - \frac{i\omega}{m}\hat{p}_x\hat{x} + \frac{\hat{p}_x^2}{m^2} \right)$$

$$= \frac{m}{2\hbar\omega} \left(\omega^2\hat{x}^2 + \frac{i\omega}{m}[\hat{x}, \hat{p}_x] + \frac{\hat{p}_x^2}{m^2} \right)$$

$$= \frac{1}{\hbar\omega} \left(\frac{1}{2}m\omega^2\hat{x}^2 + \frac{\hat{p}_x^2}{2m} \right) - \frac{1}{2}, \qquad (7.79)$$

where we have made use of the canonical commutation relation (6.148). For the reason that will be clear below, the operator $\hat{N}$ is called the *number operator*. The term in the parentheses in the last line of Eq. (7.79) is just the Hamiltonian of the quantum harmonic oscillator (7.75). Therefore, we can write

$$\hat{H} = \left(\hat{N} + \frac{1}{2} \right) \hbar\omega = \left(\hat{a}^\dagger \hat{a} + \frac{1}{2} \right) \hbar\omega. \qquad (7.80)$$

The name *number* operator is due to the fact that it represents the number of "photons" or energy quanta present in a certain system (for instance, the harmonic oscillator). Since the number of energy quanta is an observable, this explains why the number operator is so important. Its eigenvalue equation is given by

$$\hat{N}|n\rangle = n|n\rangle \quad (n = 0, 1, 2, \dots), \qquad (7.81)$$

where $|n\rangle$ is the nth (normalized) energy eigenstate and the corresponding eigenvalue n is precisely the number of energy quanta present. For the sake of notational simplicity we have adopted the further simplified (and widely used) notation $|\psi_n\rangle = |n\rangle$ for indicating the energy eigenstates. Indeed, from Eq. (7.80) it is evident that the number eigenstate $|n\rangle$ is also the energy eigenstate (as it should be), namely, we have

$$\hat{H}|n\rangle = E_n|n\rangle \quad \text{and} \quad E_n = \left(n + \frac{1}{2} \right) \hbar\omega \quad (n = 0, 1, 2, \dots).$$

$$(7.82)$$

The first few values of E_n are displayed in Fig. 7.3. On the other hand, we expect that the number operator is related to the annihilation

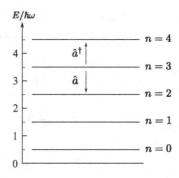

Figure 7.3 Schematic representation of the first few energy levels of a quantum harmonic oscillator as labeled by the quantum number n, which indicates the energy of the harmonic oscillator. For transition from one level to the adjacent level, an energy quantum of energy $\hbar\omega$ needs to be either emitted or be absorbed.

and creation operators since the latter determine respectively how many energy quanta are released or acquired by a certain system. It is therefore suitable to explore a little more their properties. To this purpose, let us consider the action of $\hat{N}$ on the state vector $|\hat{a}n\rangle = \hat{a}|n\rangle$

$$\hat{N}\hat{a}|n\rangle = \left(\hat{N}\hat{a} - \hat{a}\hat{N} + \hat{a}\hat{N}\right)|n\rangle$$
$$= \left([\hat{N}, \hat{a}] + \hat{a}\hat{N}\right)|n\rangle$$
$$= -\hat{a}|n\rangle + n\hat{a}|n\rangle$$
$$= (n-1)\hat{a}|n\rangle, \qquad (7.83)$$

where in the first line we have simply added and subtracted the same quantity and in the third line we have made use of both Eq. (7.81) and the following commutation relation [see Eqs. (6.151a) and (7.78)]:

$$[\hat{N}, \hat{a}] = [\hat{a}^\dagger\hat{a}, \hat{a}] = [\hat{a}^\dagger, \hat{a}]\hat{a} = -\hat{a}. \qquad (7.84)$$

From Eq. (7.83) it follows that $\hat{a}|n\rangle = |\hat{a}n\rangle$ is an eigenstate of $\hat{N}$ with eigenvalue $n-1$, that is, we have:

$$\hat{N}|\hat{a}n\rangle = (n-1)|\hat{a}n\rangle, \qquad (7.85)$$

but from Eq. (7.81) it follows that

$$\hat{N}|n-1\rangle = (n-1)|n-1\rangle. \qquad (7.86)$$

Since the eigenstates $\hat{a}|n\rangle$ and $|n - 1\rangle$ correspond to the same eigenvalue, they have to be proportional to each other. Hence we have

$$\hat{a}|n\rangle = c_n|n - 1\rangle, \tag{7.87}$$

where c_n is a proportional constant that is formally given by $c_n = \langle n - 1|\hat{a}n\rangle$. To find c_n, we left multiply both sides of Eq. (7.87) by $\langle \hat{a}n|$, which is tantamount to taking the norm squared of the state vector $\hat{a}|n\rangle$, to obtain

$$|c_n|^2 = \langle n - 1|\hat{a}n\rangle\langle \hat{a}n|n - 1\rangle = \langle n|\hat{a}^\dagger\hat{a}|n\rangle = \langle n|\hat{N}|n\rangle = n. \tag{7.88}$$

Therefore, we find $c_n = \sqrt{n}$ up to a global phase factor, which can be absorbed into $|n - 1\rangle$. In other words, we have

$$\hat{a}|n\rangle = \sqrt{n}|n - 1\rangle, \tag{7.89}$$

A similar analysis shows that the state vector $\hat{a}^\dagger|n\rangle$ is an eigenstate of $\hat{N}$ with eigenvalue $(n + 1)$ and that the following relation holds [see Problem 7.11]:

$$\hat{a}^\dagger|n\rangle = \sqrt{n + 1}|n + 1\rangle. \tag{7.90}$$

Equations (7.89) and (7.90) represent final mathematical justification of why $\hat{a}$ and $\hat{a}^\dagger$ are called annihilation (lowering) and creation (raising) operators, respectively. We may then interpret the set of the harmonic oscillator eigenstates as rungs of a ladder, which the oscillator may "climb" a level through $\hat{a}^\dagger$ or "descend" a level through $\hat{a}$. The minimum allowed value for n is zero (no photons are present) and therefore we must have

$$\hat{a}|0\rangle = 0, \tag{7.91}$$

where $|0\rangle$ is the lowest energy eigenstate, called the *ground state*. Higher energy eigenstates can be obtained from the ground state by repeatedly applying the creation operator on the latter, according to Eq. (7.90), so that the general formula can be written as [see Problem 7.12]

$$|n\rangle = \frac{(\hat{a}^\dagger)^n}{\sqrt{n!}}|0\rangle. \tag{7.92}$$

In the position representation we may write the energy eigenfunctions as $\psi_n(x) = \langle x|n\rangle$. From Eqs. (7.77a) and (7.91) we find that the energy eigenfunction for the ground state $\psi_0(x) = \langle x|0 \rangle$ is determined by the equation [see Problem 7.13]

$$\frac{d\psi_0(x)}{dx} + \frac{m\omega}{\hbar} x\psi_0(x) = 0. \tag{7.93}$$

While the above differential equation is not of the constant coefficient types that we know how to solve, a quick change of variables will do the trick. Let us define the dimensionless variable (that is, a quantity without an associated physical dimension or a "pure" number, which conventionally always has the dimension of 1)

$$\xi = \frac{m\omega}{\hbar} x^2. \tag{7.94}$$

Using the chain rule (6.75), we have

$$x\frac{d\psi_0(x)}{dx} = 2\xi\frac{d\psi_0(\xi)}{d\xi}, \tag{7.95}$$

hence we can perform a change of variables and rewrite Eq. (7.93) in terms of the new variable ξ as

$$\psi_0'(\xi) + \frac{1}{2}\psi_0(\xi) = 0, \tag{7.96}$$

where $\psi_0'(\xi) = d\psi_0(\xi)/d\xi$. The solution to the above equation is given by [see Box 6.8 and Eq. (6.130)]

$$\psi_0(\xi) = \mathcal{N}e^{-\frac{\xi}{2}}, \tag{7.97}$$

or equivalently expressed in terms of the original variable x as

$$\psi_0(x) = \mathcal{N}e^{-\frac{m\omega}{2\hbar}x^2}, \tag{7.98}$$

where $\mathcal{N}$ is a normalization constant which can be determined by imposing the usual normalization condition [see Problem 7.14]

$$\int_{-\infty}^{+\infty} dx |\psi_0(x)|^2 = 1. \tag{7.99}$$

After normalization we finally obtain the ground state energy eigenfunction for the harmonic oscillator

$$\psi_0(x) = \left(\frac{m\omega}{\pi\hbar}\right)^{\frac{1}{4}} e^{-\frac{m\omega}{2\hbar}x^2}. \tag{7.100}$$

The simplest way to determine all the excited state wave functions $\psi_n(x)$ for $n > 0$ is by repeated application of the creation operator

---

Box 7.2 Excited state eigenfunctions of the harmonic oscillator

By repeated application of the creation operator $\hat{a}^{\dagger}$ [Eqs. (7.90) and (7.92)] to the ground state wave function (7.100) we can obtain the harmonic oscillator's excited state eigenfunctions. We formulate here the first two excited state wave functions. We start from $\psi_0(x)$ and obtain $\psi_1(x)$ as

$$\psi_1(x) = \sqrt{\frac{m}{2\hbar\omega}} \left(\omega x - \frac{\hbar}{m} \frac{\partial}{\partial x} \right) \psi_0(x)$$

$$= \frac{1}{\sqrt{2}} \left(\frac{m\omega}{\pi\hbar} \right)^{\frac{1}{4}} 2\sqrt{\frac{m\omega}{\hbar}} x\, e^{-\frac{m\omega}{2\hbar}x^2}, \qquad (7.101)$$

from which $\psi_2(x)$ is obtained as

$$\psi_2(x) = \sqrt{\frac{m}{4\hbar\omega}} \left(\omega x - \frac{\hbar}{m} \frac{\partial}{\partial x} \right) \psi_1(x)$$

$$= \frac{1}{\sqrt{8}} \left(\frac{m\omega}{\pi\hbar} \right)^{\frac{1}{4}} \left(4\frac{m\omega}{\hbar}x^2 - 2 \right) e^{-\frac{m\omega}{2\hbar}x^2}, \qquad (7.102)$$

etc. In general, we may find $\psi_n(x)$ if $\psi_{n-1}(x)$ is known thanks to the relation

$$\psi_n(x) = \sqrt{\frac{m}{2\hbar n\omega}} \left(\omega x - \frac{\hbar}{m} \frac{\partial}{\partial x} \right) \psi_{n-1}(x). \qquad (7.103)$$

$\hat{a}^{\dagger}$ to the ground state wave function $\psi_0(x)$, as shown in Box 7.2. The energy of the ground state and the excited states can be obtained from Eq. (7.82). In particular, the ground state energy is called the *zero-point energy* of the harmonic oscillator and is given by

$$E_0 = \frac{1}{2}\hbar\omega. \qquad (7.104)$$

It is noted that the ground state energy E_0 is different from zero, which has been taken as the minimum value of the potential energy. From the above discussion, it appears clearly that the quantity $\hbar\omega$ may be interpreted as the energy quantum for a harmonic oscillator of angular frequency ω. The oscillator may "jump" from level j to level k only when $|j - k|$ energy quanta are either absorbed $(j < k)$ or emitted $(j > k)$ [see Fig. 7.3].

Equation (7.100) shows that the wave function for the ground state of a quantum harmonic oscillator is a pure Gaussian (or a bell-shaped curve), for which the uncertainty relation (6.171) is said to be saturated, i.e., we have

$$(\Delta \hat{p}_x \Delta \hat{x})_{\psi_0} = \frac{\hbar}{2}. \qquad (7.105)$$

Moreover, as the Gaussian extends (but falls off quickly) towards plus and minus infinity, it follows that there is a finite (albeit small) probability of finding the oscillator appearing in regions far away from the equilibrium point. This is in sharp contrast to the classical case in which the oscillator is always confined in the region where its total energy is greater than or equal to the potential energy [see Fig. 7.2]. The region in which the classical oscillator cannot appear is called the *classically forbidden region*. Therefore, the phenomenon that a quantum particle appears in the classically forbidden region is purely quantum mechanical and is referred to as *quantum tunneling*. The phenomenon of tunneling has many important applications. For instance, it describes a type of radioactive decay (called alpha decay) in which a nucleus emits an alpha particle (a helium nucleus). It is also the basic principle used in the scanning tunneling microscope (or STM for short), which is an instrument for imaging surfaces at the atomic level.

Problem 7.10 Prove Eq. (7.78). (*Hint*: Use the fact that any observable commutes with itself and the property (6.151a).)

Problem 7.11 Derive Eq. (7.90). (*Hint*: Try to build a proof analogous to the derivation of Eq. (7.89).)

Problem 7.12 Derive Eq. (7.92). (*Hint*: Assuming that the relation holds for a given n, prove that it holds for $n + 1$ as well.)

Problem 7.13 Derive Eq. (7.93). (*Hint*: Take advantage of Eqs. (6.52) and (6.117).)

Problem 7.14 The reader who has already acquired a certain competency may try to derive the normalization constant $\mathcal{N}$ in Eq. (7.100). (*Hint*: Use the mathematical formula $\int_{-\infty}^{+\infty} dy\, e^{-ay^2} = \sqrt{\frac{\pi}{a}}$.)

Problem 7.15 (a) Evaluate the commutators $[\hat{x}, \hat{H}]$ and $[\hat{p}_x, \hat{H}]$ in the Schrödinger picture for the harmonic oscillator. (b) Using the results of (a), find the time evolution of the position and momentum operators $\hat{x}(t)$ and $\hat{p}_x(t)$ in the Heisenberg picture with the initial conditions $\hat{x}(0) = \hat{x}$ and $\hat{p}_x(0) = \hat{p}_x$. Part (b) is addressed to more experienced readers. (*Hint*: Take advantage of Box 7.1 and in particular Eq. (7.72).)

7.8 Density Matrix

A very useful formalism for treating many of the subsequent problems is that of the *density matrix*, which can be considered a generalization of the concept of projector. In particular, we are now interested in the description of *many* quantum systems (for instance, photons) prepared in some state. Let us first consider the elementary example of photon polarization [see Section 3.1]. Suppose that we have prepared a large number of photons and all of the photons are in the same polarization state. Denoting horizontal polarization by $|h\rangle$ and vertical polarization by $|v\rangle$, we can write the polarization state of the photons as a two-state superposition:

$$|\psi\rangle = c_h|h\rangle + c_v|v\rangle, \tag{7.106}$$

where c_h and c_v are probability amplitudes satisfying $|c_h|^2 + |c_v|^2 = 1$. The density matrix that describes the system, which in this case coincides with the corresponding projector [see Eq. (4.44)], is given by

$$\begin{aligned}
\hat{\rho} &= |\psi\rangle\langle\psi| \\
&= \left(c_h|h\rangle + c_v|v\rangle\right)\left(c_h^*\langle h| + c_v^*\langle v|\right) \\
&= |c_h|^2\,|h\rangle\langle h| + |c_v|^2\,|v\rangle\langle v| + c_v^*c_h|h\rangle\langle v| + c_v c_h^*|v\rangle\langle h|.
\end{aligned} \tag{7.107}$$

Note that the third and fourth term in the third line are expression of quantum features [see Section 4.8]. In the basis $\{|h\rangle, |v\rangle\}$, $\hat{\rho}$ takes the matrix form

$$\hat{\rho} = \begin{bmatrix} |c_h|^2 & c_h c_v^* \\ c_h^* c_v & |c_v|^2 \end{bmatrix}. \tag{7.108}$$

What these equations tell us is that *all* of the photons being described are in the superposition state (7.106). Consider now the following situation that is quite common in physics. Suppose that we have prepared a large number of photons, with some of the photons in horizontal polarization and the others in vertical polarization. The state of the photons in this system is totally different from the superposition state $|\psi\rangle$ given by Eq. (7.106) and we need to represent it with two distinct projectors, namely, one for describing the set of photons in horizontal polarization and the other for describing the set of photons in vertical polarization. The corresponding density matrix is given by

$$\hat{\rho}' = \wp_h \hat{P}_h + \wp_v \hat{P}_v, \tag{7.109}$$

where $\wp_h + \wp_v = 1$ with $\wp_h$ and $\wp_v$ being the respective probabilities that we expect to find each photon in horizontal and vertical polarizations. We note that the density matrix is no longer a projector, but rather a *weighted sum* of projectors (where $\wp_h$ and $\wp_v$ are those weights). In matrix form $\hat{\rho}'$ is given by

$$\hat{\rho}' = \begin{bmatrix} \wp_h & 0 \\ 0 & \wp_v \end{bmatrix}, \tag{7.110}$$

where only diagonal elements are nonzero. In other words, quantum features are now absent. Quantum states described by density matrices of the type in Eq. (7.107) are called *pure states*, while those described by density matrices of the form in Eq. (7.109) are called *mixed states* (or *mixtures*). In general, suppose a quantum system may be found in the pure states $|\psi_j\rangle$ with the probability $\wp_j$, where j is some index labeling the pure states. Then, the density matrix for this mixture is defined by

$$\hat{\rho} = \sum_j \wp_j |\psi_j\rangle\langle\psi_j| = \sum_j \wp_j \hat{P}_j, \tag{7.111}$$

where

$$\sum_j \wp_j = 1. \tag{7.112}$$

We stress that the pure states $|\psi_j\rangle$ in which the system may be found do *not* have to be mutually orthogonal. It it evident that pure states are a special case of mixed states in which the system is found in one pure state with probability one.

The expectation value for any observable $\hat{O}$ in the mixed state $\hat{\rho}$ is given by [see Section 6.8]

$$\langle \hat{O} \rangle_{\hat{\rho}} = \text{Tr}(\hat{\rho}\hat{O})$$
$$= \sum_j \wp_j \langle \psi_j | \hat{O} | \psi_j \rangle. \qquad (7.113)$$

To put it in another way, the expectation value of $\hat{O}$ in the mixed state $\hat{\rho}$ is the sum of the expectation values of $\hat{O}$ in each of the pure states $|\psi_j\rangle$, weighted by the probabilities $\wp_j$. In the above expressions the symbol Tr denotes the *trace* of a (square) matrix, i.e., the sum of the diagonal elements of a matrix. In mathematical terms, we have

$$\text{Tr}\,\hat{A} = \sum_j A_{jj}, \qquad (7.114)$$

where $\hat{A}$ is a matrix. In addition, the trace satisfies the following basic properties:

$$\text{Tr}(\hat{A} + \hat{B}) = \text{Tr}\,\hat{A} + \text{Tr}\,\hat{B}, \qquad (7.115a)$$

$$\text{Tr}(c\hat{A}) = c\,\text{Tr}\,\hat{A}, \qquad (7.115b)$$

$$\text{Tr}(\hat{A}\hat{B}) = \text{Tr}(\hat{B}\hat{A}), \qquad (7.115c)$$

where c is a constant. The third property (7.115c) is called the *cyclic property* of the trace, from which it follows that

$$\text{Tr}(\hat{U}^\dagger \hat{A}\hat{U}) = \text{Tr}(\hat{U}\,\hat{U}^\dagger \hat{A})$$
$$= \text{Tr}(\hat{A}\hat{U}\,\hat{U}^\dagger)$$
$$= \text{Tr}\,\hat{A}, \qquad (7.116)$$

where $\hat{U}$ is a unitary matrix. Therefore, the trace of a matrix (or an operator) is invariant under any unitary transformation. The density matrix satisfies the following important properties:

$$\hat{\rho}^\dagger = \hat{\rho}, \qquad (7.117a)$$

$$\text{Tr}\,\hat{\rho} = 1, \qquad (7.117b)$$

$$\langle \psi | \hat{\rho} | \psi \rangle \geq 0 \text{ for all } |\psi\rangle, \qquad (7.117c)$$

$$\hat{\rho}_{\text{pure}}^2 = \hat{\rho}_{\text{pure}}. \qquad (7.117d)$$

The first property means that the density matrix is Hermitian. This is evident as the probabilities $\wp_j$ are real (and non-negative) while the projectors $\hat{P}_j$ are Hermitian. The proof of the second property is left to the reader [see Problem 7.17]. The third property means that the density matrix is positive semidefinite. This is also evident as the expectation value $\langle \psi | \hat{\rho} | \psi \rangle$ is precisely the probability of finding the system in the pure state $| \psi \rangle$, which is non-negative. The last property provides a criterion to distinguish pure states from mixtures [see Problem 7.18]. Since the density matrix for a pure state is a projector, the square of a projector is the projector itself [see Eqs. (3.61)].

The dynamics of a quantum system can also be described by making use of density matrices. In the Schrödinger picture, the state vectors $| \psi_j(t) \rangle$ carry time dependence, and so does the density matrix. By explicitly writing out the time dependence, we have

$$\hat{\rho}(t) = \sum_j \wp_j | \psi_j(t) \rangle \langle \psi_j(t) |. \tag{7.118}$$

It is noted that the density matrix can be considered a peculiar observable in that the usual observables in the Schrödinger picture do not carry time dependence. We will discuss in detail about this subtle point below. Using the Schrödinger equation (7.12), we find that the time derivative of the density matrix is given by

$$\frac{d}{dt}\hat{\rho}(t) = \sum_j \wp_j \frac{d}{dt} \left(| \psi_j(t) \rangle \langle \psi_j(t) | \right)$$

$$= -\frac{i}{\hbar} \hat{H} \sum_j \wp_j | \psi_j(t) \rangle \langle \psi_j(t) | + \frac{i}{\hbar} \sum_j \wp_j | \psi_j(t) \rangle \langle \psi_j(t) | \hat{H}$$

$$= \frac{i}{\hbar} \left[\hat{\rho}(t) \hat{H} - \hat{H} \hat{\rho}(t) \right], \tag{7.119}$$

which implies

$$i\hbar \frac{d}{dt}\hat{\rho}(t) = \left[\hat{H}, \hat{\rho}(t) \right], \tag{7.120}$$

where use has been made of the antisymmetric property of the commutator [see Eq. (6.151a)]. The above equation is known as the *von Neumann equation*. Just as the Schrödinger equation describes how a pure state evolves in time, the von Neumann equation describes how a density matrix evolves in time. For the case of a time-independent Hamiltonian that we have implicitly assumed, the solution to von Neumann equation (7.120) is given by

$$\hat{\rho}(t) = \hat{U}_t(t)\hat{\rho}(0)\hat{U}_t^\dagger(t), \qquad (7.121)$$

where $\hat{U}_t(t)$ is the time evolution operator given by Eq. (7.17) and $\hat{\rho}(0)$ is the initial density matrix of the system at time $t_0 = 0$. The above result agrees with what will be obtained from direct substitution of Eq. (7.16) into Eq. (7.118).

We note that while the von Neumann equation (7.120) and the Heisenberg equation (7.55) are formally alike, an important difference between the two equations is that the order of the Hamiltonian and the observable under the commutator is inverted. Another difference is that the eigenvalues of a density matrix are the probabilities to obtain measurement outcomes in that basis. The reason for these differences is that the density matrix is a very peculiar observable, representing a quantum state. This seems to contradict what we have said in Section 6.9. There, we have affirmed that one cannot acquire the whole information contained in a quantum state. Here, we are saying that a density matrix is an observable, what seems to imply that, when measured, it provides us with such an information. The crucial point to understand is that the density matrix is a peculiar observable precisely because it represents a quantum state. As we have seen, it either represents a pure state (as a projector) or a mixture (as a weighted sum of projectors). If it is a projector, by measuring it we are not able to determine the state, since such a measurement corresponds to a test through which we ask the system *whether* or *not* it is in that state (or if at least it has a non-zero overlap with that state). We are not asking *what* is the state of the system, which is precisely the kind of question that would allow us to extract the whole information from the system's state. For instance, let us consider the projectors given by Eq. (3.49). It is easy to see that the projector $\hat{P}_a$ has two eigenvectors (which are its two columns); the first one is the vector $|a\rangle$ itself and the other one is the zero vector. They correspond to the two possible answers of this test: (i) *yes*, and here the eigenvalue is 1; (ii) *no*, and here the eigenvalue is 0. Similarly for the projector $\hat{P}_{a'}$ whose columns are the zero vector and the ket $|a'\rangle$. To understand this difference a classical example could be useful. It is much easier to ascertain whether or not a certain object is an apple (the answer is here either yes or not) than to give a full description of the object (here the answer takes the form of a list of properties). On the other hand, if the density matrix describes a mixture, to measure a single system would mean again to measure something described by one of the projectors summed in the mixture. The outcome of this measurement would only provide partial information. We have therefore evidence that the state of a quantum system cannot be determined through a single measurement, and, as a consequence, we cannot distinguish between non-orthogonal states with a single measurement. This is an issue closely related with the no-cloning theorem, that will be dealt with in the third part of the book.[a] In general, when measuring a certain observable, we only obtain one of its eigenvalues but *not* the probability distribution of its possible properties (connected with its eigenstates) in the state being measured. The latter remains unknown to us through this *single* outcome.

Problem 7.16 Show how we can derive the matrix form of the density matrix $\hat{\rho}$ given by Eq. (7.108) from Eq. (7.107).

Problem 7.17 Prove the second property (7.117b), i.e., $\mathrm{Tr}\,\hat{\rho} = 1$ for both pure and mixed states.

Problem 7.18 By making use of the expression (7.110), show that in the case of mixtures it is not true that $\hat{\rho}^2 = \hat{\rho}$. Show that it holds true for the pure state given by (7.108).

7.9 Composite Systems

The formalism of the density matrix is very useful when dealing with composite systems (or compound systems). Let us suppose to have a system of two photons, both of which can be in horizontal or vertical polarization (or in any superposition of the two polarization states). Let $\mathcal{H}_1$ and $\mathcal{H}_2$ be the Hilbert spaces for the polarization states of

---

[a]See, for instance, (D'Ariano/Yuen, 1996), where the authors proved that to measure the state of single system with a single measurement would contradict the unitarity of the quantum mechanical time evolution.

photon 1 and photon 2, respectively, and $\mathcal{H}$ the Hilbert space for the composite system of the two photons. In this case we have two orthonormal bases

$$\{|v\rangle_1, |h\rangle_1\} \quad \text{and} \quad \{|v\rangle_2, |h\rangle_2\}, \tag{7.122}$$

where the first is a basis in $\mathcal{H}_1$ and the second a basis in $\mathcal{H}_2$. It is noted that while $\mathcal{H}_1$ and $\mathcal{H}_2$ are identical two-dimensional Hilbert spaces, they are in fact two independent Hilbert spaces. When we consider the two systems together, we need to deal with a basis in the Hilbert space $\mathcal{H}$ of the composite system that is given by a certain "product" of these two bases. This can be accomplished by means of what is called the *direct products* of the basis vectors in $\mathcal{H}_1$ and $\mathcal{H}_2$. A convenient basis in $\mathcal{H}$ can be obtained by taking the direct product of the basis vectors in Hilbert spaces $\mathcal{H}_1$ and $\mathcal{H}_2$:

$$\{|h\rangle_1 \otimes |h\rangle_2, \ |h\rangle_1 \otimes |v\rangle_2, \ |v\rangle_1 \otimes |h\rangle_2, \ |v\rangle_1 \otimes |v\rangle_2\}, \tag{7.123}$$

where the symbol $\otimes$ denotes the direct product. Note that each basis vector to the left of the symbol $\otimes$ pertains to the Hilbert space $\mathcal{H}_1$, while each basis vector to the right pertains to the Hilbert space $\mathcal{H}_2$.

The Hilbert space $\mathcal{H}$ of the composite system is defined to be the vector space spanned by the direct product basis (7.123), namely, an arbitrary state $|\Psi\rangle_{12}$ in $\mathcal{H}$ can be written as

$$|\Psi\rangle_{12} = c_{hh}|h\rangle_1 \otimes |h\rangle_2 + c_{hv}|h\rangle_1 \otimes |v\rangle_2 + c_{vh}|v\rangle_1 \otimes |h\rangle_2 + c_{vv}|v\rangle_1 \otimes |v\rangle_2. \tag{7.124}$$

Here and henceforth, whenever the notation becomes unwieldy we shall drop the symbol $\otimes$ between the kets, condense the kets into a single one, or both. For instance, the following expressions $|h\rangle_1 \otimes |h\rangle_2$, $|h\rangle_1|h\rangle_2$, $|h_1 \otimes h_2\rangle$, $|h_1, h_2\rangle$, and $|h_1 h_2\rangle$ all represent the same direct product state. Likewise, here and henceforth, we shall use the state vectors with uppercase greek letters (like $|\Psi\rangle$, $|\Phi\rangle$, etc.) to denote the states of a composite system. Moreover, the scalar product of state vectors in $\mathcal{H}$ is defined in terms of those in $\mathcal{H}_1$ and $\mathcal{H}_2$ by

$$\langle \psi_1 \otimes \psi_2 | \phi_1 \otimes \phi_2 \rangle = \langle \psi|\phi\rangle_1 \langle \psi|\phi\rangle_2. \tag{7.125}$$

Again for the sake of notational simplicity, here and henceforth, we shall partially suppress the indexes 1 and 2 that respectively denote systems 1 and 2 whenever no confusion may arise. With the above definition, it is evident that the direct product basis (7.123) is an orthonormal basis. Therefore, the corresponding basis vectors can be written in component form as

$$|h\rangle_1 \otimes |h\rangle_2 = \begin{pmatrix} 1 \\ 0 \\ 0 \\ 0 \end{pmatrix}, \quad |h\rangle_1 \otimes |v\rangle_2 = \begin{pmatrix} 0 \\ 1 \\ 0 \\ 0 \end{pmatrix},$$

$$|v\rangle_1 \otimes |h\rangle_2 = \begin{pmatrix} 0 \\ 0 \\ 1 \\ 0 \end{pmatrix}, \quad |v\rangle_1 \otimes |v\rangle_2 = \begin{pmatrix} 0 \\ 0 \\ 0 \\ 1 \end{pmatrix}, \qquad (7.126)$$

which in turn implies that $\mathcal{H}$ is a four-dimensional Hilbert space. In general, if $\dim(\mathcal{H}_1) = m$ and $\dim(\mathcal{H}_2) = n$, then we have $\dim(\mathcal{H}) = mn$, where the symbol dim denotes the dimension of a Hilbert space. The resultant Hilbert space is referred to as the direct product Hilbert space of the two Hilbert spaces $\mathcal{H}_1$ and $\mathcal{H}_2$. In mathematical terms, we have $\mathcal{H} = \mathcal{H}_1 \otimes \mathcal{H}_2$.

We are interested in ascertaining which kind of relation, if any, can exist between these two systems. Let us first suppose that the two systems are *separated* [see Section 1.1] and each is in the state of a polarization at $45°$ (or $\frac{\pi}{4}$ in radians) relative to the horizontal polarization, that is, we have

$$|\psi\rangle_1 = \frac{1}{\sqrt{2}}(|h\rangle_1 + |v\rangle_1), \quad |\psi\rangle_2 = \frac{1}{\sqrt{2}}(|h\rangle_2 + |v\rangle_2). \qquad (7.127)$$

Here by separated we mean that no action whatsoever performed locally on one of the systems can have effects on the other one. This implies that every possible combination of the allowed measurement outcomes of the two systems can occur or that the probability distributions of the possible measurement outcomes of the two systems are statistically independent. If there are no additional constraints on the two systems, we shall expect that, in measurements of a sufficient large number of identically prepared systems, the composite system on average will have equal probability of $\frac{1}{4}$ to be found in the states described by the four product basis vectors (7.123). Therefore, we may rewrite the latter state as

$$
\begin{aligned}
|\Psi\rangle_{12} &= |\psi\rangle_1 \otimes |\psi\rangle_2 \\
&= \frac{1}{\sqrt{2}}(|h\rangle_1 + |v\rangle_1) \otimes \frac{1}{\sqrt{2}}(|h\rangle_2 + |v\rangle_2) \\
&= \frac{1}{2}\left(|h\rangle_1 \otimes |h\rangle_2 + |h\rangle_1 \otimes |v\rangle_2 + |v\rangle_1 \otimes |h\rangle_2 + |v\rangle_1 \otimes |v\rangle_2\right).
\end{aligned}
$$
(7.128)

This implies, for instance, that if we find that system 1 is, say, in the state of vertical polarization, the system 2 can be either vertically or horizontally polarized, both possibilities having equal probability. In general, if system 1 and system 2 are described by the state vectors $|\psi\rangle_1$ and $|\psi\rangle_2$, respectively, and

$$
|\psi\rangle_1 = c_h|h\rangle_1 + v_c|v\rangle_1, \quad |\psi\rangle_2 = c_h'|h\rangle_2 + v_c'|v\rangle_2, \tag{7.129}
$$

then the composite system is described by the state vector $|\Psi\rangle$ that is the direct product of $|\psi\rangle_1$ and $|\psi\rangle_2$, namely, we have

$$
\begin{aligned}
|\Psi\rangle_{12} &= |\psi\rangle_1 \otimes |\psi\rangle_2 \\
&= c_h c_h'|h\rangle_1 \otimes |h\rangle_2 + c_h c_v'|h\rangle_1 \otimes |v\rangle_2 \\
&\quad + c_v c_h'|v\rangle_1 \otimes |h\rangle_2 + c_v c_v'|v\rangle_1 \otimes |v\rangle_2,
\end{aligned}
$$
(7.130)

which can be considered a reformulation of Eq. (7.124). We now consider a completely different situation. Suppose that the composite system at a certain time is described by the state vector

$$
|\Phi\rangle_{12} = \frac{1}{\sqrt{2}}(|h\rangle_1 \otimes |h\rangle_2 + |v\rangle_1 \otimes |v\rangle_2). \tag{7.131}
$$

We immediately notice something extraordinary. According to the above equation, when system 1 is in the state of horizontal polarization so is system 2, and vice versa; similarly, when system 1 is in the state of vertical polarization, the same occurs for system 2, and vice versa. In this case, the two systems are no longer separated. As a matter of fact, they can no longer be considered independent since a certain measurement outcome on system 1 allows a prediction about system 2, and vice versa. However, this state tells us even something more, namely, a measurement on system 1 will *project* system 2 in one of the two alternative states. Indeed, the state $|\Phi\rangle_{12}$ tells us that system 2 can be in a state of vertical or horizontal polarization *before* the polarization on system 1 is measured. But *after* a measurement on system 1 is carried out, system 2 has to be in either one or the other state according to the measurement outcome on system 1. In other words, it seems that a local operation on system 1 can have *non-local* effects on system 2, and vice versa. We call such a state *entangled*. Such an extraordinary situation, which is even puzzling from the point of view of special relativity since it seems to imply the instantaneous transmission of signals, will be further discussed in the third part of the book. By now, let us consider the density matrix for the entangled state $|\Phi\rangle_{12}$ in Eq. (7.131), namely,

$$
\begin{aligned}
\hat{\rho}_{12} &= |\Phi\rangle\langle\Phi|_{12} \\
&= \frac{1}{2}(|h\rangle\langle h|_1 \otimes |h\rangle\langle h|_2 + |v\rangle\langle v|_1 \otimes |v\rangle\langle v|_2 \\
&\quad + |h\rangle\langle v|_1 \otimes |h\rangle\langle v|_2 + |v\rangle\langle h|_1 \otimes |v\rangle\langle h|_2).
\end{aligned}
\tag{7.132}
$$

Note that the components $|h\rangle\langle v|_1 \otimes |h\rangle\langle v|_2$ and $|v\rangle\langle h|_1 \otimes |v\rangle\langle h|_2$ represent features or the interference terms, not between states of different systems but between different states *of the same* system. Therefore, entanglement can be understood to a certain extent as an extension of the concept of superposition to a composite system. However, there is an important difference: superposition is relative to *the basis* [see Section 4.5] while entanglement is intrinsic in *the state* of the system and is therefore independent of the basis used. As a matter of fact, entanglement combines the typical quantum superposition with correlations that have a classical root. *Both* these kind of classical correlations and quantum superposition factors constitute entanglement [see Fig. 7.4].

Our conclusion could be put in question by pointing out that the density matrix $\hat{\rho}_{12}$ in Eq. (7.132) has a particular form. As a matter of fact, if take another kind of entanglement, e.g.,

$$
\begin{aligned}
\hat{\rho}'_{12} &= \frac{1}{2}(|h\rangle\langle h|_1 \otimes |v\rangle\langle v|_2 + |v\rangle\langle v|_1 \otimes |h\rangle\langle h|_2 \\
&\quad + |h\rangle\langle v|_1 \otimes |v\rangle\langle h|_2 + |v\rangle\langle h|_1 \otimes |h\rangle\langle v|_2),
\end{aligned}
\tag{7.133}
$$

where for the sake of simplicity we have put in the first row the terms having a classical analogue while in the second row the cross terms expressing features, things may appear different. It is

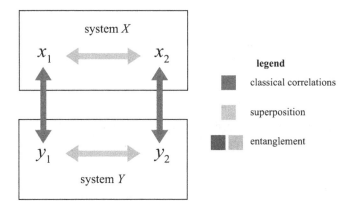

Figure 7.4 Particular case in which the entanglement of two degrees of freedom is between two subsystems and each observable has only two eigenstates. Although a particular case, the figure is very enlightening and therefore also easily generalizable.

clear now that each time we have photon 1 polarized vertically, photon 2 is polarized horizontally and vice versa. Therefore, the two photon show a kind of anti-parallelism. However, as before also the crossed term follow this rule as far as the relations between the two subsystems are considered. On the contrary, when the single subsystems are considered, we have precisely the *same* cross terms in $\hat{\rho}'_{12}$ as those in $\hat{\rho}_{12}$ [see Eq. (7.132)]. Therefore, the typical quantum behavior, the quantumness of the features is here connected with the superposition of the two subsystems rather than with the relations between them.

Suppose now that we would like to describe one of the two systems without considering the other one. This happens quite commonly when we are interested only in one of the systems and consider the other as the environment. Let us therefore start with the density matrix in Eq. (7.132) and assume that we wish to omit the consideration of system 2. This can be done by performing a *partial trace* over system 2 (denoted by Tr_2), that is, by summing the diagonal elements with respect to system 2 only, or equivalently, by summing the expectation values of $\hat{\rho}_{12}$ in a complete set of orthonormal basis states of system 2. The resultant *reduced density matrix* of system 1 contains information only about system 1 and is given by

$$\hat{\rho}_1 = \text{Tr}_2 \, \hat{\rho}_{12}$$
$$= \langle h|\hat{\rho}_{12}|h\rangle_2 + \langle v|\hat{\rho}_{12}|v\rangle_2$$
$$= \frac{1}{2}(|h\rangle\langle h|_1 + |v\rangle\langle v|_1). \tag{7.134}$$

It is evident that the reduced density matrix $\hat{\rho}_1$ describes a mixture. In other words, the reduced subsystem of a composite system prepared in an entangled state is a mixture.

Problem 7.19 Compute the density matrix

$$\hat{\rho}'_{12} = |\Psi\rangle\langle\Psi|_{12},$$

where $|\Psi\rangle_{12}$ is given by Eq. (7.128). It is a long and tedious calculation that nevertheless can enlighten the meaning of separation of the two involved systems.

Problem 7.20 Show that a system in the entangled state $|\Phi\rangle_{12}$ given by Eq. (7.131) will remain as an entangled state in other basis. You may use the basis $\{|a\rangle_1 \otimes |a\rangle_2, |a\rangle_1 \otimes |a'\rangle_2, |a'\rangle_1 \otimes |a\rangle_2, |a'\rangle_1 \otimes |a'\rangle_2\}$ and the inverse of the transformation given by Eq. (4.20).

Problem 7.21 Compute the reduced density matrix $\hat{\rho}_2 = \text{Tr}_1 \, \hat{\rho}_{12}$, where the density matrix $\hat{\rho}_{12}$ is given by Eq. (7.132).

7.10 Summary

In this chapter we have

- Introduced the time-dependent Schrödinger equation that describes the time evolution of a quantum system.
- Shown how to calculate the probability amplitudes as a function of time given that we know the initial state and the Hamiltonian of a quantum system.
- Explained the precise sense in which quantum mechanics can be viewed as a deterministic theory about probabilities.
- Proved that the time evolution ruled by the Schrödinger equation is unitary (and therefore reversible).

- Presented the Schrödinger picture formulation and the Heisenberg picture formulation of quantum dynamics and proved their equivalence.
- Given a very sketchy presentation of the free particle and the harmonic oscillator in one dimension.
- Developed the density matrix formalism and distinguished between pure states and mixtures.
- Introduced the concepts of entanglement and the reduced density matrix.

Chapter 8

Angular Momentum and Spin

In this chapter we shall treat two of the most important observables of quantum mechanical systems: angular momentum and spin. We shall argue that angular momentum is the generator of rotations. In quantum mechanics, angular momentum is quantized in units of the reduced Planck constant. The quantization of angular momentum is dealt with by finding the angular momentum eigenvalues and eigenstates. We shall then apply the theory of angular momentum to study the hydrogen atom. Spin is an intrinsic angular momentum of subatomic particles and therefore corresponds to an important intrinsic degree of freedom. We shall discuss briefly how spin was discovered and many of its important properties. Finally, we shall also explore here some further subtleties of quantum theory.[a]

---

[a] This chapter is without any doubt the most mathematically difficult in the book. We suggest the reader to proceed even more carefully and to invest more time. Although difficult, this chapter will constitute a very important preparation for understanding the third part of the book. Nonetheless, the reader may skip some mathematical derivations at a first reading and try to grasp the essential results.

Quantum Mechanics for Thinkers
Gennaro Auletta and Shang-Yung Wang
Copyright © 2014 Pan Stanford Publishing Pte. Ltd.
ISBN 978-981-4411-71-4 (Hardcover), 978-981-4411-72-1 (eBook)
www.panstanford.com

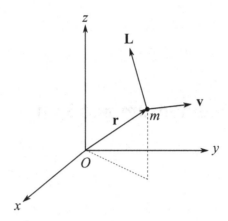

Figure 8.1 A classical particle with mass m, velocity $\mathbf{v}$, and position $\mathbf{r}$ in a certain reference frame has an angular momentum $\mathbf{L} = \mathbf{r} \times \mathbf{p} = m\mathbf{r} \times \mathbf{v}$. Note that the vector $\mathbf{L}$ is orthogonal to both vectors $\mathbf{r}$ and $\mathbf{p}$.

8.1 Angular Momentum as Generator of Rotations

When we consider quantum systems in two or three spatial dimensions, we need to take into account also the effect of rotations. In this case, we need to introduce a physical quantity called the *angular momentum*, that is, the momentum induced by or connected with rotations. In classical physics the angular momentum of a particle about a given origin is defined by [see Fig. 8.1]

$$\mathbf{L} = \mathbf{r} \times \mathbf{p}, \tag{8.1}$$

where $\mathbf{r}$ is the position vector of the particle relative to the origin, $\mathbf{p}$ is the momentum of the particle, and the symbol $\times$ denotes the cross product of two spatial vectors [see Box 8.1]. In three dimensions, the position, momentum, and angular momentum are expressed in Cartesian coordinates as three-component spatial (row) vectors $\mathbf{r} = (x, y, z)$, $\mathbf{p} = (p_x, p_y, p_z)$, and $\mathbf{L} = (L_x, L_y, L_z)$, respectively; then, according to Eq. (8.6) we have

$$L_x = yp_z - zp_y, \quad L_y = zp_x - xp_z, \quad L_z = xp_y - yp_x. \tag{8.2}$$

The angular momentum of a system of particles is the sum of the angular momenta of the individual particles

$$\mathbf{L} = \sum_{i=1}^{n} \mathbf{r}_i \times \mathbf{p}_i, \tag{8.3}$$

Box 8.1 Cross product of vectors

In mathematics, the cross product (or vector product) of two Euclidean vectors **a** and **b** is denoted by **a** × **b**. It is a vector that is perpendicular to both **a** and **b**, with a direction given by the right-hand rule and a magnitude equal to the area of the parallelogram that the vectors span. The right-hand rule states that the direction of the cross product **a** × **b** is determined by placing **a** and **b** *tail-to-tail*, flattening the right hand, extending it in the direction of **a**, and then curling the fingers in the direction that the angle **b** makes with **a**. The thumb then points in the direction of **a** × **b** [see Fig. 8.2]. In other words, the cross product **a** × **b** is defined by [see also Box 3.1]

$$\mathbf{a} \times \mathbf{b} = \|\mathbf{a}\| \, \|\mathbf{b}\| \, \sin\theta. \tag{8.4}$$

where $\|\mathbf{a}\|$ and $\|\mathbf{b}\|$ are the respectively magnitudes of vectors **a** and **b**, $0 \leq \theta \leq \pi$ is the angle between **a** and **b**, and **n** is a unit vector in the direction given by the right-hand rule. Moreover, the cross product satisfies the following properties:

$$\mathbf{a} \times \mathbf{a} = 0, \tag{8.5a}$$

$$\mathbf{a} \times \mathbf{b} = -\mathbf{b} \times \mathbf{a}, \tag{8.5b}$$

$$\mathbf{a} \times (\mathbf{b} + \mathbf{c}) = (\mathbf{a} \times \mathbf{b}) + (\mathbf{a} \times \mathbf{c}), \tag{8.5c}$$

$$(k\mathbf{a}) \times \mathbf{b} = \mathbf{a} \times (k\mathbf{b}) = k(\mathbf{a} \times \mathbf{b}), \tag{8.5d}$$

$$\mathbf{a} \times (\mathbf{b} \times \mathbf{c}) + \mathbf{b} \times (\mathbf{c} \times \mathbf{a}) + \mathbf{c} \times (\mathbf{a} \times \mathbf{b}) = 0, \tag{8.5e}$$

where k is a real number.

In Cartesian coordinates, the vector product of the vectors $\mathbf{a} = a_x \mathbf{e}_x + a_y \mathbf{e}_y + a_z \mathbf{e}_z$ and $\mathbf{b} = b_x \mathbf{e}_x + b_y \mathbf{e}_y + b_z \mathbf{e}_z$ is given in component form by

$$\mathbf{a} \times \mathbf{b} = (a_y b_z - a_z b_y)\mathbf{e}_x + (a_z b_x - a_x b_z)\mathbf{e}_y + (a_x b_y - a_y b_x)\mathbf{e}_z, \tag{8.6}$$

where $\mathbf{e}_x$, $\mathbf{e}_y$, and $\mathbf{e}_z$ are unit vectors in the x, y, and z directions, respectively. The above expression can be written in a shorthand notation by using the determinant of a formal matrix

$$\mathbf{a} \times \mathbf{b} = \begin{vmatrix} \mathbf{e}_x & \mathbf{e}_y & \mathbf{e}_z \\ a_x & a_y & a_z \\ b_x & b_y & b_z \end{vmatrix}. \tag{8.7}$$

In two dimensions, the analogue of the cross product for $\mathbf{a} = a_x \mathbf{e}_x + a_y \mathbf{e}_y$ and $\mathbf{b} = b_x \mathbf{e}_x + b_y \mathbf{e}_y$ is obtained by setting $a_z = b_z = 0$ in the above expressions.

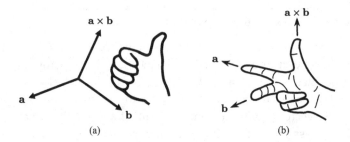

Figure 8.2 The right-hand rule. (a) The rule as defined in Box 8.1. (b) The right-hand rule for establishing a system of Cartesian coordinates: here the vectors **a** and **b** need to be orthogonal. Adapted from http://en.wikipedia.org/wiki/File:Right-hand_grip_rule.svg and http://en.wikipedia.org/wiki/File:Right_hand_rule_cross_product.svg.

where $\mathbf{r}_i$ and $\mathbf{p}_i$ are the respective position and momentum of the ith particle, and n is the number of particles in the system. Just like momentum to a moving system, angular momentum is related to the capability of a rotating system to have a certain dynamical impact on other systems. For instance, everyday experience tells us that the heavier a circular saw is, the larger its radius is, or the faster it is rotating, the more cutting power it will deliver. Indeed, the net torque applied on a system is equal to the time derivative of its angular momentum

$$\boldsymbol{\tau} = \frac{d\mathbf{L}}{dt}$$
$$= \frac{d\mathbf{r}}{dt} \times \mathbf{p} + \mathbf{r} \times \frac{d\mathbf{p}}{dt}$$
$$= \mathbf{v} \times \mathbf{p} + \mathbf{r} \times \mathbf{F}$$
$$= \mathbf{r} \times \mathbf{F}, \tag{8.8}$$

where the vector product of velocity and momentum is zero because the two vectors are parallel. The above expression is simply a reformulation of Newton's second law (1.2) [see also Eq. (6.108)].

It is very important that we understand what angular momentum means from a fundamental point of view. Angular momentum is a conserved quantity in a closed system, i.e., a system that is not subject to the action of external forces [see Section 1.1]. Moreover, since the angular momentum of a system is what describes its rotations, it can be thought of as being strictly related to the

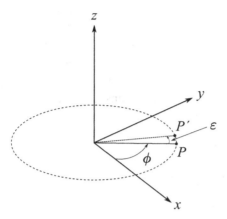

Figure 8.3 An infinitesimal counterclockwise rotation about the z axis brings a point P on the unit circle in the xy plane to a point P' on the unit circle.

generator of rotations, that is, as the dynamical parameter or operator that is associated with changes in the kinematic parameter represented by the angular displacement. In the previous two chapter we have already dealt with two other dynamical quantitates: momentum, which is associated with changes in the kinematic parameter position, and energy, which is associated with changes of the kinematic parameter time. It is a theorem of both classical and quantum mechanics (known as the Noether theorem) that tells us that changes of a kinematic parameter is ruled by a dynamical quantity that is the generator of that transformation.[a] Then, by analogy with the case of momentum (the generator of space translations), which has been discussed in Section 6.6, let us first consider an *infinitesimal* rotation about the z axis that brings a point P on the unit circle in the xy plane to a point P' on the unit circle [see Fig. 8.3]. Here and henceforth, we follow the convention that rotations occur in the sense prescribed by the right-hand rule [see Box 8.1], i.e., counterclockwise about the z axis. The coordinates of the points P and P' are related by

$$\begin{pmatrix} x \\ y \end{pmatrix} = \begin{pmatrix} \cos\phi \\ \sin\phi \end{pmatrix} \xrightarrow{\varepsilon} \begin{pmatrix} x' \\ y' \end{pmatrix} = \begin{pmatrix} \cos(\phi + \varepsilon) \\ \sin(\phi + \varepsilon) \end{pmatrix}, \qquad (8.9)$$

---

[a](Arnold, 1978, pp. 88–89), (Auletta *et al.*, 2009, Chapter 8).

where ε is an infinitesimal angle measured in radians. For the sake of presentational simplicity, in the above expression the Cartesian coordinates of points have been written as column vectors and the unaffected z components have been suppressed. Using the addition formulas (2.22) and the approximations $\sin \varepsilon \approx \varepsilon$ and $\cos \varepsilon \approx 1$ for infinitesimal ε (measured in radians) [see Eq. (6.95)], we can rewrite the above expression approximately as

$$\begin{pmatrix} x \\ y \end{pmatrix} \xrightarrow{\varepsilon} \begin{pmatrix} x' \\ y' \end{pmatrix} \approx \begin{pmatrix} \cos \phi - \varepsilon \sin \phi \\ \sin \phi + \varepsilon \cos \phi \end{pmatrix} = \begin{pmatrix} x - \varepsilon y \\ y + \varepsilon x \end{pmatrix}. \tag{8.10}$$

Now, the rotated position $\begin{pmatrix} x' \\ y' \end{pmatrix}$ can be expressed in terms of the original position $\begin{pmatrix} x \\ y \end{pmatrix}$ as

$$\begin{pmatrix} x' \\ y' \end{pmatrix} \approx \left(1 + \varepsilon \hat{R}_z\right) \begin{pmatrix} x \\ y \end{pmatrix}, \tag{8.11}$$

where the operator $(1 + \varepsilon \hat{R}_z)$ acts on each component of the column vector $\begin{pmatrix} x \\ y \end{pmatrix}$ and $\hat{R}_z$ is the generator of rotations about the z axis [see Table 6.2]

$$\hat{R}_z = x \frac{\partial}{\partial y} - y \frac{\partial}{\partial x}. \tag{8.12}$$

Again in order to go over from an infinitesimal rotation to a *finite* rotation by an angle γ about the z axis, we may think of the latter as a result of an infinite number of repeated applications of the infinitesimal rotation (8.11). Following the same procedure used in Section 6.6 by first subdividing the finite angle γ into n equal partitions, each of angle γ/n, and then taking the limit for n approaching infinity, we obtain

$$\begin{pmatrix} x' \\ y' \end{pmatrix} = e^{\gamma \hat{R}_z} \begin{pmatrix} x \\ y \end{pmatrix}, \tag{8.13}$$

where $e^{\gamma \hat{R}_z}$ is the rotation operator by an angle γ about the z axis. As we will see below, the z component of the angular momentum operator in quantum mechanics is closely related to the generator of rotations about the z axis.

Problem 8.1 Check that applying the transformation $(1 + \varepsilon \hat{R}_z)$ to x and y, where $\hat{R}_z$ is given by (8.12), we obtain $x - \varepsilon y$ and $y + \varepsilon x$, respectively.

8.2 Angular Momentum Operator

According to the correspondence principle [see Principle 7.1], the classical definition of angular momentum (8.1) can be carried over to quantum mechanics by reinterpreting position and momentum in classical mechanics as the position and momentum operators in quantum mechanics. Therefore, the angular momentum operator in quantum mechanics is defined by

$$\hat{\mathbf{L}} = \hat{\mathbf{r}} \times \hat{\mathbf{p}}. \tag{8.14}$$

As a matter of fact, there are two types of angular momenta in quantum mechanics. The type of angular momentum defined in the above equation, which is a generalization of angular momentum in classical mechanics and hence associated with the rotational (or orbital) motion of a system in real space, is referred to as the *orbital angular momentum* (or simply angular momentum whenever no confusion may arise). The other type of angular momentum which is a fundamental intrinsic property of subatomic particles and has no analogue in classical mechanics is called the *spin angular momentum* (or *spin* for short). We will first discuss the orbital angular momentum and come back to the spin angular momentum in Section 8.6.

From the expressions for the position and momentum operators in the position representation in Eqs. (6.56) and (6.137), the three Cartesian components of the angular momentum operator $\hat{L}_x$, $\hat{L}_y$, and $\hat{L}_z$ in the position representation corresponding to the classical counterparts (8.2) are respectively given by

$$\hat{L}_x = \hat{y}\hat{p}_z - \hat{z}\hat{p}_y = -i\hbar \left(y\frac{\partial}{\partial z} - z\frac{\partial}{\partial y} \right), \tag{8.15a}$$

$$\hat{L}_y = \hat{z}\hat{p}_x - \hat{x}\hat{p}_z = -i\hbar \left(z\frac{\partial}{\partial x} - x\frac{\partial}{\partial z} \right), \tag{8.15b}$$

$$\hat{L}_z = \hat{x}\hat{p}_y - \hat{y}\hat{p}_x = -i\hbar \left(x\frac{\partial}{\partial y} - y\frac{\partial}{\partial x} \right). \tag{8.15c}$$

From the above expressions, we see that the Planck constant has the dimension of angular momentum. Indeed, as will be seen below, the reduced Planck constant $\hbar$ is the quantum of angular momentum. It is also evident that the angular momentum operator is a Hermitian operator because each of its components is a Hermitian operator. Moreover, the magnitude squared of the angular momentum operator is defined by

$$\hat{\mathbf{L}}^2 = \hat{L}_x^2 + \hat{L}_y^2 + \hat{L}_z^2, \tag{8.16}$$

which is also a Hermitian operator. We are now in a position to establish the relation between the angular momentum operator and the generator of rotations. Upon comparing Eqs. (8.12) and (8.15c), we have

$$\hat{L}_z = -i\hbar \hat{R}_z, \tag{8.17}$$

which shows clearly that the z component of the angular momentum operator is the generator of rotations about the z axis (the factor $-i$ is there to ensure that as an observable the operator $\hat{L}_z$ is a Hermitian operator). Hence, the rotation operator about the z axis by an angle γ (measured in radians) can be written in terms of $\hat{L}_z$ as the unitary operator

$$\hat{U}_z(\gamma) = e^{-\frac{i}{\hbar}\gamma \hat{L}_z}, \tag{8.18}$$

which we note is understood as an active transformation [see Section 7.4]. Similarly, the operators $\hat{L}_x$ and $\hat{L}_y$ are respectively the generators of rotations about the x axis and y axis. As a consequence, the rotation operator about an arbitrary direction $\mathbf{n}$ (which is therefore a unit vector) by an angle γ is given by the unitary operator

$$\hat{U}_{\mathbf{n}}(\gamma) = e^{-\frac{i}{\hbar}\gamma \mathbf{n} \cdot \hat{\mathbf{L}}}, \tag{8.19}$$

where $\mathbf{n} \cdot \hat{\mathbf{L}} = n_x \hat{L}_x + n_y \hat{L}_y + n_z \hat{L}_z$ with $n_x^2 + n_y^2 + n_z^2 = 1$. In other words, under a rotation of angle γ about the direction $\mathbf{n}$, the state of a system $|\psi\rangle$ transforms as

$$|\psi\rangle \xrightarrow{\gamma \mathbf{n}} |\psi'\rangle = \hat{U}_{\mathbf{n}}(\gamma)|\psi\rangle. \tag{8.20}$$

Recall that the direction of a vector can be described by two angles: the angle (now denoted by β) between the vector and the z axis, and the angle (now denoted by α) of its orthogonal projection on the xy plane measured from the x axis [see Fig. 1.1]. Therefore, a rotation in three-dimensional space can be uniquely described by three angles α, β, and γ. These angles are referred to as the Euler angles [see Fig. 8.4].

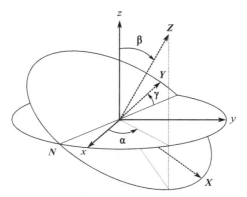

Figure 8.4 A rotation of angle γ about the Z axis can be uniquely described by three Euler angles α, β, and γ, which define the relative orientation between the fixed coordinate system xyz and the rotated coordinate system XYZ by three successive rotations. Here we use the so-called the zyz or γ convention. The line N where the xy and XY planes intersect is called the line of nodes.

Since components of the position and momentum operators satisfy the canonical commutation relations (6.150) and therefore do not in general commute, components of the angular momentum operator do not necessarily commute with one another. The commutation relations between components of the angular momentum operator can be calculated by using the canonical commutation relations (6.150) and the properties of commutators (6.151). For instance, using the expressions (8.15a) and (8.15b), we have

$$
\begin{aligned}
\left[\hat{L}_x, \hat{L}_y\right] &= [\hat{y}\hat{p}_z - \hat{z}\hat{p}_y, \hat{z}\hat{p}_x - \hat{x}\hat{p}_z] \\
&= [\hat{y}\hat{p}_z, \hat{z}\hat{p}_x] + [\hat{z}\hat{p}_y, \hat{x}\hat{p}_z] \\
&= \hat{y}[\hat{p}_z, \hat{z}]\hat{p}_x + \hat{x}[\hat{z}, \hat{p}_z]\hat{p}_y \\
&= \hat{x}[\hat{z}, \hat{p}_z]\hat{p}_y - \hat{y}[\hat{z}, \hat{p}_z]\hat{p}_x \\
&= i\hbar(\hat{x}\hat{p}_y - \hat{y}\hat{p}_x) \\
&= i\hbar\hat{L}_z.
\end{aligned}
\tag{8.21}
$$

Applying a similar but tedious procedure to the other components, we finally obtain the following non-trivial commutation relations [see Problem 8.2]

$$
\left[\hat{L}_x, \hat{L}_y\right] = i\hbar\hat{L}_z, \quad \left[\hat{L}_y, \hat{L}_z\right] = i\hbar\hat{L}_x, \quad \left[\hat{L}_z, \hat{L}_x\right] = i\hbar\hat{L}_y.
\tag{8.22}
$$

The above commutation relations imply that components of the angular momentum are not jointly measurable [see Section 6.9]. This is due to the fact that the angular momentum is the generator of rotations and that rotations in three dimensional space do not in general commute. It also can be shown that each of the components of the angular momentum operator commutes with its magnitude squared, that is, we have [see Problem 8.3]

$$[\hat{L}_j, \hat{\mathbf{L}}^2] = 0, \tag{8.23}$$

where $j = x, y, z$ and $\hat{\mathbf{L}}^2$ is defined in Eq. (8.16). The commutation relations (8.23) means that only the magnitude squared of the angular momentum and one, and only one, of its components are simultaneously measurable. This is because the magnitude squared of a spatial vector is invariant under rotations. Moreover, angular momentum in general does not commute with position and momentum either. The non-trivial commutation relations of $\hat{L}_z$ and the components of position and momentum are given by

$$[\hat{L}_z, \hat{x}] = i\hbar\hat{y}, \quad [\hat{L}_z, \hat{p}_x] = i\hbar\hat{p}_y, \quad [\hat{L}_z, \hat{y}] = -i\hbar\hat{x}, \quad [\hat{L}_z, \hat{p}_y] = -i\hbar\hat{p}_x. \tag{8.24}$$

Similar commutation relations hold true for $\hat{L}_x$ and $\hat{L}_y$. The reason for this is that while the magnitude squared of a spatial vector is invariant under rotations, as is evident from Eq. (8.9) or (8.13), the components of a spatial vector do change under rotations, also as is evident from Eq. (8.9) or (8.13), even though they obey certain transformation rules.

Problem 8.2 Using the canonical commutation relations, derive the commutation relations $[\hat{L}_y, \hat{L}_z] = i\hbar\hat{L}_x$ and $[\hat{L}_z, \hat{L}_x] = i\hbar\hat{L}_y$.

Problem 8.3 Using Eq. (8.22), derive the commutation relation $[\hat{L}_j, \hat{\mathbf{L}}^2] = 0$.

Problem 8.4 Check the validity of Eq. (8.24) and derive the corresponding expressions for $\hat{L}_x$ and $\hat{L}_y$.

8.3 Quantization of Angular Momentum

In quantum mechanics, angular momentum is quantized, that is, it cannot vary continuously but can only take discrete values.

Quantization of angular momentum was first postulated by Niels Bohr in his model of the atom [see Section 4.2].[a] Starting from his angular momentum quantum rule, Bohr was able to calculate the energies of the allowed orbits of the hydrogen atom and other hydrogen-like atoms and ions. We now proceed to find the eigenvalues and eigenstates of the angular momentum operator. Recall that, as we have discussed in the previous section, only the magnitude squared of the angular momentum and one, and only one, of its components are simultaneously measurable. The convention is to choose this component as the z component of the angular momentum operator, $\hat{L}_z$. Mathematically, the commutation relation $[\hat{L}_z, \hat{\mathbf{L}}^2] = 0$ implies that the operators $\hat{L}_z$ and $\hat{\mathbf{L}}^2$ have common eigenstates. Let us denote their common eigenstates by $|l, m\rangle$, where l and m are (discrete) quantum numbers that are respectively related to the eigenvalues of the operators $\hat{L}_z$ and $\hat{\mathbf{L}}^2$.

Let us postulate that the eigenvalue equation for $\hat{L}_z$ has the following form

$$\hat{L}_z|l, m\rangle = m\hbar|l, m\rangle, \qquad (8.25)$$

that is, the states $|l, m\rangle$ are eigenstates of $\hat{L}_z$ and the corresponding eigenvalues are $m\hbar$. The number m is called the *magnetic quantum number*, whose value will be determined below. Since the expectation value of the square of an Hermitian operator is non-negative [see Section 6.8], we have

$$m^2\hbar^2 = \langle l, m|\hat{L}_z^2|l, m\rangle = \langle l, m|(\hat{\mathbf{L}}^2 - \hat{L}_x^2 - \hat{L}_y^2)|l, m\rangle \le \langle l, m|\hat{\mathbf{L}}^2|l, m\rangle, \qquad (8.26)$$

where use has been made of Eq. (8.16) and the fact that the eigenstates $|l, m\rangle$ are normalized. The above inequality implies that m^2 is bounded by the eigenvalues of $\hat{\mathbf{L}}^2$, hence we have

$$-l \le m \le l, \qquad (8.27)$$

where the eigenvalue l is the maximum value of m and is called the *azimuthal quantum number*.

Our task now is to find the possible values of m and l. Since we have a range of value of m for each fixed l, it proves convenient to introduce the raising and lowering operators

$$\hat{l}_+ = \hat{l}_x + i\hat{l}_y, \quad \hat{l}_- = \hat{l}_x - i\hat{l}_y, \qquad (8.28)$$

---

[a](Bohr, 1913).

respectively, where $\hat{l}_j = \hat{L}_j/\hbar$ (with $j = x, y, z$). Obviously, they satisfy

$$\hat{l}_+ = \hat{l}_-^\dagger, \quad \hat{l}_- = \hat{l}_+^\dagger. \tag{8.29}$$

We note that the operators $\hat{l}_+$ and $\hat{l}_-$ are not observables because they are not Hermitian operators but each is the Hermitian conjugate of the other, as it happens for the annihilation and creation operators $\hat{a}$ and $\hat{a}^\dagger$ of the harmonic oscillator [see Eqs. (7.77)]. Let us now consider the action of $\hat{L}_z$ on the state vector $\hat{l}_+|l, m\rangle$

$$\begin{aligned} \hat{L}_z\hat{l}_+|l, m\rangle &= \left(\hat{L}_z\hat{l}_+ - \hat{l}_+\hat{L}_z + \hat{l}_+\hat{L}_z\right)|l, m\rangle \\ &= \left([\hat{L}_z, \hat{l}_+] + \hat{l}_+\hat{L}_z\right)|l, m\rangle \\ &= \hbar\hat{l}_+|l, m\rangle + \hat{l}_+\hat{L}_z|l, m\rangle \\ &= \hbar\hat{l}_+|l, m\rangle + m\hbar\hat{l}_+|l, m\rangle \\ &= (m + 1)\hbar\hat{l}_+|l, m\rangle, \end{aligned} \tag{8.30}$$

where we have made use of Eq. (8.25) and of the commutation relation [see Eq. (8.22)]

$$\begin{aligned} [\hat{L}_z, \hat{l}_+] &= \frac{1}{\hbar}\left[\hat{L}_z, \hat{L}_x + i\hat{L}_y\right] \\ &= \frac{1}{\hbar}\left[\hat{L}_z, \hat{L}_x\right] + \frac{i}{\hbar}\left[\hat{L}_z, \hat{L}_y\right] \\ &= i\hat{L}_y + \hat{L}_x \\ &= +\hbar\hat{l}_+. \end{aligned} \tag{8.31}$$

From Eq. (8.30) it follows that $\hat{l}_+|l, m\rangle$ is an eigenstate of $\hat{L}_z$ with eigenvalue $(m + 1)\hbar$. Since the eigenstates $\hat{l}_+|l, m\rangle$ and $|l, m + 1\rangle$ correspond to the same eigenvalue of $\hat{L}_z$, they have to be proportional to each other. Hence we have

$$\hat{l}_+|l, m\rangle = c_{lm}^+|l, m + 1\rangle, \tag{8.32}$$

where c_{lm}^+ is a proportional constant [see Problem 8.5]. A similar analysis shows that the state vector $\hat{l}_-|l, m\rangle$ is an eigenstate of $\hat{L}_z$ with eigenvalue $(m - 1)\hbar$ [see Problem 8.6]. Therefore, we have

$$\hat{l}_-|l, m\rangle = c_{lm}^-|l, m - 1\rangle, \tag{8.33}$$

where c_{lm}^- is a proportional constant [see Problem 8.7]. Now, recalling that l and $-l$ are respectively the maximum and minimum eigenvalues of the magnetic quantum number m, we must also have [see also Eq. (7.91)]

$$\hat{l}_+|l, l\rangle = 0, \quad \hat{l}_-|l, -l\rangle = 0. \tag{8.34}$$

Hence, we can start with the eigenstate $|l, l\rangle$ of $\hat{L}_z$ and apply repeatedly the lowering operator $\hat{l}_-$ on that eigenstate to obtain the other eigenstates $|l, l-1\rangle, |l, l-2\rangle, \ldots$ of $\hat{L}_z$. The process has to stop after finite steps because $\hat{l}_-|l, -l\rangle = 0$. It is then possible to order the eigenstates of $\hat{L}_z$ in descending order of m as

$$|l, l\rangle, \ |l, l-1\rangle, \ \ldots, \ |l, -l+1\rangle, \ |l, -l\rangle. \tag{8.35}$$

Note that the same result can be obtained by starting with the eigenstate $|l, -l\rangle$ of $\hat{L}_z$ and applying repeatedly the raising operator $\hat{l}_+$ on that eigenstate. Hence, for each possible value of l, there are $2l + 1$ possible eigenstates of $\hat{L}_z$. Moreover, since $2l + 1$ must be a positive integer, the value of l should be either a non-negative integer or a positive half-integer (i.e., a number of the form $n+\frac{1}{2}$, where n is a non-negative integer). However, as we will see below, for the orbital angular momentum currently under discussion that is associated with the rotational or orbital motion of a system in real space, the azimuthal quantum number l can only take integer values. The half-integer values are exclusive for the spin quantum number that will be discussed in Section 8.6. As a result, the possible values of the azimuthal quantum number l are given by

$$l = 0, \ 1, \ 2, \ \ldots. \tag{8.36}$$

Therefore, the magnetic quantum number m takes integer values as well. For a given l, the possible values of m are given in descending order by

$$m = l, \ l-1, \ \ldots, \ -l+1, \ -l. \tag{8.37}$$

The group of the $2l + 1$ states in Eq. (8.35) that are related by the raising and lowering operators $\hat{l}_+$ and $\hat{l}_-$ are called a multiplet. Having determined the possible values of the quantum numbers l and m, we can write down the orthonormal conditions and completeness relation for the angular momentum eigenstates $|l, m\rangle$ as [see Eqs. (6.11) and (6.12)]

$$\langle l, m|l', m'\rangle = \delta_{ll'}\delta_{mm'} \quad \text{and} \quad \sum_{l=0}^{\infty}\sum_{m=-l}^{l} |l, m\rangle\langle l, m| = \hat{1}, \quad (8.38)$$

respectively.

In order to find the eigenvalues of $\hat{\mathbf{L}}^2$, we use the definitions of the raising and lowering operators in Eq. (8.28) to write down the following product:

$$\hat{l}_-\hat{l}_+ = \left(\hat{l}_x - i\hat{l}_y\right)\left(\hat{l}_x + i\hat{l}_y\right)$$
$$= \hat{l}_x^2 + \hat{l}_y^2 + i\left[\hat{l}_x, \hat{l}_y\right]$$
$$= \hat{\mathbf{l}}^2 - \hat{l}_z^2 - \hat{l}_z. \quad (8.39)$$

Taking into account that (i) from the first identity in Eq. (8.34) and the previous result we have

$$\left[\hat{\mathbf{l}}^2 - \hat{l}_z\left(\hat{l}_z + 1\right)\right]|l, l\rangle = \hat{l}_-\hat{l}_+|l, l\rangle = 0, \quad (8.40)$$

and (ii) from Eq. (8.25) it follows that

$$\hat{l}_z\left(\hat{l}_z + 1\right)|l, l\rangle = l(l + 1)\hbar^2|l, l\rangle, \quad (8.41)$$

where we do not need to consider global phase factor (recalling the relation between lower-case and upper-case operators), we can finally obtain

$$\left[\hat{\mathbf{L}}^2 - l(l + 1)\hbar^2\right]|l, l\rangle = 0, \quad (8.42)$$

or equivalently,

$$\hat{\mathbf{L}}^2|l, l\rangle = l(l + 1)\hbar^2|l, l\rangle. \quad (8.43)$$

Using the fact that $\hat{\mathbf{L}}^2$ commutes with $\hat{l}_-$, i.e., $[\hat{\mathbf{L}}^2, \hat{l}_-] = 0$, we can generalize the above result to other values of m and obtain

$$\hat{\mathbf{L}}^2|l, m\rangle = l(l + 1)\hbar^2|l, m\rangle, \quad (8.44)$$

which is the eigenvalue equation for the operator $\hat{\mathbf{L}}^2$ that we were looking for. Therefore, the eigenvalue of the operator $\hat{\mathbf{L}}^2$ is $l(l + 1)\hbar^2$ instead of $l^2\hbar^2$ as one would have expected. This is a peculiarity of quantum mechanics in that the eigenvalue of $\hat{\mathbf{L}}^2$ is not the square of the maximum eigenvalue of $\hat{l}_z$. As can seen from Eq. (8.39), it is a direct consequence of the fact that components of the angular

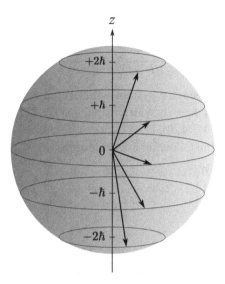

Figure 8.5 Quantization of angular momentum for the case of $l = 2$. The vectors from top to bottom illustrate respectively the angular momentum with $L_z = +2\hbar$, $+\hbar$, 0, $-\hbar$, $-2\hbar$. The magnitude of the angular momentum is given by $L = \sqrt{2(2+1)}\hbar = \sqrt{6}\hbar$.

momentum operator do not commute with one another. Indeed, if they would commute, the last term in the parentheses on the right-hand side of Eq. (8.40) would be absent and the eigenvalue of $\hat{\mathbf{L}}^2$ would be equal to $l^2\hbar^2$. The above discussion on the quantization of angular momentum is illustrated in Fig. 8.5 for the case of $l = 2$. In this case, there are five angular momentum eigenstates in the multiplet, corresponding to $L_z = +2\hbar$, $+\hbar$, 0, $-\hbar$, $-2\hbar$, respectively, and the magnitude of the angular momentum is given by $L = \sqrt{2(2+1)}\hbar = \sqrt{6}\hbar$. Therefore, the angular momentum and its z component can never be aligned (a consequence of the uncertainty principle). Before ending this section, we note that since it is up to us which space direction we call the z axis, the eigenvalues of the component of the angular momentum in an arbitrary direction are always quantized in the same way as in Eq. (8.37).

Problem 8.5 Using Eq. (8.39), show that
$$\hat{l}_+|l, m\rangle = \sqrt{l(l+1) - m(m+1)}\,|l, m+1\rangle.$$
Problem 8.6 Show that $[\hat{L}_z, \hat{l}_-] = -\hbar\hat{l}_-$, then use the commutation relation to derive Eq. (8.33).

Problem 8.7 Express $\hat{l}_+\hat{l}_-$ in terms of $\hat{\mathbf{L}}^2$ and $\hat{L}_z$, then use the result to show that

$$\hat{l}_-|l, m\rangle = \sqrt{l(l+1) - m(m-1)}\,|l, m-1\rangle.$$

Problem 8.8 Consider a system with orbital angular momentum quantum number $l = 1$, which implies $m = 1, 0, -1$. Show that in the basis $\{|1, 1\rangle, |1, 0\rangle, |1, -1\rangle\}$, the operators $\hat{L}_x$, $\hat{L}_y$, and $\hat{L}_z$ are represented by the following 3×3 matrices:

$$\hat{L}_x = \frac{\hbar}{\sqrt{2}}\begin{bmatrix} 0 & 1 & 0 \\ 1 & 0 & 1 \\ 0 & 1 & 0 \end{bmatrix}, \quad \hat{L}_y = \frac{\hbar}{\sqrt{2}}\begin{bmatrix} 0 & -i & 0 \\ i & 0 & -i \\ 0 & i & 0 \end{bmatrix}, \quad \hat{L}_z = \hbar\begin{bmatrix} 1 & 0 & 0 \\ 0 & 0 & 0 \\ 0 & 0 & -1 \end{bmatrix}.$$

$$(8.45)$$

8.4 Angular Momentum Eigenfunctions

While algebraically straightforward, the above approach to the eigenvalues and eigenstates of the angular momentum is somewhat conceptually abstract since we have not yet derived the specific forms of the eigenstates according to particular values of l and m. It therefore seems desirable to take a more concrete approach to the same problem by considering eigenfunctions of the angular momentum operator. However, a full treatment of the angular momentum eigenfunctions is a mathematically challenging subject to be dealt with here.[a] In the present context, we will limit ourselves to a very sketchy presentation and bypass most of the details.

The three-dimensional angular momentum eigenfunctions $\psi_{lm}(\mathbf{r})$ are the angular momentum eigenstates $|l, m\rangle$ represented in the position eigenbasis $\{|\mathbf{r}\rangle\}$, that is,

$$\psi_{lm}(\mathbf{r}) = \langle\mathbf{r}|l, m\rangle. \tag{8.46}$$

When dealing with angular momentum, because of the rotational nature of the problem it is convenient to use spherical coordinates instead of the usual Cartesian (rectangular) coordinates. In spherical coordinates the position of a point P in three-dimensional space is specified by three numbers (r, θ, ψ), where r is the radial distance from the origin O to P (with $0 \leq r < \infty$), θ is the polar angle of the

---

[a]The interested reader may have a look at (Auletta *et al.*, 2009, Chapter 6).

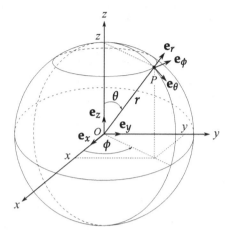

Figure 8.6 Relation between the Cartesian (rectangular) coordinates (x, y, z) and the spherical coordinates (r, θ, ϕ) of a point P in three-dimensional space. See text for details.

line segment OP measured from the z axis (with $0 \leq \theta \leq \pi$), and ϕ is the azimuthal angle of the line segment OP measured from the x axis to its orthogonal projection on the xy plane (with $0 \leq \phi < 2\pi$) [see Figs. 1.1 and 8.6]. The spherical coordinates (r, θ, ϕ) of a point are related to its Cartesian coordinates (x, y, z) by

$$r = \sqrt{x^2 + y^2 + z^2}, \quad \theta = \arctan \frac{\sqrt{x^2 + y^2}}{z}, \quad \phi = \arctan \frac{y}{x}. \tag{8.47}$$

In the above expressions arctan denotes the inverse tangent function. We recall that the tangent function is the ratio of the sine function to the cosine function [see Box 2.3]. Just as the square root function $y = \sqrt{x}$ is defined such that $y^2 = x$, the function $y = \arctan x$ is defined such that $\tan y = x$. Conversely, the Cartesian coordinates (x, y, z) can be obtained from the spherical coordinates (r, θ, ϕ) by

$$x = r \sin\theta \cos\phi, \quad y = r \sin\theta \sin\phi, \quad z = r \cos\theta. \tag{8.48}$$

We note that relations like those in the above two expressions that describe conversions between two coordinate systems of the same space are called coordinate transformations. Some elementary properties of the coordinate transformation between Cartesian and spherical coordinates can be found in Boxes 8.2 and 8.3.

Box 8.2 Spherical coordinates I

Ir is well known that in Cartesian coordinates the basis vectors $\mathbf{e}_x$, $\mathbf{e}_y$, and $\mathbf{e}_z$, i.e., the unit vectors in the x, y, and z directions are position independent [see Boxes 1.1 and 3.1]. However, this in general is not true in spherical coordinates. The spherical basis $\{\mathbf{e}_r, \mathbf{e}_\theta, \mathbf{e}_\phi\}$ can be expressed in terms of the Cartesian basis $\{\mathbf{e}_x, \mathbf{e}_y, \mathbf{e}_z\}$ as [see Fig. 8.6]

$$\mathbf{e}_r = \sin\theta\cos\phi\,\mathbf{e}_x + \sin\theta\sin\phi\,\mathbf{e}_y + \cos\theta\,\mathbf{e}_z, \tag{8.49a}$$

$$\mathbf{e}_\theta = \cos\theta\cos\phi\,\mathbf{e}_x + \cos\theta\sin\phi\,\mathbf{e}_y - \sin\theta\,\mathbf{e}_z, \tag{8.49b}$$

$$\mathbf{e}_\phi = -\sin\phi\,\mathbf{e}_x + \cos\phi\,\mathbf{e}_y. \tag{8.49c}$$

It is easy to check that $\{\mathbf{e}_r, \mathbf{e}_\theta, \mathbf{e}_\phi\}$ is indeed an orthonormal basis. The inverse of the above transformation can be written as

$$\mathbf{e}_x = \sin\theta\cos\phi\,\mathbf{e}_r + \cos\theta\cos\phi\,\mathbf{e}_\theta - \sin\phi\,\mathbf{e}_\phi, \tag{8.50a}$$

$$\mathbf{e}_y = \sin\theta\sin\phi\,\mathbf{e}_r + \cos\theta\sin\phi\,\mathbf{e}_\theta + \cos\phi\,\mathbf{e}_\phi, \tag{8.50b}$$

$$\mathbf{e}_z = \cos\theta\,\mathbf{e}_r - \sin\theta\,\mathbf{e}_\theta. \tag{8.50c}$$

Note that the cross products among the Cartesian basis vectors are [see Box 8.1]

$$\mathbf{e}_x \times \mathbf{e}_y = \mathbf{e}_z, \quad \mathbf{e}_y \times \mathbf{e}_z = \mathbf{e}_x, \quad \mathbf{e}_z \times \mathbf{e}_x = \mathbf{e}_y, \tag{8.51}$$

while those among the spherical basis vectors are

$$\mathbf{e}_r \times \mathbf{e}_\theta = \mathbf{e}_\phi, \quad \mathbf{e}_\theta \times \mathbf{e}_\phi = \mathbf{e}_r, \quad \mathbf{e}_\phi \times \mathbf{e}_r = \mathbf{e}_\theta. \tag{8.52}$$

Using Eqs. (8.49), we find that the coordinate derivatives of the basis vectors $\mathbf{e}_r$, $\mathbf{e}_\theta$, and $\mathbf{e}_\phi$ are given by

$$\frac{\partial\mathbf{e}_r}{\partial r} = 0, \quad \frac{\partial\mathbf{e}_r}{\partial\theta} = \mathbf{e}_\theta, \quad \frac{\partial\mathbf{e}_r}{\partial\phi} = \sin\theta\,\mathbf{e}_\phi, \tag{8.53a}$$

$$\frac{\partial\mathbf{e}_\theta}{\partial r} = 0, \quad \frac{\partial\mathbf{e}_\theta}{\partial\theta} = -\mathbf{e}_r, \quad \frac{\partial\mathbf{e}_\theta}{\partial\phi} = \cos\theta\,\mathbf{e}_\phi, \tag{8.53b}$$

$$\frac{\partial\mathbf{e}_\phi}{\partial r} = 0, \quad \frac{\partial\mathbf{e}_\phi}{\partial\theta} = \mathbf{e}_\theta, \quad \frac{\partial\mathbf{e}_\phi}{\partial\phi} = -(\sin\theta\,\mathbf{e}_r + \cos\theta\,\mathbf{e}_\theta). \tag{8.53c}$$

In other words, $\mathbf{e}_r$, $\mathbf{e}_\theta$, and $\mathbf{e}_\phi$ are invariant under radial translation (the first column in the above equations), but not under rotation (the second and third columns in the above equations).

Box 8.3 Spherical coordinates II

Here we derive the useful relations between the differential operators in Cartesian coordinates and those in spherical coordinates. From the transformation (8.47), we have [see Problem 8.9]

$$\frac{\partial r}{\partial x} = \sin\theta\cos\phi, \quad \frac{\partial\theta}{\partial x} = \frac{1}{r}\cos\theta\cos\phi, \quad \frac{\partial\phi}{\partial x} = -\frac{1}{r}\frac{\sin\phi}{\sin\theta},$$
$$(8.54a)$$

$$\frac{\partial r}{\partial y} = \sin\theta\sin\phi, \quad \frac{\partial\theta}{\partial y} = \frac{1}{r}\cos\theta\sin\phi, \quad \frac{\partial\phi}{\partial y} = \frac{1}{r}\frac{\cos\phi}{\sin\theta},$$
$$(8.54b)$$

$$\frac{\partial r}{\partial z} = \cos\theta, \quad \frac{\partial\theta}{\partial z} = -\frac{1}{r}\sin\theta, \quad \frac{\partial\phi}{\partial z} = 0, \qquad (8.54c)$$

where use has been made of the formula

$$\frac{d}{dx}\arctan x = \frac{1}{1+x^2} \qquad (8.55)$$

and the inverse transformation (8.48). Using the chain rule for partial derivatives [see Eq. (6.75)], we can write $\partial/\partial x$, $\partial/\partial y$, and $\partial/\partial z$ in terms of $\partial/\partial r$, $\partial/\partial\theta$, and $\partial/\partial\phi$ as

$$\frac{\partial}{\partial x} = \frac{\partial r}{\partial x}\frac{\partial}{\partial r} + \frac{\partial\theta}{\partial x}\frac{\partial}{\partial\theta} + \frac{\partial\phi}{\partial x}\frac{\partial}{\partial\phi}$$
$$= \sin\theta\cos\phi\frac{\partial}{\partial r} + \frac{\cos\theta\cos\phi}{r}\frac{\partial}{\partial\theta} - \frac{\sin\phi}{r\sin\theta}\frac{\partial}{\partial\phi}, \quad (8.56a)$$

$$\frac{\partial}{\partial y} = \frac{\partial r}{\partial y}\frac{\partial}{\partial r} + \frac{\partial\theta}{\partial y}\frac{\partial}{\partial\theta} + \frac{\partial\phi}{\partial y}\frac{\partial}{\partial\phi}$$
$$= \sin\theta\sin\phi\frac{\partial}{\partial r} + \frac{\cos\theta\sin\phi}{r}\frac{\partial}{\partial\theta} + \frac{\cos\phi}{r\sin\theta}\frac{\partial}{\partial\phi}, \quad (8.56b)$$

$$\frac{\partial}{\partial z} = \frac{\partial r}{\partial z}\frac{\partial}{\partial r} + \frac{\partial\theta}{\partial z}\frac{\partial}{\partial\theta} + \frac{\partial\phi}{\partial z}\frac{\partial}{\partial\phi}$$
$$= \cos\theta\frac{\partial}{\partial r} - \frac{\sin\theta}{r}\frac{\partial}{\partial\theta}. \qquad (8.56c)$$

The inverse relations can be obtained by solving the above equations for $\partial/\partial r$, $\partial/\partial\theta$, and $\partial/\partial\phi$. The relations (8.56) are of great use in expressing the Cartesian components of the orbital angular momentum $\hat{L}_x$, $\hat{L}_y$, and $\hat{L}_z$ in terms of the spherical coordinates r, θ, and ϕ [see Problem 8.10].

In spherical coordinates the angular momentum eigenfunctions are given by

$$\psi_{lm}(r, \theta, \phi) = \langle r, \theta, \phi | l, m \rangle. \tag{8.57}$$

In order to find $\psi_{lm}(r, \theta, \phi)$ we need to solve the eigenvalue equations (8.25) and (8.44) in spherical coordinates (which is a kind of position representation). From the expressions (8.15) for the angular momentum components $\hat{L}_x$, $\hat{L}_y$, and $\hat{L}_z$ in Cartesian coordinates and the relations (8.56), after some tedious algebra [see Problem 8.10] we can rewrite $\hat{L}_z$ and $\hat{\mathbf{L}}^2$ in spherical coordinates as

$$\hat{L}_z = -i\hbar \frac{\partial}{\partial \phi}, \tag{8.58a}$$

$$\hat{\mathbf{L}}^2 = -\hbar^2 \left[\frac{1}{\sin\theta} \frac{\partial}{\partial \theta} \left(\sin\theta \frac{\partial}{\partial \theta} \right) + \frac{1}{\sin^2\theta} \frac{\partial^2}{\partial \phi^2} \right], \tag{8.58b}$$

where extensive use has been made of the chain rule of partial derivatives. Note that the above equations only depend on the angles θ and ϕ. Indeed, the use of spherical coordinates is advantageous because it allows the factorization of the angular momentum eigenfunction $\psi_{lm}(r, \theta, \phi)$ into the radial (depending on r) and angular (depending on both θ and ϕ) parts as

$$\psi_{lm}(r, \theta, \phi) = f(r) Y_{lm}(\theta, \phi). \tag{8.59}$$

Collecting the results (8.58) and (8.59), we find that the eigenvalue equations (8.25) and (8.44) in spherical coordinates reduce to

$$\hat{L}_z Y_{lm}(\phi, \theta) = m\hbar Y_{lm}(\phi, \theta), \tag{8.60a}$$

$$\hat{\mathbf{L}}^2 Y_{lm}(\phi, \theta) = l(l + 1)\hbar^2 Y_{lm}(\phi, \theta), \tag{8.60b}$$

where it is noted that the radial part $f(r)$ has been canceled out in the above expressions. Therefore, angular momentum eigenfunctions are actually independent of the radial coordinate r, which in turn allows us to rewrite the eigenfunctions in a simpler form as

$$Y_{lm}(\phi, \theta) = \langle \theta, \phi | l, m \rangle. \tag{8.61}$$

We note that such a simplification is not unexpected, as we have seen at the end of Section 8.2 that the magnitude squared of a spatial vector (or, equivalently, the radial distance of a point) is invariant under rotations.

Since the eigenfunctions $Y_{lm}(\phi, \theta)$ represent the probability amplitudes whose square moduli yield the angular probability distributions of finding the system under consideration, the fact that the latter should take unique values means that $Y_{lm}(\phi, \theta)$ be single-valued functions of θ and ϕ. Mathematically, this is tantamount to the requirement that l and m are integers. A detailed analysis of the eigenvalue equations (8.60) shows the possible values of l and m are given by

$$l = 0, \ 1, \ 2, \ \ldots \quad \text{and} \quad m = l, \ l - 1, \ \ldots, \ -l + 1, \ -l, \quad (8.62)$$

which are the same results we have obtained in the previous section. In mathematical terms the angular momentum eigenfunctions $Y_{lm}(\phi, \theta)$ are known as the *spherical harmonics* [see Box 8.4], which are important in many theoretical and practical applications in physics and engineering.

The names s–states (or s–waves), p–states (or p–waves), d–states (or d–waves), and f–states (or f–waves) are used to refer to angular momentum eigenstates with quantum number $l = 0$, 1, 2 and 3, respectively. The symbols s, p, d, and f stand for the terms *sharp, principal, diffuse,* and *fundamental,* respectively, which are derived from the characteristics of their spectroscopic lines. The rest angular momentum eigenstates with higher azimuthal quantum numbers are named in subsequent alphabetical order.

Problem 8.9 The advanced reader may try to derive transformations (8.54). (*Hint:* This derivation is not easy. We suggest here a simplified method that allows to derive the derivatives in Eqs. (8.54) and (8.56). To this purpose, make use of derivatives of the form

$$\frac{\partial}{\partial r_j} = \frac{\partial r}{\partial r_j} \frac{\partial}{\partial r} + \frac{\partial \cos\theta}{\partial r_j} \frac{\partial}{\partial \cos\theta} + \frac{\partial \tan\phi}{\partial r_j} \frac{\partial}{\partial \tan\phi},$$

where $j = x, y, z$ with r_x corresponding to x, r_y to y, and r_z to z, and we have taken advantage of the following expressions $r = \sqrt{x^2 + y^2 + z^2}$, $\cos\theta = z/\sqrt{x^2 + y^2 + z^2}$, and $\tan\phi = y/x$, which can be obtained from Eqs. (8.47) and (8.48).)

Problem 8.10 The advanced reader may try to derive Eqs. (8.58). (*Hint:* Start with the relations (8.56) and try to express the Cartesian components of the orbital angular momentum $\hat{L}_x$, $\hat{L}_y$, and $\hat{L}_z$ in terms of the spherical coordinates r, θ, and ϕ.)

Box 8.4 Spherical harmonics

The spherical harmonics are defined by

$$Y_{lm}(\theta, \phi) = (-1)^m \sqrt{\frac{(2l+1)(l-m)!}{4\pi(l+m)!}} \, P_l^m(\cos\theta) \, e^{im\phi}, \quad (8.63)$$

where $P_l^m(x)$ are the associated Legendre polynomials and defined in terms of the Legendre polynomials $P_l(x)$ by

$$P_l^m(x) = (1 - x^2)^{m/2} \frac{d^m}{dx^m} P_l(x), \quad P_l(x) = \frac{1}{2^l l!} \frac{d^l}{dx^l} (x^2 - 1)^l. \quad (8.64)$$

From the definition, we find the first few spherical harmonics after some straightforward algebra [see Problem 8.11]. For $l = 0$ we have

$$Y_{0,0}(\theta, \phi) = \frac{1}{\sqrt{4\pi}}. \quad (8.65)$$

For $l = 1$ we have

$$Y_{1,0}(\theta, \phi) = \sqrt{\frac{3}{4\pi}} \cos\theta, \quad Y_{1,\pm1}(\theta, \phi) = \mp\sqrt{\frac{3}{8\pi}} \sin\theta \, e^{\pm i\phi}. \quad (8.66)$$

Final, for $l = 2$ we have

$$Y_{2,0}(\theta, \phi) = \sqrt{\frac{5}{16\pi}} (3\cos^2\theta - 1), \quad (8.67a)$$

$$Y_{2,\pm1}(\theta, \phi) = \mp\sqrt{\frac{15}{8\pi}} \sin\theta \cos\theta \, e^{\pm i\phi}, \quad (8.67b)$$

$$Y_{2,\pm2}(\theta, \phi) = \sqrt{\frac{15}{32\pi}} \sin^2\theta \, e^{\pm 2i\phi}. \quad (8.67c)$$

It is noted that the use of the plus–minus sign ($\pm$) and the minus–plus sign ($\mp$) in Eqs. (8.66) and (8.67) is a shorthand notation to present *two* equations in one expression, in which the upper and lower signs are interrupted separately and independently.

We summarize here some of the basic properties satisfied by the spherical harmonics:

$$Y_{lm}(\pi - \theta, \pi + \phi) = (-1)^l Y_{lm}(\theta, \phi), \tag{8.68a}$$

$$\int_0^{2\pi} \int_0^\pi Y_{lm}^*(\theta, \phi) Y_{l'm'}(\theta, \phi) \sin\theta d\theta d\phi = \delta_{ll'}\delta_{mm'}, \tag{8.68b}$$

$$\sum_{l=0}^\infty \sum_{m=-l}^l Y_{lm}(\theta, \phi) Y_{lm}^*(\theta', \phi') = \delta(\cos\theta - \cos\theta')\delta(\phi - \phi'). \tag{8.68c}$$

It is noted that the last two equations above are respectively a reformulation of the orthonormal conditions and completeness relation for the angular momentum eigenstates [see Eqs. (8.38)].

Problem 8.11 The advanced reader may try to derive the first few spherical harmonics presented in Box 8.4.

8.5 Central Potential and the Hydrogen Atom

So far we have discussed (in Chapter 7) only examples of quantum systems in one dimension. Equipped with the notion of orbital angular momentum, we are now in a position to discuss models of concrete quantum systems in three dimensions. An instructive model is represented by a particle subject to a central potential. In specific, we will consider an electron moving in the attractive Coulomb potential of the proton in a hydrogen atom. What we would like to do here is to give the reader a feeling of what happens in the conceptual laboratory of a theory, how *ad hoc* approximations are made in order that concrete physical problems are treatable at all, how very abstract formulations can be translated in capability to make true predictions, and how complex are the problems faced by science. Obviously, this is only a feeling. We cannot present all the very difficult mathematical developments that are employed here. Therefore, we limit ourselves to a sketchy presentation as we already did before. However, here, when dealing with the hydrogen atom things acquire a new light and we hope that the reader will finally

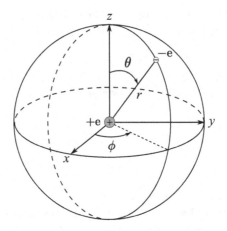

Figure 8.7 Pictorial representation of an electron (white circle) with negative charge −e orbiting a proton (gray circle) with a positive charge +e in a hydrogen atom, where e is the elementary positive charge. The electron wave function is expressed in spherical coordinates as $\psi(r, \theta, \phi)$, where r represents the distance of the electron from the proton, whose position is chosen as the origin, θ is the polar angle, and ϕ is the azimuthal angle [see Fig. 8.6].

also enjoy the mathematical apparatus as a way to understand what science is and does.

A potential energy that depends only on the distance between particles is called a central potential. In other words, a central potential for a system of two particles has the form $V = V(r)$, where r is the distance between the two particles. The hydrogen atom is a bound state of a proton and an electron. Since the electron is far less massive than the proton, the relative motion of the latter can be largely neglected. This is tantamount to choosing the position of the proton as the origin of the coordinate system [see Fig. 8.7]. Hence the distance between the electron and the proton, r, is simply the radial distance of the electron and the central potential $V(r)$ possesses spherical symmetry. Because of this spherical symmetry, it will be convenient to solve the problem in spherical coordinates [see Eqs. (8.47)]. We recall the general expression for the Hamiltonian (7.8) in three dimensions, in which the Laplacian operator $\mathbf{V}^2$ [see Eq. (6.102)] in spherical coordinates (r, θ, ϕ) can be rewritten as

(see Problem 8.12, here we spare the reader the lengthy and tedious derivation)

$$\nabla^2 = \frac{1}{r^2}\left[\frac{\partial}{\partial r}\left(r^2\frac{\partial}{\partial r}\right) + \frac{1}{\sin\theta}\frac{\partial}{\partial\theta}\left(\sin\theta\frac{\partial}{\partial\theta}\right) + \frac{1}{\sin^2\theta}\frac{\partial^2}{\partial\phi^2}\right].$$

(8.69)

It follows from Eq. (8.58b) that up to a factor of $\hbar^2$ the angular part of the Laplacian is the minus of the magnitude squared of the angular momentum operator. As a result, the three-dimensional Hamiltonian for a central potential can be written as [see again Eq. (7.8)]

$$\hat{H} = -\frac{\hbar^2}{2m}\nabla^2 + V(r)$$

$$= -\frac{1}{2m}\left[\frac{\hbar^2}{r^2}\frac{\partial}{\partial r}\left(r^2\frac{\partial}{\partial r}\right) - \frac{1}{r^2}\hat{\mathbf{L}}^2\right] + V(r)$$

$$= -\frac{\hbar^2}{2mr^2}\frac{\partial}{\partial r}\left(r^2\frac{\partial}{\partial r}\right) + \frac{\hat{\mathbf{L}}^2}{2mr^2} + V(r).$$

(8.70)

It is straightforward to check that the operators $\hat{H}$, $\hat{\mathbf{L}}^2$, and $\hat{L}_z$ are mutually commuting and hence they have common eigenfunctions (which represents another case of degenerate energy eigenvalues, as discussed in Section 7.6). Together with the fact that the angular momentum eigenfunctions are the spherical harmonics $Y_{lm}(\theta, \phi)$, this allows us to write the energy eigenfunction $\psi(r, \theta, \phi)$ in terms of the spherical harmonics as

$$\psi(r, \theta, \phi) = f(r)Y_{lm}(\theta, \phi),$$

(8.71)

where we need now to consider the radial part $f(r)$ of the energy eigenfunction (that decoupled in the previous section). This is quite understandable, since the energy levels obviously depend on the distance r from the nucleus. However, the above expression allows us still to factorize the time-independent Schrödinger equation (7.29) into the angular and radial parts. According to what is previously said, the angular part of the Schrödinger equation is the eigenvalue equation for $\hat{\mathbf{L}}^2$ in spherical coordinates given by Eq. (8.60b), while the radial part of the Schrödinger equation takes the form

$$-\frac{\hbar^2}{2mr^2}\frac{\partial}{\partial r}\left(r^2\frac{\partial}{\partial r}\right)f(r) + \frac{l(l+1)\hbar^2}{2mr^2}f(r) + V(r)f(r) = Ef(r),$$

(8.72)

where the three terms on the left-hand side correspond to the three parts in the last line of Eq. (8.70). It may be noted that the square of the angular momentum contribution (the second term) takes the form of its eigenvalues $l(l + 1)\hbar^2$ times the parameter $1/2mr^2$. Obviously, since the above expression represents the eigenvalue equation of the Hamiltonian, we have the energy eigenvalue E on the right-hand side. By performing now a change of variables

$$f(r) = \frac{\xi(r)}{r} \quad (r > 0), \tag{8.73}$$

we can obtain a further simplification of the first term on the left-hand side of the above equation:

$$
\begin{aligned}
\frac{1}{r^2} \frac{\partial}{\partial r} \left(r^2 \frac{\partial}{\partial r} \right) \frac{\xi(r)}{r} &= \frac{1}{r^2} \frac{\partial}{\partial r} \left(r^2 \frac{\partial}{\partial r} \frac{\xi(r)}{r} \right) \\
&= \frac{1}{r^2} \frac{\partial}{\partial r} \left(r\xi'(r) - \xi(r) \right) \\
&= \frac{1}{r} \xi''(r) + \frac{1}{r^2} \xi'(r) - \frac{1}{r^2} \xi'(r) \\
&= \frac{1}{r} \xi''(r),
\end{aligned}
\tag{8.74}
$$

where $\xi'(r) = d\xi(r)/dr$, $\xi''(r) = d^2\xi(r)/dr^2$, and use has been made of the product rule (6.73) and the quotient rule (6.74) for derivatives. This result allows us to simplify Eq. (8.72) to

$$-\frac{\hbar^2}{2m} \xi''(r) + \frac{\hbar^2 l(l + 1)}{2mr^2} \xi(r) + V(r)\xi(r) = E\xi(r). \tag{8.75}$$

We note that in term of $\xi(r)$ the radial equation (8.75) is formally identical to the one-dimensional Schrödinger equation (7.28) for a particle moving in an effective potential given by

$$V_{\text{eff}}(r) = V(r) + \frac{l(l + 1)\hbar^2}{2mr^2}. \tag{8.76}$$

As mentioned, the additional term in the effective potential originates from the orbital angular momentum of the electron. It is called the *centrifugal potential barrier* and, being repulsive, prevents the electron from reaching the center of the potential.

The previous formalism is fundamental when dealing with the hydrogen atom. In this case, the parameter m is the election mass

$m_e = 9.109 \times 10^{-31}$ kg and the potential energy $V(r)$ is the celebrated Coulomb potential

$$V(r) = -\frac{1}{4\pi\epsilon_0}\frac{e^2}{r}, \qquad (8.77)$$

which represents the electrostatic interaction between a proton with charge e and an electron with charge $-e$ at a distance r from the proton. In the above expression, while e is the elementary charge, i.e., the electric charge carried by a proton, ϵ_0 is the so-called vacuum permittivity. Their values in SI units are given by

$$e = 1.602 \times 10^{-19}\ \text{C} \quad \text{and} \quad \epsilon_0 = 8.854 \times 10^{-12}\ \text{F} \cdot \text{m}^{-1}, \qquad (8.78)$$

where C is the symbol for the electric charge unit coulomb and F the symbol for the capacitance unit farad. Using the Coulomb potential, the radial equation (8.75), after rearrangement of the terms, can be rewritten as[a]

$$\xi''(r) + \frac{2m_e}{\hbar^2}\left[E + \frac{1}{4\pi\epsilon_0}\frac{e^2}{r} - \frac{l(l+1)\hbar^2}{2m_e r^2}\right]\xi(r) = 0. \qquad (8.79)$$

Fig. 8.8(a) shows that the effective potential resulting from the sum of the attractive Coulomb potential and the repulsive centrifugal potential has a well shape. For $E < 0$ there are discrete bound states, corresponding to a hydrogen atom in the ground state (i.e., the lowest energy state) and the various excited states. A look at Fig. 8.8 shows that the ground state is when the principal quantum $n = 1$, otherwise we have excited states [see also Section 7.7]. Moreover, for $E > 0$ there is a continuum of unbound states (that is, the electron has escaped the attractive Coulomb potential becoming a free particle [see Section 7.6]), corresponding to the ionized hydrogen atom.

In physics it is always advisable to make equations dimensionless (not depending on specific physical quantities but pure numbers) by introducing natural units characteristic to the problem. In the

---

[a]Our analysis here assumes that the proton is infinitely massive with respect to the electron. For actual proton mass $m_p \simeq 1836\, m_e$, the electron mass m_e that appears in the Hamiltonian (8.70) should be replaced by the so-called reduced mass

$$m = \frac{m_e m_p}{m_e + m_p} \simeq 0.995\, m_e,$$

which to an excellent approximation can be identified with the electron mass.

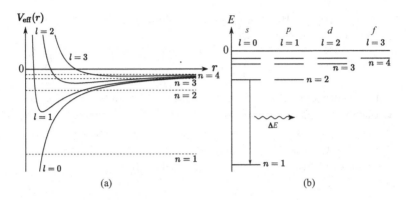

Figure 8.8 (a) The effective potential $V_{\text{eff}}(r)$ resulting from the sum of the attractive Coulomb potential and the repulsive centrifugal potential is plotted schematically for azimuthal quantum number $l = 0$, 1, 2, and 3. For energy below zero there are discrete bound states, corresponding to a hydrogen atom in the ground state and the various excited states. The energies of the lowest four bound states (the ground state and the first three excited states) are indicated by horizontal dashed lines for reference. For energy above zero there is a continuum of unbound states, corresponding to the ionized hydrogen atom. (b) Schematic representation of the first four energy levels of the hydrogen atom as labeled by the principal quantum numbers n (indicating the energy of the electron) and the azimuthal quantum number l (indicating the orbital angular momentum of the electron). The vertical line with an arrow represents an electron transition from the $n = 2$ state to the $n = 1$ state, with emission of a photon (illustrated by a wiggly line) of energy $\Delta E = E_2 - E_1 = -R_\infty/4 + R_\infty = 3R_\infty/4 = 10.2$ eV. Adapted from (Auletta, 2011a, p. 162).

present problem, let us first introduce two natural units that can be built from the physical constants m_e, $\hbar$, e, and ϵ_0: the natural unit of length is the *Bohr radius* a_0, and the natural unit of energy is the *Rydberg unit of energy* R_∞. They are respectively defined by

$$a_0 = \frac{4\pi\epsilon_0\hbar^2}{m_e e^2} = 5.29\times10^{-11}\,\text{m} \quad \text{and} \quad R_\infty = \frac{m_e e^4}{32\pi^2\epsilon_0^2\hbar^2} = 13.6\,\text{eV},$$

(8.80)

where eV stands for electron volt and is a unit of energy equal to approximately $1.60217653 \times 10^{-19}$ J. The Bohr radius is equal to the most probable distance between the proton and the electron in a hydrogen atom in its ground state, while the Rydberg constant measures the ionization energy of a hydrogen atom in its ground state and hence is related to the ground state energy of the hydrogen atom. This allows us now to define the dimensionless variables $\tilde{r}$ and the dimensionless constant $\tilde{E}$:

$$\tilde{r} = \frac{r}{a_0}, \quad \tilde{E} = -\frac{E}{R_\infty}. \tag{8.81}$$

In other words, these two quantities are multiples of the above natural parameters that are therefore considered units of measure. In what follows we limit ourselves to bound states for which $E < 0$ and therefore $\tilde{E} > 0$, thus the choice of the minus sign. Using the expression

$$r_0^2 \frac{d^2}{dr^2} \xi(r) = \frac{d^2}{d\tilde{r}^2} \xi(\tilde{r}), \tag{8.82}$$

which is a result of the chain rule (6.75), we can perform a change of variables and rewrite Eq. (8.79) in terms of the $\tilde{r}$ as

$$\left[\frac{d^2}{d\tilde{r}^2} - \frac{l(l+1)}{\tilde{r}^2} + \frac{2}{\tilde{r}} - \tilde{E} \right] \xi(\tilde{r}) = 0. \tag{8.83}$$

It is crucial in the following development to introduce the dimensionless quantity n that is related to $\tilde{E}$ by

$$n = \frac{1}{\sqrt{\tilde{E}}}. \tag{8.84}$$

We will see below that n takes only positive integer values and is the quantum number that labels the orbital levels [see Fig. 8.8]. Since it is always convenient to write the unknown function in a differential equation in terms of an exponential multiplied by another function, here we adopt the assumption that $\xi(\tilde{r})$ is of the form[a]

$$\xi(\tilde{r}) = \tilde{r}^{l+1} e^{-\frac{\tilde{r}}{n}} W(\tilde{r}), \tag{8.85}$$

from which we can obtain a reformulation of the radial equation in terms of the function $W(\tilde{r})$ as [see Problem 8.14]

$$\tilde{r} W''(\tilde{r}) + 2 \left(l + 1 - \frac{\tilde{r}}{n} \right) W'(\tilde{r}) + \frac{2(n - l - 1)}{n} W(\tilde{r}) = 0. \tag{8.86}$$

With another change of variables

$$\eta = \frac{2\tilde{r}}{n}, \tag{8.87}$$

---

[a]We note that the form of $\xi(\tilde{r})$ in Eq. (8.85) is actually an informed guess that is based on the asymptotic behavior of $\xi(\tilde{r})$ for both small $\tilde{r}$ (i.e., $\tilde{r} \to 0$) and large $\tilde{r}$ (i.e., $\tilde{r} \to \infty$).

we finally obtain the desired equation

$$\eta W''(\eta) + [2(l + 1) - \eta]W'(\eta) + (n - l - 1)W(\eta) = 0, \quad (8.88)$$

where the derivatives of $W(\eta)$ are now taken with respect to η. In mathematically terms, the above equation is called the *associated Laguerre equation*.

From the physical requirement that energy eigenfunction $\psi(r, \theta, \phi)$ should satisfy the normalization condition [see Eqs. (8.71) and (8.68b) as well as Section 6.3]

$$\int |\psi(r, \theta, \phi)|^2 d^3r = \int_0^\infty |f(r)|r^2 dr = 1, \quad (8.89)$$

we find that $W(\eta)$ cannot grow faster than $e^{\frac{\eta}{2}}$ as $\eta \to \infty$, which in turn implies that $n - l - 1$ has to be a non-negative integer and that $W(\eta)$ is in fact a polynomial in η [see Box 8.5]. The resultant polynomials are called the *associated Laguerre polynomials*. Since the azimuthal quantum number l is a non-negative integer, it follows that n must be a positive integer, i.e., $n = 1, 2, \ldots$, and that for n fixed the allowed values of l cannot be greater than $n - 1$, i.e., we have $l \leq n - 1$.

Then, given the definitions (8.81) and (8.84), for each value of n the corresponding energy eigenvalue E_n is given by

$$E_n = -\frac{R_\infty}{n^2} = -\frac{13.6 \text{ eV}}{n^2} \quad (n = 1, 2, 3, \ldots). \quad (8.90)$$

The number n is called the *principal quantum number*, which as can be seen from the above equation determines the energy eigenvalues of the hydrogen atom. Therefore, the radial energy eigenfunction are categorized by the quantum numbers n and l and denoted by $f_{nl}(r)$, while the full energy eigenfunction are categorized by the quantum numbers n, l, and m and denoted by $\psi_{nlm}(r, \theta, \phi)$. In physical terms, the quantum numbers n, l, and m correspond to the electron's energy, angular momentum, and angular momentum component in an arbitrary direction (conventionally called the z direction), respectively. We note that the energy eigenvalue E_n does *not* depend on the azimuthal quantum number l and the magnetic quantum number m. In other words, there may be energy eigenstates with the same value of n but different values of l and m that have the same energy eigenvalue E_n. These

---

Box 8.5 Radial wave functions of the hydrogen atom

To find the solution of the associated Laguerre equation (8.88), we assume that the solution $W(\eta)$ can be represented by a power series of the form

$$W(\eta) = \sum_{j=0}^{\infty} c_j \eta^j, \tag{8.91}$$

where c_j are coefficients to be determined. The derivatives of $W(\eta)$ with respect to η are given by

$$W'(\eta) = \sum_{j=1}^{\infty} j c_j \eta^{j-1}, \quad W''(\eta) = \sum_{j=2}^{\infty} j(j-1) c_j \eta^{j-2}. \tag{8.92}$$

Substituting the above expressions into Eq. (8.88), we find the following recurrence relation between successive coefficients:

$$c_{j+1} = \frac{j+l+1-n}{(j+1)[j+2(l+1)]} c_j \quad (j = 0, 1, 2, \dots). \tag{8.93}$$

For instance, the first three coefficients are given by

$$c_0, \quad c_1 = \frac{l+1-n}{2(l+1)} c_0, \quad c_2 = \frac{l+2-n}{2(2l+3)} c_1. \tag{8.94}$$

If we arbitrarily fix c_0, then all c_j can be obtained from c_0. For $j \gg 1$ (i.e., j is much greater than 1), we have

$$c_{j+1} \simeq \frac{1}{j} c_j \quad \text{or} \quad c_j \simeq \frac{1}{j!} c_0, \tag{8.95}$$

which leads to [see Eq. (6.93)]

$$W(\eta) \simeq \sum_{j=1}^{\infty} \frac{1}{j!} \eta^j \simeq c_0 e^{\eta}. \tag{8.96}$$

This implies that for $\eta \to \infty$ we have [see Eq. (8.85)]

$$\xi(\eta) \simeq c_0 e^{-\eta/2} e^{\eta} \simeq c_0 e^{\eta/2}, \tag{8.97}$$

which fails to satisfy the normalization condition for $\psi(r, \theta, \phi)$ [see Eq. (8.89)]. The only way to avoid the exponential divergence is to have the series (8.91) terminate at some integer $j = k$, which could happen only if $n = l + 1 + k$, where k is a non-negative integer. In order to get an idea how the polynomial solution is obtained, we consider the simplest case $n = 1$ and $l = 0$. From Eq. (8.93), we have $c_j = 0$ for $j \geq 1$, which leads to $W(\eta) = c_0$ and $\xi(\tilde{r}) = c_0 \tilde{r} e^{-\tilde{r}}$ [see Eqs. (8.91) and (8.85)] Therefore, we obtain $f_{10}(r) = c e^{-r/a_0}$ [see Eq. (8.73)], where the constant c is fixed by the normalization condition (8.89). The solutions for other values of n and l can be obtained in a similar manner [see Problem 8.16].

energy eigenstates are referred to as *degenerate energy eigenstates* and we say that the hydrogen atom's energy levels have degeneracy (as already anticipated). Moreover, since for fixed n the allowed values of l are given by $l = 0, 1, \ldots, n - 1$ and since for fixed l there are $2l + 1$ possible values for m [see Eq. (8.37)], it follows that there are n^2 degenerate energy eigenstates corresponding to the energy eigenvalue E_n [see Problem 8.15].

The energy eigenfunctions $\psi_{nlm}(r, \theta, \phi)$ are usually called the atomic orbitals. Orbitals with the same value of n are said to comprise an electron shell. Hence the quantity n^2 indicates the number of orbitals in the nth shell. The ground state (i.e., the lowest energy state) of the hydrogen atom is categorized by $n = 1$ and $l = 0$ and is non-degenerate. The first excited state is categorized by $n = 2$ and $l = 0, 1$ and has four-fold degeneracy. The energy level diagram of the hydrogen atom is schematically depicted in Fig. 8.8(b). Unlike the energy levels of a harmonic oscillator, which are equally spaced, the energy levels of the hydrogen atom become more and more dense as n goes to infinity, i.e., as the energy approaches to the limit $E_\infty = 0$ This limiting value separates the discrete bound states (with $E < 0$) of the hydrogen atom and the continuum unbound states (with $E > 0$) of the ionized hydrogen atom, for which the electron has escaped the electrostatic Coulomb force field of the proton as it happens, for instance, with the photoelectric effect [see Section 4.2]. The electron in a hydrogen atom can make a transition from a higher energy state n to a lower energy state n' by emitting a photon of energy $\Delta E = E_n - E_{n'}$. The emitted electromagnetic radiation constitutes the emission spectrum of the hydrogen atom. The fact that only discrete quanta are allowed in such transitions is precisely what ensures the stability of the hydrogen atom by avoiding the Larmor effect [see again Section 4.2]. Likewise, by absorbing a photon of energy $\Delta E = E_n - E_{n'}$ the electron can make a transition from the lower energy state n' to the high energy state n. The absorbed electromagnetic radiation constitutes the absorption spectrum of the hydrogen atom.

To give some concrete examples we write down the first few radial eigenfunctions of the hydrogen atom:

$$f_{10}(r) = 2 \left(\frac{1}{a_0} \right)^{\frac{3}{2}} e^{-\frac{r}{a_0}}, \tag{8.98a}$$

$$f_{20}(\tilde{r}) = \frac{1}{\sqrt{2}} \left(\frac{1}{a_0} \right)^{\frac{3}{2}} \left(1 - \frac{r}{2a_0} \right) e^{-\frac{r}{2a_0}}, \tag{8.98b}$$

$$f_{21}(\tilde{r}) = \frac{1}{2\sqrt{6}} \left(\frac{1}{a_0} \right)^{\frac{3}{2}} \frac{r}{a_0} e^{-\frac{r}{2a_0}}. \tag{8.98c}$$

A sketch of the derivation of the ground state radial eigenfunction $f_{10}(r)$ is given in Box 8.5. It is important to understand that we are considering here only the radial part and not the total eigenfunction. We recall that the latter has the form given by Eq. (8.71), which now can be written as

$$\psi_{nlm}(r, \theta, \phi) = f_{nl}(r) Y_{lm}(\theta, \phi). \tag{8.99}$$

For $n = 1$ and $l = m = 0$, we obtain the ground state eigenfunction [see Eq. (8.65)]

$$\psi_{100}(r, \theta, \phi) = \frac{1}{\sqrt{\pi}} \left(\frac{1}{a_0} \right)^{\frac{3}{2}} e^{-\frac{r}{a_0}}. \tag{8.100}$$

Moreover, from the normalization condition (8.89) the probability of finding the electron in the spherical shell enclosed between two concentric spheres of radii r and $r + dr$ is $|f_{nl}(r)|^2 r^2 dr$. Hence, we deduce that the radial probability density is given by

$$\wp_{nl}(r) = |f_{nl}(r)|^2 r^2. \tag{8.101}$$

From $\wp_{nl}(r)$ we can determine for each orbital at which distance from the proton the angular averaged probability of finding the electron is maximum. The first few radial probability densities of the hydrogen atom as a function of r are plotted in Fig. 8.9. As can

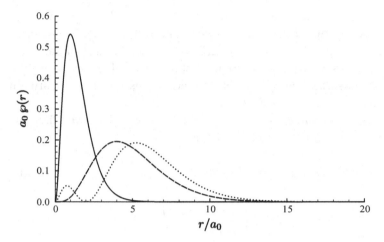

Figure 8.9 Plot of the first few radial probability densities of the hydrogen atom as a function of $\tilde{r}$: $\wp_{10}(r)$ (solid line), $\wp_{20}(r)$ (dotted line), and $\wp_{21}(r)$ (dashed line).

be seen in the figure, the most probable distance between the proton and the electron in a ground-state hydrogen atom is indeed the Bohr radius.

Problem 8.12 The advanced reader may try to derive Eq. (8.69), i.e., the Laplacian operator in spherical coordinates. (*Hint*: Start with Eq. (6.102) and the relations (8.56).)

Problem 8.13 Express the orbital angular momentum $\hat{\mathbf{L}}$ in spherical coordinates, that is, find the spherical components of the orbital angular momentum $\hat{L}_r$, $\hat{L}_\theta$, and $\hat{L}_\phi$.

Problem 8.14 The reader who has truly assimilated the results so far can try to derive Eq. (8.86).

Problem 8.15 Show that for the hydrogen atom there are n^2 degenerate energy eigenstates corresponding to the energy eigenvalue E_n, where $n = 1, 2, 3, \ldots$ is the principal quantum number.

Problem 8.16 The reader may try to derive the radial eigenfunctions of the hydrogen atom $f_{20}(r)$ and $f_{21}(r)$ up to a normalization constant.

8.6 Spin Angular Momentum

One of the most amazing discoveries in quantum mechanics was the purely quantum mechanical quantity called *spin*, which is an intrinsic angular momentum of subatomic particles (like the electron, proton, neutron, etc.) and therefore also plays an important role in the hydrogen as well as in other atoms.[a] The experimental evidence of spin was first discovered in a series of experiments performed in 1922 by Stern and Gerlach[b] in spite of the fact that they did not realize it was the quantity that we call today spin that they have observed. In these experiments, a beam of silver atoms is sent through a strongly inhomogeneous magnetic field oriented in such a way that the gradient of the magnetic field [see Section 6.5] is perpendicular to the beam axis. The silver atoms have a single electron in their outermost shell, and thus the magnetic moment of the atoms is essentially that of the unpaired outer electron. As the atoms passed through the region of the inhomogeneous magnetic field, the gradient of the field causes a deflection of the atoms according to the orientation of their magnetic moments. The silver atoms are then deposited on a collector plate perpendicular to the incoming beam axis [see Fig. 8.10]. The result showed that the field separates the beam into two distinct parts. This is an experimental evidence for quantization of angular momentum, indicating that the angular momentum and the associated magnetic moment of the electron had two possible projections along the direction of the external field. Not knowing the orbital angular momentum of the silver atom is actually zero, Stern and Gerlach erroneously attributed the twofold splitting to a quantized orbital angular momentum of magnitude $\hbar$ as presumed in the Bohr model. In 1925, Goudsmit and Uhlenbeck postulated that the electron had an intrinsic angular momentum, called the spin angular momentum, which is independent of orbital characteristics and assume only two discrete values.[c] This, together with the discovery in 1927 that the

---

[a]We should have indeed already considered spin in our discussion of the hydrogen atom, but a full treatment of the spin effects goes far beyond the scope of this book. Nevertheless, a highly simplified treatment of the electron spin is provided in p. 239.
[b](Gerlach/Stern, 1922a), (Gerlach/Stern, 1922b), (Gerlach/Stern, 1922c).
[c](Uhlenbeck/Goudsmit, 1925), (Uhlenbeck/Goudsmit, 1926).

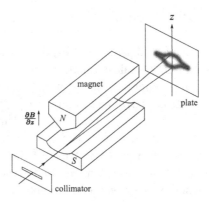

Figure 8.10 Schematic setup of the Stern–Gerlach experiment. A beam of silver atoms is sent through a strongly inhomogeneous magnetic field with the gradient $\partial B / \partial z$ perpendicular to the beam axis. The beam is separated into two distinct parts on the collector plate due to quantization of the spin angular momentum and the associated magnetic moment of the electron. See text for details.

orbital angular momentum and the associated magnetic moment of the silver atom is zero, eventually led to a reinterpretation of the double splitting as really due to spin.[a]

The spin operator is denoted by $\hat{\mathbf{S}}$ and its Cartesian components by $\hat{S}_x$, $\hat{S}_y$, and $\hat{S}_z$. Being an angular momentum, the spin operator $\hat{\mathbf{S}}$ satisfies the commutation relations that are also valid for the orbital angular momentum operator $\hat{\mathbf{L}}$ [see Eqs. (8.22)], namely,

$$\left[\hat{S}_x, \hat{S}_y\right] = i\hbar\hat{S}_z, \quad \left[\hat{S}_y, \hat{S}_z\right] = i\hbar\hat{S}_x, \quad \left[\hat{S}_z, \hat{S}_x\right] = i\hbar\hat{S}_y. \quad (8.102)$$

Similarly, the magnitude squared of the spin operator, defined by

$$\hat{\mathbf{S}}^2 = \hat{S}_x^2 + \hat{S}_y^2 + \hat{S}_z^2, \quad (8.103)$$

commutes with each of the components of the spin operator, that is, we have

$$\left[\hat{\mathbf{S}}^2, \hat{S}_j\right] = 0, \quad (8.104)$$

where $j = x, y, z$. Analogous to the case of the orbital angular momentum, let the common eigenstates of the operators $\hat{\mathbf{S}}^2$ and $\hat{S}_z$

---

[a]For a very readable historical account of the Stern–Gerlach experiment, see (Friedrich/Herschbach, 2003).

be denoted by $|s, m_s\rangle$, where s is the *spin quantum number* and m_s is the *spin magnetic quantum number*. Following the same analysis presented in Section 8.3 by introducing the spin raising and lowering operators

$$\hat{S}_{\pm} = \hat{S}_x \pm i\hat{S}_y, \tag{8.105}$$

we find that the eigenvalue equations for $\hat{S}_z$ and $\hat{\mathbf{S}}^2$ are respectively given by

$$\hat{S}_z|s, m_s\rangle = m_s\hbar|s, m_s\rangle, \tag{8.106a}$$

$$\hat{\mathbf{S}}^2|s, m_s\rangle = s(s+1)\hbar^2|s, m_s\rangle, \tag{8.106b}$$

where

$$s = 0, \frac{1}{2}, 1, \frac{3}{2}, \dots \quad \text{and} \quad m_s = s, s-1, \dots, -s+1, -s. \tag{8.107}$$

For the electron we have $s = \frac{1}{2}$ and $m_s = \pm\frac{1}{2}$, corresponding the two possible spin states observed in the Stern–Gerlach experiment.

The spin is an important intrinsic degree of freedom of subatomic particles and it enters in a huge number of experimental and theoretical problems. In the spin $\frac{1}{2}$ case (i.e., for $s = \frac{1}{2}$), it is convenient to write the spin eigenstates $|\frac{1}{2}, \frac{1}{2}\rangle$ and $|\frac{1}{2}, -\frac{1}{2}\rangle$ respectively as $|\uparrow_z\rangle$ and $|\downarrow_z\rangle$, namely, we have

$$\hat{S}_z|\uparrow_z\rangle = \frac{\hbar}{2}|\uparrow_z\rangle, \quad \text{and} \quad \hat{S}_z|\downarrow_z\rangle = -\frac{\hbar}{2}|\downarrow_z\rangle. \tag{8.108}$$

They are usually referred to as the *spin up* and *spin down* states in the z direction, respectively. They also constitute an orthonormal basis of the two-dimensional Hilbert space, called the spin $\frac{1}{2}$ space, and therefore can be written as

$$|\uparrow_z\rangle = \begin{pmatrix} 1 \\ 0 \end{pmatrix} \quad \text{and} \quad |\downarrow_z\rangle = \begin{pmatrix} 0 \\ 1 \end{pmatrix}. \tag{8.109}$$

In the following we shall often omit the subscript referring to the direction whenever no confusion may arise. In the basis $\{|\uparrow_z\rangle, |\downarrow_z\rangle\}$, the operators $\hat{S}_x$, $\hat{S}_y$, and $\hat{S}_z$ are represented by

$$\hat{S}_x = \frac{\hbar}{2}\hat{\sigma}_x, \quad \hat{S}_y = \frac{\hbar}{2}\hat{\sigma}_y, \quad \hat{S}_z = \frac{\hbar}{2}\hat{\sigma}_z. \tag{8.110}$$

The matrices $\hat{\sigma}_x$, $\hat{\sigma}_y$, and $\hat{\sigma}_z$ are called the *Pauli matrices* (or Pauli operators)[a]

$$\hat{\sigma}_x = \begin{bmatrix} 0 & 1 \\ 1 & 0 \end{bmatrix}, \quad \hat{\sigma}_y = \begin{bmatrix} 0 & -i \\ i & 0 \end{bmatrix}, \quad \hat{\sigma}_z = \begin{bmatrix} 1 & 0 \\ 0 & -1 \end{bmatrix}. \tag{8.111}$$

These matrices are both Hermitian and unitary, and therefore satisfy the following important properties:

$$\hat{\sigma}_x^2 = \hat{\sigma}_y^2 = \hat{\sigma}_z^2 = \hat{I}, \tag{8.112a}$$

$$\hat{\sigma}_j \hat{\sigma}_k + \hat{\sigma}_k \hat{\sigma}_j = 2\delta_{jk} \hat{I}, \tag{8.112b}$$

$$[\hat{\sigma}_x, \hat{\sigma}_y] = 2i\hat{\sigma}_z, \quad [\hat{\sigma}_y, \hat{\sigma}_z] = 2i\hat{\sigma}_x, \quad [\hat{\sigma}_z, \hat{\sigma}_x] = 2i\hat{\sigma}_y, \tag{8.112c}$$

where $j, k = x, y, z$ and $\hat{I}$ is the 2×2 identity matrix. It is noted that Eqs. (8.112c) are a restatement of the commutation relations (8.102) for $s = \frac{1}{2}$ [see Problem 8.18]. The eigenvalues and eigenstates of $\hat{S}_x$ and $\hat{S}_y$ satisfy the eigenvalue equations

$$\hat{S}_x |\uparrow_x\rangle = \frac{\hbar}{2} |\uparrow_x\rangle, \quad \hat{S}_x |\downarrow_x\rangle = -\frac{\hbar}{2} |\downarrow_x\rangle, \tag{8.113a}$$

$$\hat{S}_y |\uparrow_y\rangle = \frac{\hbar}{2} |\uparrow_y\rangle, \quad \hat{S}_y |\downarrow_y\rangle = -\frac{\hbar}{2} |\downarrow_y\rangle, \tag{8.113b}$$

where

$$|\uparrow_x\rangle = \frac{1}{\sqrt{2}} (|\uparrow_z\rangle + |\downarrow_z\rangle), \quad |\downarrow_x\rangle = \frac{1}{\sqrt{2}} (|\uparrow_z\rangle - |\downarrow_z\rangle), \tag{8.114a}$$

$$|\uparrow_y\rangle = \frac{1}{\sqrt{2}} (|\uparrow_z\rangle + i|\downarrow_z\rangle), \quad |\downarrow_y\rangle = \frac{1}{\sqrt{2}} (|\uparrow_z\rangle - i|\downarrow_z\rangle). \tag{8.114b}$$

When taking into account of the spin degree of freedom, the state vector of (e.g., a spin $\frac{1}{2}$) particle is generalized to

$$|\psi\rangle = |\psi_\uparrow\rangle \otimes |\uparrow_z\rangle + |\psi_\downarrow\rangle \otimes |\downarrow_z\rangle, \tag{8.115}$$

or in component form [see Eq. (8.109)]

$$|\psi\rangle = \begin{pmatrix} |\psi_\uparrow\rangle \\ |\psi_\downarrow\rangle \end{pmatrix}. \tag{8.116}$$

The corresponding wave function in position space $\psi(\mathbf{r}) = \langle \mathbf{r} | \psi \rangle$ can be expressed as a two-component *spinor* of the form

$$\psi(\mathbf{r}) = \begin{pmatrix} \psi_\uparrow(\mathbf{r}) \\ \psi_\downarrow(\mathbf{r}) \end{pmatrix}, \tag{8.117}$$

---

[a](Pauli, 1927).

where $\psi_\uparrow(\mathbf{r}) = \langle \mathbf{r}|\psi_\uparrow\rangle$ and $\psi_\downarrow(\mathbf{r}) = \langle \mathbf{r}|\psi_\downarrow\rangle$ are the (spatial) wave functions associated with the spin up and spin down states, respectively. As a simple example let us go back to the problem of the hydrogen atom discussed in the previous section. Since the Coulomb potential is independent of spin, so is the energy eigenvalue E_n [see Eq. (8.90) and comments]. Therefore, the energy eigenfunctions now take the form [see Eq. (8.99)]

$$\psi_{nlm\sigma}(r, \theta, \phi) = f_{nl}(r)Y_{lm}(\theta, \phi)\chi_\sigma, \qquad (8.118)$$

where $\sigma = \frac{1}{2}, -\frac{1}{2}$ is the spin magnetic quantum number and χ_σ are the spinors (which coincide with the vectors in Eq. (8.109))

$$\chi_{\frac{1}{2}} = \begin{pmatrix} 1 \\ 0 \end{pmatrix} \quad \text{and} \quad \chi_{-\frac{1}{2}} = \begin{pmatrix} 0 \\ 1 \end{pmatrix}. \qquad (8.119)$$

It is easy to see that for the hydrogen atom the inclusion of the electron spin doubles the degeneracy of the energy eigenvalue E_n from n^2 to $2n^2$. In other words, the electron in the hydrogen atom may flip its spin without changing its energy.[a]

Just as the orbital angular momentum is the generator of rotations in real space, the spin angular momentum is the generator of rotation in spin space. By analogy with Section 8.2, the rotation operator in spin space about a direction $\mathbf{n}$ (which is a unit vector) by an angle ϕ is given by the unitary operator [see Eq. (8.19)]

$$\hat{U}_{\mathbf{n}}^{(s)}(\phi) = e^{-\frac{i}{\hbar}\phi\mathbf{n}\cdot\hat{\mathbf{S}}}, \qquad (8.120)$$

where s in the superscript denotes the spin quantum number and $\mathbf{n}\cdot\hat{\mathbf{S}} = n_x\hat{S}_x + n_y\hat{S}_y + n_z\hat{S}_z$ with $n_x^2 + n_y^2 + n_z^2 = 1$. For the spin $\frac{1}{2}$ case, we have

$$\hat{U}_{\mathbf{n}}^{(1/2)}(\phi) = e^{-\frac{i}{2}\phi\mathbf{n}\cdot\hat{\sigma}}$$
$$= \cos\frac{\phi}{2}\hat{I} - i\sin\frac{\phi}{2}\mathbf{n}\cdot\hat{\sigma}, \qquad (8.121)$$

---

[a]We remind the reader that the actual energy levels of the hydrogen atom are much more complicated and the degeneracy is "lifted" by effects due to the relativistic motion of the electron, the proton spin, etc.

where $\hat{\sigma} = (\hat{\sigma}_x, \hat{\sigma}_y, \hat{\sigma}_z)$ and

$$\mathbf{n} \cdot \hat{\sigma} = n_x \hat{\sigma}_x + n_y \hat{\sigma}_y + n_z \hat{\sigma}_z$$

$$= n_x \begin{bmatrix} 0 & 1 \\ 1 & 0 \end{bmatrix} + n_y \begin{bmatrix} 0 & -i \\ i & 0 \end{bmatrix} + n_z \begin{bmatrix} 1 & 0 \\ 0 & -1 \end{bmatrix}$$

$$= \begin{bmatrix} n_z & n_x - in_y \\ n_x + in_y & -n_z \end{bmatrix}. \tag{8.122}$$

In obtaining the second equality in Eq. (8.121), use has been made of Eq. (2.18) and the identity

$$(\mathbf{a} \cdot \hat{\sigma})(\mathbf{b} \cdot \hat{\sigma}) = (\mathbf{a} \cdot \mathbf{b}) \hat{I} + i\hat{\sigma} \cdot (\mathbf{a} \times \mathbf{b}), \tag{8.123}$$

where $\mathbf{a}$ and $\mathbf{b}$ are spatial vectors. It is noted that this identity can be derived from the properties (8.112a) and (8.112b). For a rotation of angle ϕ about the z axis we have

$$\hat{U}_z^{(1/2)}(\phi) = \cos \frac{\phi}{2} \hat{I} - i \sin \frac{\phi}{2} \hat{\sigma}_z$$

$$= \begin{bmatrix} \cos \frac{\phi}{2} - i \sin \frac{\phi}{2} & 0 \\ 0 & \cos \frac{\phi}{2} + i \sin \frac{\phi}{2} \end{bmatrix}$$

$$= \begin{bmatrix} e^{-\frac{i}{2}\phi} & 0 \\ 0 & e^{\frac{i}{2}\phi} \end{bmatrix}, \tag{8.124}$$

while for rotations about the x and y axes we have

$$\hat{U}_x^{(1/2)}(\phi) = \cos \frac{\phi}{2} \hat{I} - i \sin \frac{\phi}{2} \hat{\sigma}_x = \begin{bmatrix} \cos \frac{\phi}{2} & -i \sin \frac{\phi}{2} \\ -i \sin \frac{\phi}{2} & \cos \frac{\phi}{2} \end{bmatrix}, \tag{8.125a}$$

$$\hat{U}_y^{(1/2)}(\phi) = \cos \frac{\phi}{2} \hat{I} - i \sin \frac{\phi}{2} \hat{\sigma}_y = \begin{bmatrix} \cos \frac{\phi}{2} & -\sin \frac{\phi}{2} \\ \sin \frac{\phi}{2} & \cos \frac{\phi}{2} \end{bmatrix}, \tag{8.125b}$$

where we have used the Pauli matrices (8.111). Moreover, for $\phi = \pi$ we find from Eq. (8.121) that

$$\hat{U}_x^{(1/2)}(\pi) = -i\hat{\sigma}_x, \quad \hat{U}_y^{(1/2)}(\pi) = -i\hat{\sigma}_y, \quad \hat{U}_z^{(1/2)}(\pi) = -i\hat{\sigma}_z. \tag{8.126}$$

Therefore, for spin $\frac{1}{2}$, up to a global phase factor the Pauli matrices $\hat{\sigma}_x$, $\hat{\sigma}_y$, and $\hat{\sigma}_z$ are the respective rotation operators by an angle π about the x, y, and z axis. For later convenience, the actions of the Pauli matrices on the states $| \uparrow_z \rangle$ and $| \downarrow_z \rangle$ are tabulated in Table 8.1.

Table 8.1 The actions of the Pauli matrices on the spin up and spin down states in the z direction

$\hat{\sigma}_x\lvert\uparrow_z\rangle = \lvert\downarrow_z\rangle$	$\hat{\sigma}_x\lvert\downarrow_z\rangle = \lvert\uparrow_z\rangle$
$\hat{\sigma}_y\lvert\uparrow_z\rangle = i\lvert\downarrow_z\rangle$	$\hat{\sigma}_y\lvert\downarrow_z\rangle = -i\lvert\uparrow_z\rangle$
$\hat{\sigma}_z\lvert\uparrow_z\rangle = \lvert\downarrow_z\rangle$	$\hat{\sigma}_z\lvert\downarrow_z\rangle = -\lvert\uparrow_z\rangle$

The rotation operator allows us to obtain the spin states in an arbitrary direction **n** from those in the z axis by rotations that align the z axis with **n**. A convenient choice that is consistent with the Euler angles is to perform two successive rotations, namely, a rotation about the y axis followed by a rotation about the z axis. Let **n** be the unit vector in the direction specified by the angles θ and ϕ in Fig. 8.6 (with $\mathbf{e}_r = \mathbf{n}$), i.e., in component form we have $\mathbf{n} = (\sin\theta\cos\phi, \sin\theta\sin\phi, \cos\theta)$. Then the combined rotation that aligns the z axis with **n** is given by a rotation of the angle θ about the y axis followed by a rotation of the angle ϕ about the z axis, i.e.,

$$
\begin{aligned}
\hat{D}^{(1/2)}(\theta, \phi) &= \hat{U}_z^{(1/2)}(\phi)\hat{U}_y^{(1/2)}(\theta) \\
&= \begin{bmatrix} e^{-\frac{i}{2}\phi} & 0 \\ 0 & e^{\frac{i}{2}\phi} \end{bmatrix} \begin{bmatrix} \cos\frac{\theta}{2} & -\sin\frac{\theta}{2} \\ \sin\frac{\theta}{2} & \cos\frac{\theta}{2} \end{bmatrix} \\
&= \begin{bmatrix} e^{-\frac{i}{2}\phi}\cos\frac{\theta}{2} & -e^{-\frac{i}{2}\phi}\sin\frac{\theta}{2} \\ e^{\frac{i}{2}\phi}\sin\frac{\theta}{2} & e^{\frac{i}{2}\phi}\cos\frac{\theta}{2} \end{bmatrix},
\end{aligned}
\tag{8.127}
$$

which is called the *rotation matrix* for spin $\frac{1}{2}$. Hence, the spin up and spin down states in the direction **n** are respectively given by

$$
\begin{aligned}
\lvert\uparrow_\mathbf{n}\rangle &= \hat{D}^{(1/2)}(\theta, \phi)\lvert\uparrow_z\rangle \\
&= \begin{bmatrix} e^{-\frac{i}{2}\phi}\cos\frac{\theta}{2} & -e^{-\frac{i}{2}\phi}\sin\frac{\theta}{2} \\ e^{\frac{i}{2}\phi}\sin\frac{\theta}{2} & e^{\frac{i}{2}\phi}\cos\frac{\theta}{2} \end{bmatrix} \begin{pmatrix} 1 \\ 0 \end{pmatrix} \\
&= \begin{pmatrix} e^{-\frac{i}{2}\phi}\cos\frac{\theta}{2} \\ e^{\frac{i}{2}\phi}\sin\frac{\theta}{2} \end{pmatrix},
\end{aligned}
\tag{8.128a}
$$

$$| \downarrow_\mathbf{n} \rangle = \hat{D}^{(1/2)}(\theta, \phi)| \downarrow_z \rangle$$

$$= \begin{bmatrix} e^{-\frac{1}{2}\phi} \cos \frac{\theta}{2} & -e^{-\frac{1}{2}\phi} \sin \frac{\theta}{2} \\ e^{\frac{1}{2}\phi} \sin \frac{\theta}{2} & e^{\frac{1}{2}\phi} \cos \frac{\theta}{2} \end{bmatrix} \begin{pmatrix} 0 \\ 1 \end{pmatrix}$$

$$= \begin{pmatrix} -e^{-\frac{1}{2}\phi} \sin \frac{\theta}{2} \\ e^{\frac{1}{2}\phi} \cos \frac{\theta}{2} \end{pmatrix}. \tag{8.128b}$$

It is straightforward to check that the states $| \uparrow_\mathbf{n} \rangle$ and $| \downarrow_\mathbf{n} \rangle$ are eigenstates of the operator

$$\hat{S}_\mathbf{n} = \hat{\mathbf{S}} \cdot \mathbf{n} = \frac{\hbar}{2} \begin{bmatrix} \cos \theta & e^{-i\phi} \sin \theta \\ e^{i\phi} \sin \theta & -\cos \theta \end{bmatrix}, \tag{8.129}$$

and the corresponding eigenvalues are given by $\hbar/2$ and $-\hbar/2$, respectively.

We end this section by showing that spin superposition states can be built experimentally from the basis states $| \uparrow_z \rangle$ and $| \downarrow_z \rangle$. Let us consider a Mach–Zehnder type interferometer experiment shown in Fig. 8.11. A beam of spin $\frac{1}{2}$ particles (say, neutrons) in the spin up state in the z direction is split into two beams at the first beam splitter BS1, a variable one with the respective transmission and reflection coefficients given by T $= \cos \frac{\theta}{2}$ and R $= \sin \frac{\theta}{2}$ (where $0 \le \theta \le \pi$ is some parameter) [see Section 5.5]. Then, a spin flipper SF flips the spin of the upper beam to the spin down state in the z direction. After reflections at two mirrors M1 and M2, a phase shifter PS shifts the phase of the upper beam by a relative phase ϕ (where $0 \le \phi < 2\pi$). Finally, the two beams are recombined at the second beam splitter BS2, which is a 50–50 one, and subsequently detected at two detectors D1 and D2. The initial state of the particles may be described by [see Eq. (8.109)]

$$|i\rangle = |d\rangle \otimes | \uparrow_z \rangle, \tag{8.130}$$

where $|d\rangle$ describes the spatial part (as that for the photon in the usual Mach–Zehnder interferometer) and $| \uparrow_z \rangle$ the spin part of the particle's state. The action of the first beam splitter BS1, the spin

---

[a](Summhammer *et al.*, 1983).

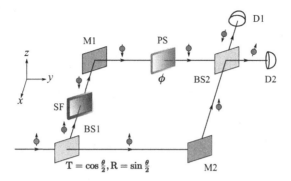

Figure 8.11 Schematic setup for realization of spin superposition through a Mach–Zehnder type interferometer. See text for details. For $\theta = \frac{\pi}{2}$ and $\phi = 0$, the spin of the outgoing particles is in the x direction as shown here. In the actual experimental setup, spin superposition is realized through single-crystal neutron interferometry.[a]

flipper SF, and the phase shifter PS can be written as [see Eq. (5.8)]

$$|i\rangle \xrightarrow{\text{BS1}} \left(\cos\frac{\theta}{2}|d\rangle + \sin\frac{\theta}{2}|u\rangle \right) \otimes |\uparrow_z\rangle$$

$$\xrightarrow{\text{SF}} \cos\frac{\theta}{2}|d\rangle \otimes |\uparrow_z\rangle + \sin\frac{\theta}{2}|u\rangle \otimes |\downarrow_z\rangle$$

$$\xrightarrow{\text{PS}} \cos\frac{\theta}{2}|d\rangle \otimes |\uparrow_z\rangle + e^{i\phi}\sin\frac{\theta}{2}|u\rangle \otimes |\downarrow_z\rangle. \tag{8.131}$$

The effect of the second beam splitter BS2 is to recombine the two beams and therefore create superposition. From the transformations (2.26) and (2.27) on $|d\rangle$ and $|u\rangle$ induced by BS2, we find that the final state of the particles after they leave BS2 is given by

$$|f\rangle = \frac{1}{\sqrt{2}}\left[\cos\frac{\theta}{2}\left(|1\rangle + |2\rangle\right) \otimes |\uparrow_z\rangle + e^{i\phi}\sin\frac{\theta}{2}\left(|1\rangle - |2\rangle\right) \otimes |\downarrow_z\rangle \right],$$

$$= \frac{1}{\sqrt{2}}\left[|1\rangle \otimes \left(\cos\frac{\theta}{2}|\uparrow_z\rangle + e^{i\phi}\sin\frac{\theta}{2}|\downarrow_z\rangle \right) \right.$$

$$\left. + |2\rangle \otimes \left(\cos\frac{\theta}{2}|\uparrow_z\rangle - e^{i\phi}\sin\frac{\theta}{2}|\downarrow_z\rangle \right) \right], \tag{8.132}$$

where in the second equality we have rewritten the result in terms of the spin superposition states. It is instructive to consider some representative values of the parameter θ and the phase shift ϕ. For $\theta = \frac{\pi}{2}$ and $\phi = 0$ the final state is given by

$$|f\rangle_{\theta=\frac{\pi}{2},\phi=0} = \frac{1}{\sqrt{2}} \left[|1\rangle \otimes \frac{1}{\sqrt{2}} (|\uparrow_z\rangle + |\downarrow_z\rangle) + |2\rangle \otimes \frac{1}{\sqrt{2}} (|\uparrow_z\rangle - |\downarrow_z\rangle) \right]$$

$$= \frac{1}{\sqrt{2}} (|1\rangle \otimes |\uparrow_x\rangle + |2\rangle \otimes |\downarrow_x\rangle), \tag{8.133}$$

where we have use Eq. (8.114a). Similarly, for $\theta = \phi = \frac{\pi}{2}$ we have

$$|f\rangle_{\theta=\phi=\frac{\pi}{2}} = \frac{1}{\sqrt{2}} \left[|1\rangle \otimes \frac{1}{\sqrt{2}} (|\uparrow_z\rangle + i|\downarrow_z\rangle) + |2\rangle \otimes \frac{1}{\sqrt{2}} (|\uparrow_z\rangle - i|\downarrow_z\rangle) \right]$$

$$= \frac{1}{\sqrt{2}} (|1\rangle \otimes |\uparrow_y\rangle + |2\rangle \otimes |\downarrow_y\rangle), \tag{8.134}$$

where use has been made of Eq. (8.114b). Indeed, it can be seen from Eqs. (8.128) that by tuning the parameter θ and the phase shift ϕ, we are able to create spin superposition states (up to a global phase) such that the spin of the outgoing particles is along an arbitrary direction **n**.

Problem 8.17 Show that in the basis $\{|\uparrow_z\rangle, |\downarrow_z\rangle\}$, the operator $\hat{\mathbf{S}}^2$ is given in matrix form by

$$\hat{\mathbf{S}}^2 = \frac{3\hbar^2}{4} \begin{bmatrix} 1 & 0 \\ 0 & 1 \end{bmatrix}.$$

Problem 8.18 Using the Pauli matrices given by Eq. (8.111), verify the commutation relations (8.112c).

8.7 Addition of Angular Momenta

Very often we need to consider the total orbital angular momentum $\hat{\mathbf{L}}_1 + \hat{\mathbf{L}}_2$ of a system of two particles (1 and 2), the total spin angular momentum $\hat{\mathbf{S}}_1 + \hat{\mathbf{S}}_2$ of two particles (1 and 2), or the total angular momentum $\hat{\mathbf{L}} + \hat{\mathbf{S}}$ of a particle with spin. In these situations we shall deal with the problem of *how to add two angular momenta*. The formalism of angular momentum addition is very general, so we will follow the usual convention to denote a generic angular momentum by $\hat{\mathbf{J}}$, which could be orbital ($\hat{\mathbf{L}}$), spin ($\hat{\mathbf{S}}$), or some combined quantity (e.g., $\hat{\mathbf{L}} + \hat{\mathbf{S}}$). Consider the addition of two angular momenta $\hat{\mathbf{J}}_1$ and $\hat{\mathbf{J}}_2$:

$$\hat{\mathbf{J}} = \hat{\mathbf{J}}_1 + \hat{\mathbf{J}}_2, \tag{8.135}$$

where the respective angular momentum quantum numbers (j_1, m_1) and (j_2, m_2) associated with $\hat{\mathbf{J}}_1$ and $\hat{\mathbf{J}}_2$ are known (they can therefore mean azimuthal quantum number and magnetic quantum number, spin quantum number and spin magnetic quantum number, or also any combination pair by pair). The operators $\hat{\mathbf{J}}_1$ and $\hat{\mathbf{J}}_2$ commute as they refer to two independent particles or two different properties of the same particle. This implies that the components of $\hat{\mathbf{J}}$ obey the commutation relations (8.22) [see Problem 8.19]. Let us denote the quantum numbers associated with $\hat{\mathbf{J}}$ by (j, m), then our goal is to find the possible values of j and m in terms of $j_1, m_1, j_2,$ and m_2.

We first note that the common eigenstates of $\hat{\mathbf{J}}_1^2, \hat{J}_{1z}, \hat{\mathbf{J}}_2^2,$ and $\hat{J}_{2z}$ are the product states

$$|j_1, m_1; j_2, m_2\rangle = |j_1, m_1\rangle \otimes |j_2, m_2\rangle. \tag{8.136}$$

From Eq. (8.135), the z component of $\hat{\mathbf{J}}$ is given by

$$\hat{J}_z = \hat{J}_{1z} + \hat{J}_{2z}, \tag{8.137}$$

which upon acting on $|j_1, m_1; j_2, m_2\rangle$ yields

$$\begin{aligned} \hat{J}_z|j_1, m_1; j_2, m_2\rangle &= (\hat{J}_{1z} + \hat{J}_{2z})|j_1, m_1\rangle \otimes |j_2, m_2\rangle \\ &= (m_1 + m_2)\hbar|j_1, m_1\rangle \otimes |j_2, m_2\rangle \\ &= (m_1 + m_2)\hbar|j_1, m_1; j_2, m_2\rangle. \end{aligned} \tag{8.138}$$

It is noted that $\hat{J}_{1z}$ acts only on $|j_1, m_1\rangle$ and $\hat{J}_{2z}$ acts only on $|j_2, m_2\rangle$. Hence, the state $|j_1, m_1; j_2, m_2\rangle$ is an eigenstate of $\hat{J}_z$ with the eigenvalue given by $(m_1 + m_2)\hbar$. The generalization of this examination is shown in Box 8.6.

Here, we consider the simplest (but a very important) case $j_1 = j_2 = \frac{1}{2}$. This case corresponds to the total spin of two spin $\frac{1}{2}$ particles, and therefore we will change the notation to that commonly used for spins. With the possible values of m_{s_1} and m_{s_2} given by $\frac{1}{2}, -\frac{1}{2}$, we find that the total spin magnetic quantum number m_s can take four possible values, namely, $m_s = 1, 0, 0, -1$, where we note that the value zero appears twice [see Table 8.2]. This

Table 8.2 Possible values of m for $j_1 = j_2 = \frac{1}{2}$

m_1	m_2	m
$\frac{1}{2}$	$\frac{1}{2}$	1
$\frac{1}{2}$	$-\frac{1}{2}$	0
$-\frac{1}{2}$	$\frac{1}{2}$	0
$-\frac{1}{2}$	$-\frac{1}{2}$	-1

in turn means that the possible values of the total spin quantum number s are given indeed by $s = 1, 0$. For $s = 1$ we have $m_s = 1, 0, -1$ and the common eigenstates $|s, m_s\rangle$ of $\hat{\mathbf{S}}^2$ and $\hat{S}_z$ are given by [see Eq. (8.141)]

$$|s = 1, m_s = 1\rangle = |s_1, m_{s_1}; s_2, m_{s_2}\rangle = \left| \frac{1}{2}, \frac{1}{2}; \frac{1}{2}, \frac{1}{2} \right\rangle$$

$$= |\uparrow_z\rangle_1 \otimes |\uparrow_z\rangle_2, \qquad (8.139a)$$

$$|s = 1, m_s = 0\rangle = \frac{1}{\sqrt{2}} \left(|s_1, m_{s_1}; s_2, m_{s_2}\rangle + |s_1, m_{s_1}; s_2, m_{s_2}\rangle \right)$$

$$= \frac{1}{\sqrt{2}} \left(\left| \frac{1}{2}, \frac{1}{2}; \frac{1}{2}, -\frac{1}{2} \right\rangle + \left| \frac{1}{2}, -\frac{1}{2}; \frac{1}{2}, \frac{1}{2} \right\rangle \right)$$

$$= \frac{1}{\sqrt{2}} \left(|\uparrow_z\rangle_1 \otimes |\downarrow_z\rangle_2 + |\downarrow_z\rangle_1 \otimes |\uparrow_z\rangle_2 \right),$$

$$(8.139b)$$

$$|s = 1, m_s = -1\rangle = |s_1, m_{s_1}; s_2, m_{s_2}\rangle = \left| \frac{1}{2}, -\frac{1}{2}; \frac{1}{2}, -\frac{1}{2} \right\rangle$$

$$= |\downarrow_z\rangle_1 \otimes |\downarrow_z\rangle_2. \qquad (8.139c)$$

For $s = 0$ we have $m_s = 0$ and the common eigenstate $|s, m_s\rangle$ of $\hat{\mathbf{S}}^2$ and $\hat{S}_z$ is [see Eq. (8.141)]

$$|s = 0, m_s = 0\rangle = \frac{1}{\sqrt{2}} \left(|s_1, m_{s_1}; s_2, m_{s_2}\rangle - |s_1, m_{s_1}; s_2, m_{s_2}\rangle \right)$$

$$= \frac{1}{\sqrt{2}} \left(\left| \frac{1}{2}, \frac{1}{2}; \frac{1}{2}, -\frac{1}{2} \right\rangle - \left| \frac{1}{2}, -\frac{1}{2}; \frac{1}{2}, \frac{1}{2} \right\rangle \right)$$

$$= \frac{1}{\sqrt{2}} \left(|\uparrow_z\rangle_1 \otimes |\downarrow_z\rangle_2 - |\downarrow_z\rangle_1 \otimes |\uparrow_z\rangle_2 \right). \qquad (8.140)$$

Box 8.6 Clebsch–Gordan coefficients

Let us denote the common eigenstates of $\hat{J}^2$ and $\hat{J}_z$ by $|j, m\rangle$.[a] From Eq. (8.138) and the superposition principle [see Principle 2.1] we can express $|j, m\rangle$ as

$$|j, m\rangle = \sum_{\substack{-j_1 \leq m_1 \leq j_1 \\ -j_2 \leq m_2 \leq j_2}}^{\prime} C^{j, j_1, j_2}_{m, m_1, m_2} |j_1, m_1; j_2, m_2\rangle, \qquad (8.141)$$

where the constants $C^{j, j_1, j_2}_{m, m_1, m_2}$ are called *Clebsch–Gordan coefficients* and a prime on the summation symbol means the summation over m_1 and m_2 is subject to the constraint $m_1 + m_2 = m$. So the relation between m_1, m_2, and m is given by

$$m = m_1 + m_2. \qquad (8.142)$$

This result implies that the number of possible values of m is the product of those of m_1 and m_2, i.e., $(2j_1 + 1)(2j_2 + 1)$. Since j is defined to be the maximum of m, one would expect that $j = j_1 + j_2$. This however is not correct because if it were the case then the number of possible values of m would be $2j + 1 = 2(j_1 + j_2) + 1$, instead of the actual number $(2j_1+1)(2j_2+1)$. A detailed analysis shows that the possible values of j are given by

$$j = j_1 + j_2, \ j_1 + j_2 - 1, \ \dots, \ |j_1 - j_2| + 1, \ |j_1 - j_2|, \qquad (8.143)$$

and to each value of j there are $2j + 1$ possible values of m given by

$$m = j, \ j - 1, \ \dots, \ -j + 1, \ -j. \qquad (8.144)$$

The total number of possible values of m is then given by

$$\sum_{j=|j_1-j_2|}^{j_1+j_2} (2j + 1) = (2j_1 + 1)(2j_2 + 1), \qquad (8.145)$$

which agrees with the result we obtained above.

---

[a]More precisely, we should consider the common eigenstates $|j_1, j_2, j, m\rangle$ of $\hat{J}_1^2$, $\hat{J}_2^2$, $\hat{J}^2$, and $\hat{J}_z$. This however does not change our discussion and conclusions. Therefore, $|j, m\rangle$ can be thought of as a shorthand for $|j_1, j_2, j, m\rangle$.

For obvious reasons, the state with total spin $s = 1$ is called a *triplet state*, while the state with total spin $s = 0$ is called a *singlet state*. It is noted that the states $|1, 0\rangle$ and $|0, 0\rangle$ are entangled states [see Section 7.9]. The spin entanglement is revealed by the fact that when particle 1 is found in the spin up state, particle 2 is in the spin down state and vice versa. These entangled states are of great importance in the measurement problem and quantum information.

Problem 8.19 Using the fact that the two angular momentum operators $\hat{J}_1$ and $\hat{J}_2$ commute, show that the total angular momentum operator $\hat{J} = \hat{J}_1 + \hat{J}_2$ qualifies as a true angular momentum because its components obey the commutation relations $[\hat{J}_x, \hat{J}_y] = i\hbar\hat{J}_z$, $[\hat{J}_y, \hat{J}_z] = i\hbar\hat{J}_x$, and $[\hat{J}_z, \hat{J}_x] = i\hbar\hat{J}_y$.

Problem 8.20 Consider a system of two spin $\frac{1}{2}$ particles that is in the spin singlet state $|\Psi\rangle$ given by Eq. (8.140). An observer, called Alice, will measure the spin of particle 1, while another observer, called Bob, will measure the spin of particle 2.

(a) What is the probability for Alice to obtain $S_{1z} = \frac{\hbar}{2}$ when Bob makes no measurement?
(b) Repeat (a) for for Alice to obtain $S_{1x} = \frac{\hbar}{2}$.
(c) Bob measures the spin of particle 2 and obtains $S_{2z} = \frac{\hbar}{2}$, then Alice measures S_{1z} of particle 1. What can we conclude about the outcome of Alice's measurement?
(d) Repeat (c) if Alice measures $S_{1x} = \frac{\hbar}{2}$?

8.8 Identical Particles and Spin

Until now, apart from the last section our focus has been on the study of quantum mechanics of a single particle. For instance, we have considered the dynamics of a free particle, a harmonic oscillator, and the only electron in a hydrogen atom. However, most physical systems involve interaction of many particles, e.g., electrons in a solid, atoms in a gas, etc. In quantum mechanics, particles of the same type cannot be distinguished from one another, even in principle. This is because they have the same intrinsic physical properties, such as mass, electric charge, and spin, and according to Heisenberg's uncertainty principle [see Principle 6.1], it is not possible to ascribe a definite trajectory to a quantum particle. The indistinguishable particles are therefore referred to as *identical particles*.

An important consequence of particle indistinguishability is that the state vectors representing physical states of a system of identical particles have to satisfy certain symmetry requirements. For the sake of simplicity let us consider a system of two noninteracting identical particles. Suppose that one particle is in the state $|a\rangle$ and the other is in the state $|b\rangle$. Since the particles are indistinguishable, we do not know which particle is in the state $|a\rangle$ and which particle is in the state $|b\rangle$. Hence the two-particle system could be in the state $|a\rangle_1 \otimes |b\rangle_2$ *or* in the state $|b\rangle_1 \otimes |a\rangle_2$, where the subscripts 1 and 2 denote particle 1 and particle 2, respectively. From the superposition principle [see Principle 2.1], it follows that the state of the system $|\psi\rangle$ is a superposition state of the form

$$|\psi\rangle = \alpha|a\rangle_1 \otimes |b\rangle_2 + \beta|b\rangle_1 \otimes |a\rangle_2, \qquad (8.146)$$

where α and β are coefficients with $|\alpha|^2 + |\beta|^2 = 1$. Now consider the situation that we could interchange the two particles and denote the resultant state of the system by $|\psi'\rangle$, i.e.,

$$|\psi\rangle \xrightarrow{\hat{P}_{12}} |\psi'\rangle = \alpha|b\rangle_1 \otimes |a\rangle_2 + \beta|a\rangle_1 \otimes |b\rangle_2, \qquad (8.147)$$

where $\hat{P}_{12}$ is the particle exchange operator which exchanges the coordinates and spins of particle 1 and particle 2. Since identical particles are indistinguishable, the exchange of them cannot be detected by measurements. To put it another way, interchange of identical particles does not alter the physical states of the system. Therefore, the states vectors $|\psi\rangle$ and $|\psi'\rangle$ differ only by a global phase ϕ [see Section 5.2]. So we have

$$|\psi'\rangle = e^{i\phi}|\psi\rangle, \qquad (8.148)$$

or equivalently,

$$\alpha = e^{i\phi}\beta \quad \text{and} \quad \beta = e^{i\phi}\alpha. \qquad (8.149)$$

The above coupled equations implies that

$$\alpha = \pm\beta. \qquad (8.150)$$

For $\alpha = +\beta$, we have the *symmetric* state vector

$$|\psi\rangle_S = \frac{1}{\sqrt{2}}(|a\rangle_1 \otimes |b\rangle_2 + |b\rangle_1 \otimes |a\rangle_2), \qquad (8.151)$$

which does not change sign under exchange of particles and is an eigenstate of the exchange operator $\hat{P}_{12}$ with eigenvalue $+1$. For $\alpha = -\beta$, we obtain the *antisymmetric* state vector

$$|\psi\rangle_A = \frac{1}{\sqrt{2}}(|a\rangle_1 \otimes |b\rangle_2 - |b\rangle_1 \otimes |a\rangle_2), \qquad (8.152)$$

which changes sign under exchange of particles and is an eigenstate of the exchange operator $\hat{P}_{12}$ with eigenvalue -1.

The above analysis can be generalized to systems of N identical particles. Following the distinction between symmetric and antisymmetric state vectors, particles can be categorized into two main groups according to their spin:

(i) Bosons, derived from the name of the Indian physicist Bose,[a] are described by symmetric state vectors and have integer spin.

(ii) Fermions, derived from the name of the Italian physicist Fermi,[b] are characterized by antisymmetric state vectors and have half-integer spin.

All observed elementary and composite particles are either bosons or fermions. Elementary bosons include the photon, gluons, and W and Z bosons, and there are two types of elementary fermions: quarks (of which protons and neutrons are composed) and leptons (which include the electron and electron neutrino). To date, all known matter is made up of elementary fermions while the fundamental forces are mediated by elementary bosons [see Box 8.7].

Composite particles can either be bosons or fermions depending on their spins. From the addition rule of angular momenta [see Eq. (8.143)], it follows that composite bosons are made up of constituent bosons or an *even* number of constituent fermions, and composite fermions consist of an *odd* number of constituent fermions. For example, the proton and neutron each contains three quarks and are fermions, and the pion and kaon each contains two quarks (more precisely, one quark and one antiquark) and are

---

[a](Bose, 1924).
[b](Fermi, 1926).

Box 8.7 Elementary particles

The observed elementary particles include elementary fermions (six quarks and six leptons) and elementary bosons (three kinds). The names and symbols of the observed elementary particles are listed in the following table.

Fermions		Bosons
Quarks (q)	Leptons (l)	
up (u), down (d)	electron (e^-), electron neutrino (ν_e)	photon (γ)
charm (c), strange (s)	muon (μ^-), muon neutrino (ν_μ)	W, Z bosons ($W^\pm$, Z^0)
top (t), bottom (b)	tau lepton (τ^-), tau neutrino (ν_τ)	gluons (eight types) (g)

The photon and gluons are massless, while the other elementary particles all have mass. All leptons and quarks are spin $\frac{1}{2}$ particles, and all elementary bosons are spin 1 particles. The electron, muon, and tau lepton all have a charge of -1 (in units of the elementary charge [see Eq. (8.78)]), and all neutrinos are electrically neutral. Quarks have fractional electric charge: up, charm, and top quarks have a charge of $+\frac{2}{3}$, while down, strange, and bottom quarks have a charge of $-\frac{1}{3}$. The elementary bosons are force carriers: the photon carries electromagnetic force, the W and Z bosons carry the weak force, responsible for the radioactive decay, nuclear fission, and nuclear fusion, and the gluons carry the strong force, responsible for the stability of atomic nuclei. The photon, Z boson, and gluons are electrically neutral, while the W^+ and W^- bosons have a charge of $+1$ and -1, respectively.

For each elementary particles there is a corresponding antiparticle with the same mass and spin, but opposite electric charge. The photon, Z boson, and gluons are their own antiparticles, and the W^+ and W^- bosons are particle and antiparticle of each other. Antiparticles of leptons are known as antileptons ($\bar{l}$); antiparticles of quarks are called antiquarks ($\bar{q}$). For instance, the antiparticle of the electron is the positron (e^+), the antiparticle of the electron neutrino is the electron antineutrino ($\bar{\nu}_e$), the antiparticle of the up quark is the anti–up quark ($\bar{u}$). However, it is an open question whether the neutrinos are their own antiparticles.

An elementary boson called the Higgs boson is hypothesized to explain why other elementary particles have mass. It is theorized to be a massive particle of spin 0. A huge experimental effort has been devoted since 2009 to the search for the Higgs boson at the Large Hadron Collider (LHC), located at the European Organization for Nuclear Research (CERN) near Geneva, Switzerland. In July 2012, CERN announced an observation of a new particle consistent with the Higgs boson. Later on in March 2013, CERN confirmed that the new particle is a Higgs boson.[a] Up-to-date information on the state of Higgs boson discovery can be found at the web pages of the ATLAS[b] and CMS[c] collaborations.

---

[a]The 2013 Nobel prize in physics has been awarded to two physicists for their work on the theory of the Higgs boson.
[b]http://atlas.ch
[c]http://cms.cern.ch

---

therefore bosons. Moreover, the helium–4 (^{4}He) atom, consisting of two protons, two neutrons and two electrons is a boson, while the helium–3 (^{3}He) atom, made up of two protons, one neutron, and two electrons is a fermion. Composite bosons are important in superconductivity, superfluidity, and applications of Bose–Einstein condensates.

Let us examine the situation if two fermions are placed in the same state. By setting $|a\rangle = |b\rangle$ in Eq. (8.152), we would obtain $|\psi\rangle_A = 0$ identically, meaning that such a state cannot exist. This result is an expression of the celebrated *exclusion principle*, which was first formulated by Pauli in 1925.[a]

Principle 8.1 (Exclusion) *No two identical fermions may occupy the same quantum state.*

---

[a](Pauli, 1925).

This principle is of fundamental importance and has profound consequences in a wide variety of systems involving identical fermions, ranging from nuclear physics and atomic physics to condensed matter physics and astrophysics. As a simple illustrative example, let us consider a system of $2N$ noninteracting identical fermions of spin $\frac{1}{2}$ trapped in a harmonic potential well. The Pauli exclusion principle implies that each energy level of the harmonic oscillator can be occupied at most by two fermions, one in the spin up state and the other in the spin down state. Hence, the ground state of the system is the state in which the lowest N energy levels of the harmonic oscillator are all filled, while all higher energy levels are empty. In other words, many of the fermions are forced into higher energy levels. When this happens the fermions are said to be degenerate. The ground state energy of the degenerate fermion system is given by

$$E_0(N) = 2 \sum_{n=0}^{N-1} \left(n + \frac{1}{2} \right) \hbar\omega = N^2 \hbar\omega, \qquad (8.153)$$

where ω is the angular frequency of the harmonic oscillator. On the other hand, if the $2N$ particles were distinguishable, then the ground state of the system would be the state in which all the particles occupy the lowest energy level of the harmonic oscillator. In this case the ground state energy of the system would become

$$2N \left(\frac{1}{2} \hbar\omega \right) = N \hbar\omega, \qquad (8.154)$$

which for large N is much smaller than $E_0(N)$. This significant increase in the ground state energy for systems of noninteracting identical fermions gives rise to the so-called *degeneracy pressure*. In fact, compact stars like white dwarfs and neutron stars are supported from further gravitational collapse by degeneracy pressure of the electrons and the neutrons, respectively.

8.9 Summary

In this chapter we have

- Introduced the angular momentum as the generator of rotations.

- Derived the commutation relations for components of the angular momentum.
- Studied in a very sketchy way the eigenvalues and eigenfunctions of the angular momentum operator.
- Considered the problem of a particle in a central potential and applied the formalism to study the energy eigenvalues and eigenfunctions of the hydrogen atom.
- Introduced the new quantum observable spin and the associated quantum numbers.
- Learned rules for addition of angular momenta.
- Introduced the spin singlet state and spin triplet state.
- Discussed the relation between identical particles and spin.
- Introduced the concept of bosons and fermions, and formulated the Pauli exclusion principle.

PART III

ONTOLOGICAL ISSUES: PROPERTIES

Chapter 9

Measurement Problem

In this chapter we shall consider one of the most difficult problems of quantum mechanics, that is, how it is possible that we have a dynamics ruled by a reversible equation but we obtain random irreversible events when measuring. Many different interpretations have been provided for solving this puzzle. We will discuss some of them by starting with the von Neumann's projection postulate, one of the first attempts to address the measurement problem and its implications. The role played by the environment in the concept of measurement is then emphasized. We shall also introduce entropy as a measure of the average information content. Finally, the famous Schrödinger's cat paradox will be examined.

9.1 Statement of the Problem

As mentioned, the problem of measurement in quantum mechanics is one of the most difficult and controversial ones, and shows many subtleties that are puzzling even for the great minds of quantum physics. However, it also raises so many fundamental and conceptual questions of both physical and philosophical order that make this

Quantum Mechanics for Thinkers
Gennaro Auletta and Shang-Yung Wang
Copyright © 2014 Pan Stanford Publishing Pte. Ltd.
ISBN 978-981-4411-71-4 (Hardcover), 978-981-4411-72-1 (eBook)
www.panstanford.com

subject fascinating. To understand what is the basic problem here, let us consider a simple superposition of the kind [see Eq. (2.10)]:

$$|\psi\rangle = c_d|d\rangle + c_u|u\rangle, \tag{9.1}$$

where c_d, c_u are probability amplitudes with $|c_d|^2 + |c_u|^2 = 1$, and $|d\rangle$, $|u\rangle$ are orthonormal states

$$|d\rangle = \begin{pmatrix} 1 \\ 0 \end{pmatrix}, \quad |u\rangle = \begin{pmatrix} 0 \\ 1 \end{pmatrix}. \tag{9.2}$$

It is always possible to *prepare* a system in this state, for instance by using appropriate beam splitter or polarization filter, as shown in the first part of the book [see, for instance, Sections 2.3 and 3.2]. We now desire to measure the observable whose eigenstates (possible outcomes) are precisely $|u\rangle$ and $|d\rangle$. In other words, in the case of interferometry, we wish to know whether the photon is in either the upper or lower path, which is well possible with certain experimental setups [see Section 5.6]. We know very well that we can obtain the probabilities of finding $|u\rangle$ or $|d\rangle$ by computing the square moduli of the respective amplitudes, that is, $|c_d|^2$ and $|c_u|^2$ [see Section 2.4]. Obviously, any difference in the probabilities is due to a difference in the transmission and reflection coefficients of the beam splitter [see Section 5.5], according to which kind of experimental setup one chooses. However, we also know that the results of our measurement themselves are, in any case, random, which allows us to treat this problem in the most elementary terms. So, suppose that we get the state $|d\rangle$. The difficulty here is that there is *no* unitary transformation such that

$$|\psi\rangle \longrightarrow |d\rangle. \tag{9.3}$$

This is a serious problem since, as we have seen in Chapter 7, time evolution of quantum systems is unitary (and therefore also reversible). We could have better said that the transition (9.3) is not unitary apart from some very particular circumstances. Let us examine again the action of the beam splitter as a unitary operator [see Section 4.6]. As mentioned, we like to reduce the problem to the most simple terms and therefore we shall assume that $c_u = c_d = \frac{1}{\sqrt{2}}$. With this assumption, now let us consider what happens if we perform the inverse transformation on a superposition state given by Eq. (9.1), namely, on the state

$$|\psi\rangle = \frac{1}{\sqrt{2}} \begin{pmatrix} 1 \\ 1 \end{pmatrix}. \tag{9.4}$$

In particular, we would like to know if we are able to recover the state $|d\rangle$. To this purpose, we need to use the conjugate transpose of the unitary operator $\hat{U}_{BS}$ [see Eq. (4.35)], which here coincides with the latter (i.e., $\hat{U}_{BS}^{\dagger} = \hat{U}_{BS}$). Then, we obtain

$$\hat{U}_{BS}|\psi\rangle = \frac{1}{\sqrt{2}} \begin{bmatrix} 1 & 1 \\ 1 & -1 \end{bmatrix} \frac{1}{\sqrt{2}} \begin{pmatrix} 1 \\ 1 \end{pmatrix}$$

$$= \frac{1}{2} \begin{pmatrix} 2 \\ 0 \end{pmatrix} = \begin{pmatrix} 1 \\ 0 \end{pmatrix}$$

$$= |d\rangle. \tag{9.5}$$

Therefore, in some conditions we can obtain an eigenstate from an initial superposition state through a unitary transformation. However, we were able to do that here only because we know the initial superposition state (i.e., the coefficients of the superposition) and are therefore able to apply a suitable transformation. In the general case, when there is an arbitrary superposition (which is likely unknown, otherwise to measure the system makes no sense), we cannot obtain such a result. This is evident by the fact that we cannot use the *same* transformation $\hat{U}_{BS}$ to get the alternative output $|u\rangle$. In this case, we should have used instead the transformation

$$\hat{U}'_{BS} = \frac{1}{\sqrt{2}} \begin{bmatrix} 1 & -1 \\ 1 & 1 \end{bmatrix}, \tag{9.6}$$

which is still unitary but different from the previous one. The reason for this situation is simple. By having a look at Fig. 9.1, it is evident that to bring the state $|\psi\rangle$ to coincide with the state $|d\rangle$ we need a clockwise rotation by the angle θ. Even in the case in which the superposition is 50–50, to bring the state vector $|\psi\rangle$ to coincide with the state vector $|u\rangle$ we need a rotation with an opposite sign (a counterclockwise one). The case is even worser when the superposition is not 50–50 as it is the case for Fig. 9.1, that is, when the angle between $|\psi\rangle$ and $|d\rangle$ is different from that between $|\psi\rangle$ and $|u\rangle$. In other words, there does *not* exist a single unitary operation

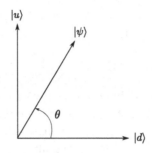

Figure 9.1 Unitary transformation of a state vector $|\psi\rangle$ making and angle θ with the state $|d\rangle$.

that transforms an arbitrary superposition state into its component states. Therefore, we are allowed to say that measurement in quantum mechanics is not ruled by a unitary transformation, i.e., not by the Schrödinger equation [see Section 7.3], and therefore its outcomes are random and unpredictable.

Problem 9.1 Which is the key factor in the chosen transformation $\hat{U}_{\text{BS}}$ that allows us to obtain the result (9.5)?

Problem 9.2 Apply the unitary operator $\hat{U}'_{\text{BS}}$ to the state $|\psi\rangle$ given by Eq. (9.4) to obtain the state $|u\rangle$.

Problem 9.3 Find the conjugate transpose of the operator $\hat{U}'_{\text{BS}}$ and show that $\hat{U}'_{\text{BS}}$ is unitary.

9.2 Density Matrix and Projectors

The previous analysis raises the question of what kind of process measurement is. Formally, the process of measurement can be expressed in a very precise way by making use of a certain operation on the density matrix. For the state vector $|\psi\rangle$ given by Eq. (9.1), the corresponding density matrix is $\hat{\rho} = |\psi\rangle\langle\psi|$. Let us represent the result of a measurement in the following way:

$$
\begin{aligned}
\hat{\rho}' &= \hat{P}_d \hat{\rho} \hat{P}_d + \hat{P}_u \hat{\rho} \hat{P}_u \\
&= |d\rangle\langle d|(c_d|d\rangle + c_u|u\rangle)\big((\langle d|c_d^* + \langle u|c_u^*)|d\rangle\langle d| \\
&\quad + |u\rangle\langle u|(c_d|d\rangle + c_u|u\rangle)\big((\langle d|c_d^* + \langle u|c_u^*)|u\rangle\langle u| \\
&= |c_d|^2|d\rangle\langle d| + |c_u|^2|u\rangle\langle u| \\
&= |c_d|^2 \hat{P}_d + |c_u|^2 \hat{P}_u, \quad\quad\quad\quad\quad\quad (9.7)
\end{aligned}
$$

where the result is clearly a mixture [see Section 7.8]. This is quite natural, since we expect interdependence (as represented by the off-diagonal terms) in the density matrix that are associated with quantum features to disappear. Indeed, in such a case the quantum mechanical problem of measurement will reduce to a treatable classical case, in which the different outcomes have a certain probability to occur [see Section 7.9]. The reason for the fact that we obtain a mixture here is that projectors are not unitary operator and, in the general case, we cannot know the initial state transformed by a projector by considering only the projected state [see again Section 7.8]. This is geometrically evident, since different state vectors can project on the same vector, i.e., on the same state [see also Fig. 3.3].

In order to go further with our examination, it is important to recall the concept of the expectation value of the observable being measured, which can also be expressed in terms of the density matrix as [see Eq. (7.113)]

$$\langle \hat{O} \rangle_{\hat{\rho}} = \mathrm{Tr}(\hat{\rho}\hat{O}). \tag{9.8}$$

The probability of obtaining the event described by the projector $\hat{P}_j$ given the initial state $\hat{\rho}$ of the system can be written as

$$\wp_j = \mathrm{Tr}(\hat{\rho}\hat{P}_j). \tag{9.9}$$

Suppose now that, given the state $\hat{\rho}$ of the system, the quantum detection event $\hat{P}_j$ actually occurs, that is, the probability $\wp_j$ of the event $\hat{P}_j$ is nonzero or $\mathrm{Tr}(\hat{\rho}\hat{P}_j) \neq 0$. According to von Neumann,[a] given that the initial state of the system is $\hat{\rho}$ and the event $\hat{P}_j$ has occurred, then the state of the system *after* the detection event is given by

$$\hat{\rho}_j = \frac{\hat{P}_j \hat{\rho} \hat{P}_j}{\mathrm{Tr}(\hat{\rho}\hat{P}_j)}. \tag{9.10}$$

Moreover, it can be shown that $\hat{\rho}_j$ is again a density matrix. From the above expression, it follows that we can rewrite Eq. (9.7) in term of $\hat{\rho}_d$ and $\hat{\rho}_u$ as [see Eq. (9.9)]

$$\hat{\rho}' = \wp_d \hat{\rho}_d + \wp_u \hat{\rho}_u. \tag{9.11}$$

---

[a](Von Neumann, 1932).

In fact, in the general case we have

$$\hat{\rho}' = \sum_j \hat{P}_j \hat{\rho} \hat{P}_j = \sum_j |c_j|^2 \hat{P}_j = \sum_j \wp_j \hat{\rho}_j, \qquad (9.12)$$

where j is an index over a complete set of projectors, $\wp_j = |c_j|^2$ are a shorthand notation for the probabilities of the kind shown before, and the $\hat{\rho}_j$'s can be computed according to Eq. (9.10). As we will see in the following section, these formulas help us understand how a final single detection event (here denoted by $\hat{P}_j$) can be obtained starting with an initial superposition state or a pure state.

Problem 9.4 Explain why the density matrix $\hat{\rho}'$ given by Eq. (9.7) represents a mixture.

9.3 Projection Postulate

In 1932, von Neumann postulated that a transformation like Eq. (9.7) or (9.10) characterizes measurement in quantum mechanics, an assumption that was called *projection postulate*.[a] In specific, von Neumann stated that there are two different time evolutions in quantum mechanics:

 (i) The unitary evolution ruled by the Schrödinger equation, which is deterministic, continuous, linear, and reversible.
(ii) The projection-like transformation occurring during a measurement, which is random, discontinuous, non-linear, and irreversible [see also Section 6.9].

Von Neumann called the latter *reduction of the wave packet* since we pass from a multicomponent superposition to one of its components, as it is again evident by Eq. (9.10) or Fig. 3.3. Since he was unable to find a physical justification for the latter, von Neumann became convinced that such a process could not happen spontaneously in nature without the involvement of the mind. The reason for assuming this was that a human observer is always present when there is a measurement process and apparently it is the only non-physical reality involved in such a process. However, von Neumann

---

[a](Von Neumann, 1932).

was not at all clear about the role of the observer's consciousness,[a] whether

(i) to attribute to it a real power of intervention on the physical reality from the outside, or

(ii) to confine its capabilities to a sort of illusory depiction of the physical world, whilst this was still ruled by the ordinary Schrödinger equation.

It is interesting to note that since the times of Descartes one has debated whether or not the mind could intervene on the physical reality from the outside. For most scholars (among them Leibniz[b]) this would clearly represent a violation of the closure of the causes in the physical world.[c] With this expression we mean the epistemological requirement that all physical phenomena must be explained with physical causes or factors, without any recurrence to causes or factors that either cannot be explained in physical terms (at least for our current state of knowledge) or pertain to scientific domains that are not ruled directly by physical laws, like psychology.

The two possible interpretations of von Neumann's postulate soon split physicists into the subjectivist and objectivist parties. There is a considerable amount of subjectivist interpretations.[d] The first consistent proposal of this kind can be found in a work written by London and Bauer.[e] These authors developed a very detailed analysis of the measurement process including a specific treatment of the apparatus and its connections with the object system. They remarked that when measuring a system prepared in a superposition state (9.1), according to the laws of quantum mechanics such a superposition should affect somehow also the apparatus, so that we should have the transformation

$$|\psi\rangle|A_0\rangle \longrightarrow c_d|d\rangle|a_d\rangle + c_u|u\rangle|a_u\rangle, \qquad (9.13)$$

where $|a_d\rangle$ and $|a_u\rangle$ are the apparatus states corresponding respectively to the system states $|d\rangle$ and $|u\rangle$. It is evident that the

---

[a] (Tarozzi, 1996).
[b] (Leibniz PS).
[c] (Kim, 1984).
[d] (Auletta, 2004c).
[e] (London/Bauer, 1939).

transformed state is an entangled state. Here, we have assumed that the apparatus is in an initial ready state $|A_0\rangle$ and, for the sake of simplicity, we have also suppressed the direct product symbol, but we stress that the apparatus kets $|a_d\rangle$ and $|a_u\rangle$ and the system kets $|d\rangle$ and $|u\rangle$ pertain to two different Hilbert spaces (or subspaces). The state obtained on the right-hand side of the transformation in Eq. (9.13) is really puzzling. On the one hand, it establishes an unequivocal connection between states of the object system and states of the apparatus. Indeed, to each measured state of the object system there is a corresponding pointer state of the apparatus. However, it also seems to imply that the apparatus is in a superposition of different pointer states [see Section 7.9], which is clearly absurd, since we expect from an apparatus to provide us with a specific and univocal answer (either u or d). Moreover, as a matter of fact, an ambiguous apparatus has never been observed. London and Bauer were perfectly aware of the circumstance that if we compute the density matrix of the transformed object system–apparatus composite system

$$\hat{\rho}_{AS} = |c_d|^2 \, |d\rangle\langle d| \otimes |a_d\rangle\langle a_d| + |c_u|^2 \, |u\rangle\langle u| \otimes |a_u\rangle\langle a_u|$$
$$+ c_d c_u^* |d\rangle\langle u| \otimes |a_d\rangle\langle a_u| + c_d^* c_u |u\rangle\langle d| \otimes |a_u\rangle\langle a_d| \qquad (9.14)$$

and take a partial trace over the object system S [see again Section 7.9], we obtain a mixture of the type given by Eq. (9.7). That is, we have

$$\hat{\rho}_A = \text{Tr}_S \, \hat{\rho}_{AS}$$
$$= \langle d|\hat{\rho}_{AS}|d\rangle + \langle u|\hat{\rho}_{AS}|u\rangle$$
$$= |c_d|^2 \, |a_d\rangle\langle a_d| + |c_u|^2 \, |a_u\rangle\langle a_u|. \qquad (9.15)$$

The crucial question for them was the physical meaning of such an operation which apparently is only a mathematical one. The global pure state contains the whole information that is in the composite system at a certain time. This information is never accessible in its totality [see Section 6.9]. When we consider the apparatus alone, we do not take into account a significant part of the information, namely, the part represented by quantum features (that contribute to the quantum interdependence between the apparatus and the object system) and therefore we only take into account a small subset of the whole information. In general, a mixture represents less information than a pure state. The part that we do not consider is precisely what is inaccessible to us and cannot be obtained under any circumstance (i.e., the quantum features). On the other hand, a mixture like $\hat{\rho}_A$ in Eq. (9.15) could correspond to the classical probability distribution between heads and tails *before* tossing a coin. Therefore, it satisfies our expectation of how an apparatus works, namely, to respond to certain measurement outcomes in an univocal way.

Summing up, we can have the desired classical result but this is obtained at a certain cost since a mixture like $\hat{\rho}_A$ in Eq. (9.15) represents a state of diminished information as compared with the entangled state on the right-hand side of the transformation (9.13). London and Bauer considered this as a very relevant problem. In a certain sense it is. However, as we shall show below, this is also the solution of the problem. We also remark that London and Bauer considered a mixture like $\hat{\rho}_A$ in Eq. (9.15) a sort of transitory (though mysterious) state and not as the final result of a measurement. Indeed, following the initial approach of von Neumann, they considered only one of the two eigenstates, either $|a_u\rangle$ or $|a_d\rangle$, to be such a final result (representing a single detection event). Therefore, they concluded that it is only the observer that is able to chose one of the two components of the initial entangled state, since it is only the observer that, when seeing one of the two apparatus states above, can say, for instance, "I perceive the apparatus as being in the state $|a_u\rangle$ and therefore the object system must be in the physical state $|u\rangle$." However, according to London and Bauer this should not be understood in solipsistic terms either, since we can communicate this experience to any member of the scientific community so that everybody is free to personally verify that the apparatus is indeed in such a state. Such an interpretation has then been further developed by Wigner[a] and even acquired subsequently a certain authority among physicists and philosophers.

The opposite orientation has been provided by the objectivist interpretation. In this case the mind only gives rise to an illusionary or partial image of reality. The first proposal in this sense goes back to a 1957 paper of Everett,[b] but this interpretation could

---

[a](Wigner, 1961).
[b](Everett, 1957).

also be considered an original development of some of Einstein's ideas that will occupy us below. The main idea of Everett is as follows. The reduction of the wave packet never actually happens but what physically occurs is only the establishment of a correlation between observed system and apparatus given by Eq. (9.13) that he called the *relative state*. The fact that we perceive only a component of the superposition as the outcome of a measurement is, according to Everett, only due to the particular perspective under which we consider the composite system. This means that for Everett all components of the entangled state on the right-hand side of the transformation (9.13) are always and equally real. B. DeWitt has developed this interpretation, giving rise to the so-called *many-worlds interpretation*, which is quite diffused in cosmological studies.[a] It asserts that actually we observe in our universe a component of some initial state of an object system, but there are other universes in each of which a counterpart of ourselves observes another component. Such a theory is reminiscent of a modal philosophy proposed by Leibniz in his correspondence with Arnould.[b] As a matter of fact, Everett's paper still left open the possibility to interpret in subjectivist terms the theory of the many components. This was done in terms of a many-minds theory of the universe.[c] It asserts that each component is observed by a different mind in a different universe or region of the same universe and every finite mind of this kind is part of an universal Mind (which could be God or a demiurge) to which the quantum state is immediately present in all its components. These are really strong speculations that seem to run against a fundamental principle of economy that has ruled natural science for centuries, namely, *entia non sunt multiplicanda praeter necessitatem*, the so-called Occam's razor.

9.4 Basis Ambiguity

To many physicists the objectivist interpretation may sound more reasonable. However, there is a specific problem occurring within

---

[a] (Dewitt, 1970).
[b] (Leibniz PS, v. II).
[c] (Lockwood, 1996).

such a context, which is known as *basis ambiguity*. Such a difficulty seems to cast doubts on the meaning of measurement as such.[a] Let us suppose that we wish to measure a generic observable [see Eq. (6.13)]

$$\hat{O} = \sum_j o_j |o_j\rangle\langle o_j|, \qquad (9.16)$$

where $|o_j\rangle$ are its eigenstates with o_j the corresponding eigenvalues. Suppose that system and apparatus as a whole evolves according to the transformation

$$|\psi\rangle|A_0\rangle \longrightarrow \sum_j c_j|o_j\rangle|a_j\rangle, \qquad (9.17)$$

where $|\psi\rangle$ is the initial state of the object system and is a superposition state of the eigenstates $|o_j\rangle$, $|A_0\rangle$ is the initially ready state of the apparatus, and $|a_j\rangle$ are the pointer states of the apparatus corresponding to $|o_j\rangle$. As we have seen, Everett calls such a transformed state a relative state and assumes that measurement is accomplished already at this stage and nothing happens thereafter. However, we know that the same state can be expanded in different bases. Indeed, using the completeness relation

$$\sum_k |a_k'\rangle\langle a_k'| = \hat{I}, \qquad (9.18)$$

we can now write the states of the apparatus as [see Section 4.6]

$$|a_j\rangle = \sum_k \langle a_k'|a_j\rangle|a_k'\rangle. \qquad (9.19)$$

Therefore, the laws of quantum mechanics do not forbid us to write the right-hand side of the transformation (9.17) as

$$\sum_j c_j|o_j\rangle|a_j\rangle = \sum_k \left(\sum_j c_j\langle a_k'|a_j\rangle|o_j\rangle\right)|a_k'\rangle$$
$$= \sum_k |o_k'\rangle|a_k'\rangle, \qquad (9.20)$$

In the above expression, the states $|o_k'\rangle$ are given by

$$|o_k'\rangle = \sum_j c_{jk}|o_j\rangle, \qquad (9.21)$$

---

[a](Zurek, 1981).

where

$$c_{jk} = c_j \langle a'_k | a_j \rangle. \tag{9.22}$$

Note that we obtain in the last line of Eq. (9.20) two different bases for both the object system and the apparatus. This seems to imply that the relative state on the right-hand side of transformation (9.17) represents not only measurement of the observable $\hat{O}$ but also of the new observable

$$\hat{O}' = \sum_j o'_j | o'_j \rangle \langle o'_j |, \tag{9.23}$$

which in general does not commute with $\hat{O}$. Therefore, there seems to be a confusion here between the formal possibility to expand the joint state of the object system and apparatus in different bases and the experimental setup for measuring a *particular* observable. If we interpret the latter in terms of the former, such a "measurement" would imply a violation of the uncertainty relations. If, on the contrary, we interpret the former in terms of the latter, this would force us to accept a definition of measurement that seems to deprive it of any significance, since any possible observables would be jointly measured, which dissolves the very act of decision that is connected with any experimental operation. In other words, measurement is by definition an interaction that deals with a certain specific observable [see Section 6.9], and this is an issue that is much more fundamental than the question whether we obtain only one of the components of a certain initial state (according to some preparation) or not. Namely, it is the step of measurement in which apparatus and object system become coupled, and is called *premeasurement* and should be kept distinguished from the initial preparation of a system in some superposition of possible outcomes. Although it is true, at least in principle, that such a premeasurement step can be made reversible[a] (which shows that up to this point Everett is right with his theory of the relative state), the fact remains that even in reversible couplings we always deal with a specific experimental setup (that selects a certain observable and therefore also a certain basis), and the latter imposes particular physical conditions that cannot be changed without doing simply another

---

[a](Wang *et al.*, 1991).

kind of experiment. Summing up, it seems that by trying to face the impossibility to get an eigenstate of a certain observable through unitary evolution, Everett and the other objectivists have incurred a far bigger problem, a confusion between mathematics and dynamics or also between preparation and premeasurement.

Problem 9.5 Prove that we arrive at a similar result by making the inverse operation, that is, starting from the expansion in Eq. (9.21) and the formula $|a'_k\rangle = \sum_j |a_j\rangle\langle a_j|a'_k\rangle$, and deriving the expansion on the right-hand side of Eq. (9.17) from the expansion in the last line of Eq. (9.20).

9.5 Role of the Environment

A different idea about the concept of measurement has been developed step by step during the 1960s and 1980s, especially as a consequence of the introduction of quantum optical devices such as lasers.[a] The so-called reduction could be the consequence of a spontaneous coupling of the object system with the apparatus and the environment, combined with the local interaction which we perform when measuring. In this case, measurement becomes a special case of a wider class of interactions between *open systems*, that is, those physical systems that are open to the environment. To this end, let us consider a composite system whose state at time $t_0 = 0$ takes the form

$$|\Psi(0)\rangle_{SAE} = (c_u|u\rangle + c_d|d\rangle)|A_0\rangle|E_?\rangle, \qquad (9.24)$$

where the object system S is *prepared* in a superposition state of $|u\rangle$ and $|d\rangle$, the apparatus A is in an initial ready state $|A_0\rangle$, and the environment E is in a unknown state $|E_?\rangle$. We have considered here a two-state system for the sake of simplicity. Now at a certain later time the apparatus and object system interact, giving rise to an entangled state. It is this coupling that is called premeasurement and, as we had seen, it is also the stage that could be still reversible. We also assume that in a subsequent time t these two systems

---

[a](Zeh, 1970), (Zurek, 1982), (Cini, 1983). See also (Auletta, 2000, Part IV), (Auletta, 2004b).

become entangled with the environment, so that [see Sections 7.3 and 7.8]

$$|\Psi(t)\rangle_{SAE} = c_u(t)|u\rangle|a_u\rangle|e_u\rangle + c_d(t)|d\rangle|a_d\rangle|e_d\rangle. \qquad (9.25)$$

We can even postulate that all quantum systems of the universe are more or less entangled. When we say *more or less* we are referring to the fact that entanglement has degrees,[a] an issue that will occupy us later.

We have also stressed that measurement is local, because it consists in some operation that we perform on a certain system with a certain apparatus in a certain place called laboratory. It is quite common that in such a case we do an extrapolation, that is, we either are not aware at all of the entanglement with the environment or decide to do not consider it. Both cases are mathematically described by a partial trace [see Section 7.9]. To this purpose, let us write down the density matrix corresponding to the state $|\Psi(t)\rangle_{SAE}$, that is,

$$\hat{\rho}_{SAE}(t) = |\Psi(t)\rangle\langle\Psi(t)|_{SAE}, \qquad (9.26)$$

and trace the environment out to obtain

$$\begin{aligned} \mathrm{Tr}_E \, \hat{\rho}_{SAE}(t) &= \langle e_u|\hat{\rho}_{SAE}(t)|e_u\rangle + \langle e_d|\hat{\rho}_{SAE}(t)|e_d\rangle \\ &= |c_u(t)|^2 \, |u\rangle\langle u| \otimes |a_u\rangle\langle a_u| + |c_d(t)|^2 \, |d\rangle\langle d| \otimes |a_d\rangle\langle a_d|. \end{aligned} \qquad (9.27)$$

This is precisely the kind of state (mixture) that we expect. It is true that this is not yet a single result, but to observe a specific outcome (which represents the final step of measurement, i.e., *detection*) starting from a state like this is quite banal from a classical point of view (however, as we shall discover, quantum mechanically there is an additional difficulty). Although the previous entanglement seems to have disappeared, quantum features, as we shall see below, are not destroyed but simply lost in the universal quantum noise background of our universe. Since what decreases by losing quantum feature is the coherence between the different components of a quantum state, this phenomenon is called *decoherence*. This should be a quite common phenomenon as it has been extensively shown that decoherence is precisely what happens

---

[a] (Vedral *et al.*, 1997a).

in controlled environments when a small quantum system is allowed to spontaneously evolve in the presence of a large system (like a bath of harmonic oscillators).[a]

There is a strong temptation to consider decoherence a pure mathematical trick, perhaps good for practical purposes but unable to solve the conceptual problem of measurement. In this case one has spoken of FAPP (For All Practical Purposes) suitable explanations of measurement.[b] We do not think that this is the case. Actually, decoherence allows us to understand a fundamental distinction between locality and non-locality that is unknown to classical physics. Indeed, at a global level (the level of our universe, ultimately) we can assume that everything happens according to the laws of quantum mechanics and therefore in a reversible and deterministic way. At this level, there are no events at all (so nothing happens in the true sense of the word) but there are only probabilities amplitudes of events ruled by those laws. At a local level, instead, random events occur that are locally irreversible. How is it possible to bring into harmony these two statements? This could well be if we can account for the concept of local irreversibility. If different local irreversible process could balance so that the whole remains unperturbed, this could represent the solution to our problem.

9.6 Entropy and Information

In order to deal with this problem, let us consider the quantum mechanical entropy, which is called the *von Neumann entropy*. Following the approach of Part II and especially the correspondence principle [see Principle 7.1], it is always opportune to take into account a classical analogue. Actually, classically we have at least two different forms of entropy. The first one is called the *Shannon entropy* (also known as the *information entropy*) and it is very relevant to the exchange of information. Let us say that we wish to find a quantitative measure of the information content of a process of

---

[a](Lindblad, 1983), (Savage/Walls, 1985).
[b](Bell, 1990).

information acquisition (reduction of incertitude) J that can yield certain possible outcomes (selections) j with the corresponding probabilities $\wp(j)$, where $\sum_j \wp(j) = 1$. The information content associated with the outcome j may be quantified by a continuous, non-negative decreasing function $I(j)$ of the probability $\wp(j)$ that the outcome j might occur. To find the function $I(j)$, we consider the following two limiting cases. When $\wp(j) = 1$, we know for certain the occurrence of the outcome j, so that the associated information is minimum and we may specify $I(j) = 0$ for $\wp(j) = 1$. When $\wp(j) \ll 1$, we know very little about the occurrence of the outcome j, so that the associated information is maximum and we may specify $I(j) \to \infty$ as $\wp(j) \to 0$.

Next we consider two of such processes of information acquisition J and K that are *independent* of each other. Then the joint probability that the outcomes $j \in J$ and $k \in K$ might occur is $\wp(j, k) = \wp(j)\wp(k)$, and the associated information of the joint outcome is $I(j, k)$. We stipulate that the information content of two independent processes be additive, namely, we have

$$I(j, k) = I(j) + I(k), \tag{9.28}$$

which is valid for all j and k. Together with the above limiting values, this condition implies that the function $I(j)$ has to take the form [see Box 9.1][a]

$$I(j) = -k \log_2 \wp(j), \tag{9.29}$$

where k is a positive constant independent of the probability and the minus sign is there to ensure that the information is a non-negative quantity. Indeed, since $0 \leq \wp(j) \leq 1$, according to Eq. (9.32c), $\log_2 \wp(j)$ would be negative except for $\wp(j) = 1$. A choice of the constant k amounts to prescribing a unit for the measure of information. Here we shall choose $k = 1$, that is,

$$I(j) = -\lg \wp(j), \tag{9.30}$$

for which the corresponding unit is called the *bit* (short for *binary digit*). The information content was called by Shannon the *surprisal* as it represents the "surprise" of observing the corresponding

---

[a]See (Khinchin, 1957) for short and effective introduction to these matters.

Box 9.1 Logarithm

The logarithm $\log_b x = y$ means that $b^y = x$, where the positive number b is called the base of the logarithm. The logarithm with base 2 is referred to as the *binary logarithm* (generally denoted by lg). It is widely used in computer science and information theory because of its close connection to the binary numeral system. Another important kind of logarithm is the *natural logarithm* (commonly denoted by ln), whose basis is the constant e (≈ 2.718281) [see Box 2.4]. It is the inverse of the exponential function [see Fig. 6.2], namely, we have

$$e^{\ln x} = x \quad (x > 0). \tag{9.31}$$

A plot of the natural logarithm and the binary logarithm is displayed in Fig. 9.2, which shows clearly that the logarithmic increase is very slow as compared with the power-law or exponential increase. The logarithm satisfies the following properties:

$$\log_b 1 = 0, \tag{9.32a}$$

$$\log_b(xy) = \log_b x + \log_b y, \tag{9.32b}$$

$$\log_b\left(\frac{x}{y}\right) = \log_b x - \log_b y, \tag{9.32c}$$

$$\log_b x^p = p \log_b x, \tag{9.32d}$$

$$\log_b x = \frac{\log_a x}{\log_a b}. \tag{9.32e}$$

Moreover, it can be shown that

$$\lim_{x \to 0^+} x \log_b x = 0, \tag{9.33}$$

where $x \to 0^+$ denotes that x approaches zero from above (i.e., x decreases in value approaching zero) and the corresponding limit is called the one-sided limit.

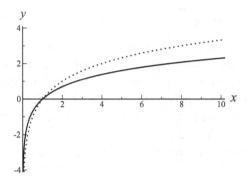

Figure 9.2 Plot of the natural logarithm ln x (solid line) and the binary logarithm lg x (dotted line). Comparing with the plot of the exponential function in Fig. 6.2, we see that the exponential growth is very fast while the logarithmic growth is very slow.

outcome. Indeed, the occurrence of a highly improbable outcome is very surprising while the occurrence of an almost sure outcome is well expected and causes little or even no surprise at all.[a]

The classical Shannon entropy associated with the information acquisition J is defined by the sum of the information contents over all the possible outcomes weighted by the corresponding probabilities, that is,

$$H(J) = -\sum_{j\in J} \wp(j) \lg \wp(j), \qquad (9.34)$$

where $\sum_j \wp(j) = 1$. In other words, it represents the averaged incertitude of all the possible outcomes in an information acquisition process (or of all possible choices that can be done in selecting certain messages). Entropy, therefore, quantifies the expected value of the information that *could be* acquired from a certain system. To put it another way, entropy quantifies the randomness of the system from which we extract information. Therefore, entropy is strictly connected with how much disorder a system displays. In particular, in a first approximation, increase in entropy means increase in disorder, while decrease in entropy means growth in order.

The other form of classical entropy is called the *Boltzmann entropy*. It appears in thermodynamics and statistical mechanics,

---

[a](Shannon, 1948).

and deals with disorder of an isolated thermodynamical system (such as an isolated gas system). In specific, the Boltzmann entropy of an isolated system of gas is defined by

$$S = k_B \ln W, \tag{9.35}$$

where k_B is the Boltzmann constant, whose value in SI units is $k_B = 1.380, 6488 \times 10^{-23}$ J/K, and W is the number of possible configurations (microstates) of the gas molecules. The Shannon and the Boltzmann entropies are closely related, especially when the maximum value of the former is considered.[a] It can be shown that the Shannon entropy reaches its maximum value when all possible outcomes are equiprobable, the situation corresponding to maximum randomness. Suppose that there are n possible states, then we have $\wp(j) = \frac{1}{n}$ for equiprobability and the maximum value of the Shannon entropy is given by [see Eq. (9.34)]

$$H^{\text{Max}} = -\sum_{j=1}^{n} \frac{1}{n} \lg \frac{1}{n} = -\lg \frac{1}{n} \sum_{j=1}^{n} \frac{1}{n} = \lg n, \tag{9.36}$$

where use has been made of relation $\sum_{j=1}^{n} \frac{1}{n} = 1$. When dealing with the Boltzmann entropy of a gas, this corresponds to setting $W = n$, that is, there are n possible configurations of the gas molecules. From the relations between the natural and the binary logarithms [see Eq. (9.32e)]

$$\ln x = \ln 2 \lg x, \tag{9.37}$$

we have

$$S = k_B \ln n$$
$$= k_B \ln 2 \lg n$$
$$= k_B \ln 2 \, H^{\text{Max}}, \tag{9.38}$$

which shows that the two quantities S and H^{Max} are equal apart from a constant factor $k_B \ln 2$, which accounts for two different units used in measuring the entropy.

---

[a]The counterpart of the Shannon entropy in statistical mechanics is the *Gibbs entropy*, which is defined by

$$S = -k_B \sum_j \wp(j) \ln \wp(j),$$

where $\wp(j)$ is the probability that the thermodynamical system is found in the microstate j and the sum is over all possible microstates of the system. For an isolated thermodynamical system, all the possible microstates are equiprobable and the Gibbs entropy reduces to the Boltzmann entropy.

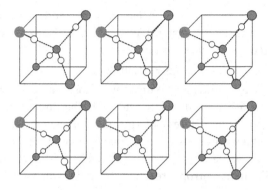

Figure 9.3 The six possible locations of hydrogen atoms (white circles) relative to oxygen atoms (gray circles) in an ice crystal. The solid lines indicate covalent bonds and the dashed lines indicate hydrogen bonds. Two of the hydrogen atoms must be near and two far away from the central oxygen atom. This incertitude about the location of the hydrogen atoms means that there is entropy even at absolute zero temperature. Adapted from (Auletta, 2011a, p. 51).

What have the Shannon and the Boltzmann entropies in common? Both measure the degree of disorder in a system. It is obviously meaningful to associate a Shannon entropy to a thermodynamic entropy (since any increase in thermodynamic entropy also means increase in the information that could be acquired), but the reverse is not necessarily the case. Unfortunately, many even identify an increase in entropy with the increase of heat, which is the thermodynamic expression of disorder. This is however not correct, for even at absolute zero temperature there is still a residual entropy given by the different possible arrangements of the atoms. This is evident for ice crystals[a] [see Fig. 9.3]. In such a case, we only deal with certain possible configurations of certain systems, but we no longer deal here with typical dynamical factors involving work and exchange of heat, although this always intervene in the generation of a certain entropic state. Therefore, the most general way to define entropy in all cases is to say that it expresses a degree of disorder in certain distributions. In the form of Boltzmann entropy, this quantification of disorder is the result

---

[a](Atkins/De Paula, 2006, pp. 609–610).

of some dynamical exchange of energy and its quality for doing work, while the Shannon entropy is connected with the degree of disorder of some sequence of signals that we receive or exchange. Having established this, entropy is not the same of information. Although any exchange of information is a dynamical process like measurement, information as such is a formal quantity. As we shall see later, information has to do with certain interdependence among the possible outputs. Indeed, while disorder is always the *result of some irreversible dynamical process* that requires some form of interaction, information can be processed in pure reversible way without any involvement of interaction (and therefore also when the system is fully isolated). What quantum systems do (in full accordance with dynamical laws like the Schrödinger equation) is precisely to process information in a reversible way. Summing up, information does not imply irreversibility while all forms of entropy deal with at least the result of irreversible processes. Although the Shannon entropy has to do with dynamical treatment of information (like that occurring in measurement), and it has to do with the disorder among signals rather than with thermodynamic quantities, ultimately it also implies thermodynamic aspects, as we shall see.

The *von Neumann entropy* is the extension of classical entropy concepts to quantum mechanics. It is formally similar to the Shannon entropy and is indeed defined by

$$H_{VN}(\hat{\rho}) = -\operatorname{Tr}(\hat{\rho} \ln \hat{\rho}), \tag{9.39}$$

where $\hat{\rho}$ is the density matrix describing the system [see Section 7.8]. As we may recall, the trace of a matrix is equivalent to the sum of the diagonal elements of that matrix, which for the density matrix represent the probabilities of obtaining the corresponding components when measuring the system, and therefore the above equation is formally similar to Eq. (9.34). To compute the von Neumann entropy $H_{VN}(\hat{\rho})$, it is convenient to first find the spectral decomposition [see Section 6.1] of density matrix $\hat{\rho}$. We have already shown that the density matrix can be considered an observable, therefore it can be diagonalized. In other words, we have the following eigenvalue equation

$$\hat{\rho}|\eta_j\rangle = \eta_j|\eta_j\rangle, \tag{9.40}$$

where η_j are the eigenvalues of $\hat{\rho}$ and $|\eta_j\rangle$ are the corresponding eigenstates. This allows us to write the spectral decomposition of $\hat{\rho}$ as

$$\hat{\rho} = \sum_j \eta_j |\eta_j\rangle\langle\eta_j|. \tag{9.41}$$

The von Neumann entropy can then be expressed as

$$H_{\text{VN}}(\hat{\rho}) = -\sum_j \eta_j \ln \eta_j. \tag{9.42}$$

Let us consider a simple two-state system. If $\hat{\rho}$ is a pure state, then the above eigenvalue equation produces two eigenvalue: $+1$ with the eigenstate, say, $|\eta_1\rangle$, and 0 with the eigenstate, say, $|\eta_2\rangle$. The specificity of the density matrix with respect to other observables is that all its eigenvalues represent probabilities. Hence, the system has a probability one to be in the state $|\eta_1\rangle$ and a probability zero to be in the orthogonal state $|\eta_2\rangle$. In such case, we have $\hat{\rho} = |\eta_1\rangle\langle\eta_1|$. This is in full accordance with what we know of projectors, for instance by considering projectors like those given by Eqs. (3.49). It is interesting to note that the von Neumann entropy is precisely zero when the system is a pure state (i.e., when the density matrix coincides with a projector) as we have seen in Eq. (9.32a) that the logarithm of one is zero. On the other hand, if $\hat{\rho}$ is a mixture then we have $\hat{\rho} = \eta_1|\eta_1\rangle\langle\eta_1| + \eta_2|\eta_2\rangle\langle\eta_2|$, where $0 < \eta_1, \eta_2 < 1$. Moreover, we have $\eta_1 + \eta_2 = 1$ because of the unit trace property [see Eq. (7.117b)]. The system has a probability η_1 in the state $|\eta_1\rangle$ and a probability η_2 in the orthogonal state $|\eta_2\rangle$. The von Neumann entropy for this mixture is given by $H_{\text{VN}}(\hat{\rho}) = -(\eta_1 \ln \eta_1 + \eta_2 \ln \eta_2)$, which is a positive quantity [see Problem 9.7].

With the help of the previous concepts, we can now understand measurement as a local and dynamical displacement of order and disorder whose net *global* balance, according to quantum mechanics, should be zero. Suppose that the global system represented by the combination of the object system, apparatus, and environment is in a zero entropy state (i.e., it is a pure state, a state such that information can be processed in a reversible way). When we locally obtain a mixtures for the object system, this means that its entropy is irreversibly and locally increased (it is a sort of symmetry break starting from the many possible configurations that we have in the initial state). However, the global system could remain fully reversible. In other words, we have found a physical meaning to the partial trace through which we obtain the mixture that is appropriate for describing the measurement process. The local increase in entropy of the object system can be expressed as determined by some loss of features into the environment during a measurement of the object system. Therefore, features are responsible for the maximum order (i.e., zero entropy) of the initial state of the object system as well as of the object system–apparatus–environment composite system, and their local loss explains the result of a local measurement.

As we have seen, a very relevant issue is that detection can be considered a kind of rupture of symmetry in which we obtain one of the components out of many possible ones in a state of zero entropy.[a] The repeatability of measurement is a necessary symptom of the wave–packet collapse as far as we expect to have the system in the same state in the subsequent trials. Now, if the system evolves unitarily (where we do not consider here the environment for the sake of simplicity) we have, for the components $|u\rangle$ and $|v\rangle$, the following unitary transfer of information from the object system to the apparatus during the premeasurement step [see also Eq. (9.13)]:

$$|u\rangle|A_0\rangle \rightarrow |u\rangle|a_u\rangle, \quad |v\rangle|A_0\rangle \rightarrow |v\rangle|a_v\rangle. \tag{9.43}$$

Since we have the preservation of the scalar product

$$\langle u|v\rangle = \langle u|v\rangle\langle a_u|a_v\rangle, \tag{9.44}$$

then, to have $\langle u|v\rangle \neq 0$ implies that $\langle a_u|a_v\rangle = 1$. This means that the states of the pointer can differ (as we expect from an apparatus) only when $\langle u|v\rangle = 0$. In other words, we can acquire information only when the states of the object system are orthogonal. Only this requirement can reconcile the linearity of the first two steps of measurement (preparation and premeasurement) with the non-linearity of the final selection. This means that we have a spontaneous emergence of the pointer states $|a_u\rangle$, $|a_v\rangle$ only when the requirement of orthogonality is satisfied. Therefore, whatever is the initial superposition of the object system, things spontaneously

---

[a](Zurek, 2007, 2013).

evolve in order that the components of the object system that are coupled with the pointer basis are orthogonal.

Problem 9.6 Show that the inclusion of outcomes that have zero probabilities (i.e., outcomes that never occur) does not change the Shannon entropy associated with a certain information acquisition process.

Problem 9.7 Suppose a source emits a stream of binary digits (0's and 1's) with the respective probabilities $\wp_0 = x$ and $\wp_1 = 1 - x$, where $0 \leq x \leq 1$. (i) What is the Shannon entropy of this source? (ii) By plotting the Shannon entropy as a function of x for $0 \leq x \leq 1$, verify that the Shannon entropy is maximum for equiprobability, i.e., when $\wp_0 = \wp_1 = \frac{1}{2}$.

9.7 Reversibility and Irreversibility

Is there any evidence for the explanation we have proposed? The ground is represented by the following theorem:

Theorem 9.1 (Reversibility) For any local irreversible transformation of a given quantum system it is always possible to choose a larger quantum system that embeds it and in which the transformation is reversible.

Suppose now that we have a mixed state (like that occurring during an irreversible measurement process [see, e.g., Eq. (9.15)]):

$$\hat{\rho}_A = \sum_j w_j |a_j\rangle\langle a_j|, \tag{9.45}$$

where $0 < w_j < 1$ with $\sum_j w_j = 1$ and $\{|a_j\rangle\}$ is a basis in the Hilbert space of the apparatus. Consider the environment (or the rest of the world) and the state vector of the composite system of the form

$$|\Psi\rangle_{AE} = \sum_j \sqrt{w_j} |a_j\rangle |e_j\rangle, \tag{9.46}$$

Table 9.1 Truth table for negation

Input	Output
p	$\neg p$
1	0
0	1

where $\{|e_j\rangle\}$ is a basis in the Hilbert space of the environment. Then, if we denote the pure state density matrix corresponding to the state $|\Psi\rangle_{AE}$ by $\hat{\rho}_{AE}$, by tracing out the environment we have

$$
\begin{aligned}
\mathrm{Tr}_E \, \hat{\rho}_{AE} &= \sum_{jk} \sqrt{w_j w_k} |a_j\rangle\langle a_k| \, \mathrm{Tr}(|e_j\rangle\langle e_k|) \\
&= \sum_{jk} \delta_{jk} \sqrt{w_j w_k} |a_j\rangle\langle a_k| \\
&= \hat{\rho}_A,
\end{aligned}
\tag{9.47}
$$

which proves the theorem. Let us now give an example. As we have mentioned, quantum systems can be considered information processors. All information processing is performed by logic gates as those that are currently active in every personal computer. This is also the reason why one of the most important developments in quantum information is quantum computing. A *logic gate* is a function of the form

$$
f : \{0, 1\}^m \rightarrow \{0, 1\}^n,
\tag{9.48}
$$

where m is the number of binary inputs and n is the number of binary outputs. In the above expression, 1 can be associated with the logical value *true* whilst 0 with the logical value *false*. For $n = 1$, the function f is called a *Boolean function*. Examples of the logical Boolean gates include NOT, AND, OR, and XOR gates.

The NOT gate is a logic gate which implements logical *negation*. It is denoted by the symbol $\neg$ and its truth table in shown in Table 9.1, in which the output is the opposite logic value of the input. The AND gate is a logic gate that implements logical *conjunction*. It is true if and only if both the inputs to the gate are true. This means that $p \wedge q = 1$ if and only if both $p = 1$ and $q = 1$, where we recall that 1 and 0 represent truth and falsity, respectively, and $\wedge$ denotes conjunction. This can be represented by means of the truth

Table 9.2 Truth table for conjunction

Input		Output
p	q	$p \wedge q$
1	1	1
1	0	0
0	1	0
0	0	0

Table 9.3 Truth table for disjunction

Input		Output
p	q	$p \vee q$
1	1	1
1	0	1
0	1	1
0	0	0

Table 9.4 Truth table for exclusive disjunction

Input		Output
p	q	$p \veebar q$
1	1	0
1	0	1
0	1	1
0	0	0

table in Table 9.2. The OR gate is a logic gate that implements logical *disjunction*. It is denoted by the symbol $\vee$ and is true if at least one of the inputs to the gate is true, as shown in the truth table in Table 9.3. Finally, the XOR gate is a logic gate that implements logical *exclusive disjunction*. It is denoted by the symbol $\veebar$ and is true if and only if the two inputs to the gate have a different truth value, as shown in the truth table in Table 9.4. Important logical rules are the so-called De Morgan's laws. They state that the negation of a conjunction is the disjunction of the negations and that the negation of a disjunction is the conjunction of the negations. That is, formally we have

$$\neg(p \wedge q) = \neg p \vee \neg q, \quad \neg(p \vee q) = \neg p \wedge \neg q. \qquad (9.49)$$

As mentioned, logic circuits are currently implemented in our computers. In other words, the electrical connections in an ordinary computer instantiate logical connections (this is what constitutes a processor). For instance, let us consider the circuit shown in Fig. 9.4. Negation can be implemented by an open gate, a disjunction by two parallel gates (that is, two gates implemented in parallel wires), and a conjunction by serial gates. Due the De Morgan's laws and the definition of implication (which we shall come back in the next chapter) any kind of logical operation can be implemented in this way.

Conjunction is a reversible operation but only when it is true, while it is not when its truth value is not known or when it is known to be false (in the latter case three different inputs map to the same output 0). If it is true, we know for certain that both p and q must be true, that is, we can infer from the truth value of the conjunction the truth value of the inputs, and vice versa. Instead, from the truth value of the disjunction or exclusive disjunction we cannot infer the truth value of the inputs, although we can infer for the former when it is known to be false. Therefore, in the general case conjunction, disjunction, and exclusive disjunction are irreversible operations. However, it is always possible to make use of some logic gate that is expanded to other inputs and that is logically equivalent to an irreversible logical operation but being reversible itself.[a] For instance, let us consider the disjunction between the negation of p and and the negation of q as shown in the truth table in Table 9.5. The reason for this truth table is that p is true when its negation $\neg p$ is false, and vice versa. Nevertheless, as we can see, the form of this truth value is similar to the truth table in Table 9.3 and therefore still irreversible. Moreover, $\neg p \vee \neg q$ is logically equivalent to $\neg(p \wedge q)$ due to De Morgan's laws (9.49). The operation $\neg(p \wedge q)$ is called logical *negated conjunction*, which is implemented by the NOT AND (or NAND) gate.

Let us consider now not just two inputs but enlarge the number of inputs to three, denoted by p, q, and r. For our purposes we shall

---

[a](Bennett/Landauer, 1985).

Table 9.5 Truth table for $\neg p \vee \neg q$

Input		Output
p	q	$\neg p \vee \neg q$
1	1	0
1	0	1
0	1	1
0	0	1

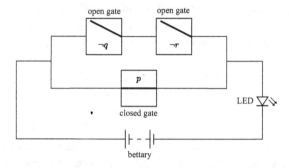

Figure 9.4 Schematic implemented of a logic circuit. Electric current flows clockwise. Gates are represented by thick lines for the sake of representation, while wires by thin lines. In order that electric current flows, it is necessary that the gates be closed. The two main wires of the circuit are parallel. In other words, current can take both paths simultaneously, which means that one path closed is sufficient for transmitting current. The wire at the top displays moreover a serial circuit, where both gates need to be closed for having a current. In specific, the circuit corresponds to the proposition $p \vee (\neg q \wedge \neg r)$.

make use of the so-called Toffoli gate,[a] whose truth table is shown in Table 9.6. We note that, according to Table 9.4, the exclusive disjunction (XOR) $r \rightarrowtail (p \wedge q)$ in the last of the two output columns on the right is true only when either (i) $r = 1$ and $p \wedge q = 0$, or (ii) $r = 0$ and $p \wedge q = 1$. It is false only when either (i) $r = 1$ and $p \wedge q = 1$, or (ii) $r = 0$ and $p \wedge q = 0$. If we take into account the fact that the conjunction between p and q is false when p or q is false [see Table 9.2], we have the following inputs represented by the three input columns on the left:

---

[a](Fredkin/Toffoli, 1982).

Table 9.6   Truth table for Toffoli gate. It is noted that p and q are the control bits and are not changed under the transformation

Input			Output	
p	q	r	$p \wedge q$	$r \rightarrowtail (p \wedge q)$
0	0	0	0	0
0	0	1	0	1
0	1	0	0	0
0	1	1	0	1
1	0	0	0	0
1	0	1	0	1
1	1	0	1	1
1	1	1	1	0

(i)   The second, forth, and sixth rows correspond to the case $r = 1$ and $p \wedge q = 0$.

(ii)   The seventh row corresponds to the case $r = 0$ and $p \wedge q = 1$.

(iii)   The latter row corresponds to $r = 1$ and $p \wedge q = 1$.

(iv)   The first, third, and fifth rows correspond to the case $r = 0$ and $p \wedge q = 0$.

Taking into account these inputs, let us consider the last of the two output columns on the right:

(i)   The values on the second, fourth, and sixth row are 1 since $r = 1$ and $p \wedge q = 0$.

(ii)   The value on the seventh row is again 1 since $r = 0$ and $p \wedge q = 1$.

(iii)   The value on the last row is 0 since $r = 1$ and $p \wedge q = 1$.

(iv)   The values on first, third, and fifth rows are 0 since $r = 0$ and $p \wedge q = 0$.

Therefore, if we only consider the cases in which both $r = 1$ and $r \rightarrowtail (p \wedge q) = 1$, we obtain precisely all the three cases in which $p \wedge q = 0$ (second, fourth, and sixth rows). This subset of the outputs is therefore determined by setting $r = 1$ and can be expressed as

$$(r = 1) \wedge \neg(p \wedge q) \quad \text{or} \quad (r = 1) \wedge (\neg p \vee \neg q). \tag{9.50}$$

It should be noted that we could obtain the same result by choosing an alternative subset, i.e., when both $r = 0$ and $r \rightarrowtail (p \wedge q) = 0$. Also this choice covers precisely all three cases in which

$p \wedge q = 0$ [see Problem 9.9]. Therefore, with both choices we have obtained the disjunction (OR) between $\neg p$ and $\neg q$, which in itself is an irreversible transformation but embedded in a larger reversible transformation. Note indeed that all outputs of the Toffoli gate are univocally mapped to the inputs: all truth values of p and q are maintained, while the values of r and $r \rtimes (p \wedge q)$ are also the same apart from the last two lines (when both p and q are true) whose truth value is inverted. This transformation can be also described quantum mechanically, as we shall see in Chapter 11.

Let us now come back to the quantum mechanical problem of measurement. When we locally have irreversible processes this means that the system under observation has lost certain features, i.e., the non-local correlations that are typical for quantum mechanics [see Section 4.8]. Actually they have not been lost but displaced into the environment or also obscured by the environment going to constitute a sort of universal background quantum noise. This is necessary if any kind of quantum dynamics must be unitary, which means that, in order to preserve a global zero-entropy state (if the state of the whole universe is a pure state), such a local increase of entropy must be counterbalanced by some reduction of entropy elsewhere, which implies constitution of order elsewhere. We experience this every day. Indeed, everywhere there are beautiful example of constitution of order, whose most striking manifestation is represented by living beings and their amazing ability to maintain and improve their order.[a] Such a universal tendency would be not completely explainable if there was not a *drive* to order given to such a compensatory effect.

In other words, if the universe as a whole obeys quantum mechanical laws (and we are inclined to think so), we would expect that the tendency to disorder is continuously balanced by an induced tendency to order, so that the net result is zero or at least fluctuates around zero [see Fig. 9.5]. In an adiabatically expanding universe the global entropy is conserved,[b] which also implies that this quantity can be conserved even when it is zero or near to zero. The conservation of both order and disorder in our

---

[a] (Ball, 1999).
[b] (Mo *et al.*, 2010, pp. 108–110 and 129–32).

Figure 9.5   A constant-entropy (perhaps a zero-entropy) universe in which local disruptive and order building processes do not affect the configuration of the whole. Adapted from (Auletta, 2011a, p. 57).

universe is likely a far more general principle than the conservation of physical quantities like mass, energy, and momentum. There are situations, at least for very short time intervals, in which (according to the energy–time uncertainty relation) the conservation of energy can be violated as in the "creation" of virtual (i.e., extremely short-living) particles in a vacuum. Moreover, quantum mechanics does not obey the principle of least action that is connected with those conservation laws. Obviously, the tendency to disorder is spontaneous and in this sense more fundamental. Leibniz and the fathers of classical mechanics considered indeed that nature always chooses the "easiest" solutions, and in the sense of spontaneity disorder is "easier" than order. This principle has a counterpart in statistical mechanics, according to which in all the accessible states of an isolated system the number of the disordered ones are much more than that of the ordered ones (for instance, there are likely infinite ways to break a cup, but only few to build it). The crucial point is that every time such a tendency manifests itself (and assuming that we are right in the previous assumptions), a compensatory tendency to order should also be produced to preserve this net balance. This second tendency can be said to be less fundamental and not spontaneous, that is, forced by the first one.

The fact that irreversible phenomena are only the result of local processes also implies that the matter structures of our universe emerge as a determination process starting from quantum mechanics. This, however, does not imply at all that these matter structures and the macroscopic world in general are illusionary phenomena, as supporters of the many-worlds interpretation [see Section 9.3] would be inclined to think. The only consequence is that local reality is relational and perspective-like[a] while laws may very well be invariant. A certain philosophical tradition has brought us to a confusion between the concepts of subjectivity and relativity. However, they should be kept carefully distinct. When we say that a phenomenon has a subjective reality we mean that without the contribution of the mind it would not be as we understand or perceive it (this is the ground of both the subjectivist and the objectivist interpretations of quantum mechanics). It is clear that in a very wide sense, everything that we perceive or understand is subjective (a truth that was often forgotten in a classical-physical framework where it was commonly assumed that our theories depict reality as it is). In a more useful and stricter sense, the term *subjective* means plus or minus what is illusionary or has at least an illusionary component. On the contrary, *relative* only means that a certain system has the property that we assign to it only in the intercourse with something else. In the case of quantum systems, they can give rise to irreversible local processes when interacting with other quantum systems open to the environment. It is not necessary at all that human beings be present in order that this happens and indeed it has happened many times in the history of our universe long before humans had appeared at all.[b] It is amazing that the objectivist interpretation of quantum mechanics went very near to this conclusion when introduced the concept of relative state. The only problem there was the inability to distinguish between local irreversibility and global reversibility and to assume that only global descriptions mirror reality.

Problem 9.8 Drawing comparative truth tables of $\neg(p \wedge q)$ and $\neg p \vee \neg q$, show that they are logically equivalent (i.e., they are both true or false with the same value assignment).

---

[a](Wheeler, 1983).
[b](Joos/Zeh, 1985).

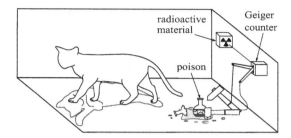

Figure 9.6 Pictorial representation of the Schrödinger's cat thought experiment. Adapted from http://en.wikipedia.org/wiki/File:Schrodingers_cat.svg.

Problem 9.9 Individuate the rows in which $r = 0$ and $r \rightarrowtail (p \wedge q) = 0$ and write down the counterpart of Eq. (9.50).

9.8 Schrödinger's Cat

A relevant point of the previous examination is that quantum effects are never totally washed out but only damped when larger systems like molecules or polymers are built. This is evident today at the chemical and molecular level, and indeed quantum chemistry became in the last years one of the most affirmed fields.[a] However, this should have some consequences also for macroscopic systems, even if the effects are here so attenuated that it is very difficult to detect them. In fact, almost a century ago such a possibility was already foreseen by Schrödinger.[b] This is the celebrated Schrödinger's cat paradox (which, as we shall discover, is not a paradox). Let us consider the situation depicted in Fig. 9.6, in which a cat is confined in a box together with a trace of radioactive material with a certain decay probability. The decayed atom would activate a Geiger counter which is connected through a relay to a hammer that may break an ampulla, thus releasing some poison, and so killing the cat. Now, according to the probabilistic character of the radioactive decay, which is a typical quantum process [see p. 186], after some time the wave function describing the system should represent a superposition of *alive cat* and *dead cat*. But this seems impossible,

---

[a] (Atkins/De Paula, 2006), (Atkins/Friedman, 2005).
[b] (Schrödinger, 1935a).

and as a matter of fact nobody has ever observed such a situation in the ordinary experience at macroscopic level. On the other hand, according to the projection postulate [see Section 9.3], we may assume that a measurement, in our case the opening of the box to observe the cat, will univocally determine the state of the cat (*either* alive *or* dead), thus eliminating the superposition. If so, we ascribe an enormous and perhaps unjustified power to the observation, that is the power to realize *ex novo* a very special physical situation, the so-called *wave function collapse* (or wave packet reduction).

We may describe the whole system composite of the radioactive material R, the hammer H, and the cat C by the superposition state:

$$|\Psi\rangle_{RHC} = c_D |d\rangle |a\rangle |D\rangle + c_A |d'\rangle |a'\rangle |A\rangle, \qquad (9.51)$$

where c_D and c_A are probability amplitudes, $|d\rangle$ and $|d'\rangle$ represent an atom decaying and not decaying, respectively, $|a\rangle$ and $|a'\rangle$ the hammer active and not active, respectively, and finally the states $|D\rangle$ and $|A\rangle$ the cat dead and alive, respectively. In other words, we have a superposition here between the component $|d\rangle |a\rangle |D\rangle$, in which there is an atom decay, the hammer works, and the poison kills the cat, and the component $|d'\rangle |a'\rangle |A\rangle$, in which there is no atom decay, the hammer is at rest, and the cat survives. Moreover, the state $|\Psi\rangle_{RHC}$ is an entangled state in which the radioactive material, the hammer, and the cat have a special entanglement. Although the above expression is correct, it is very unlikely to have a cat in animated suspension between life and death (anyway nobody could observe it by definition). However, this does not represent an impossibility as such and is rather a consequence of the huge complexity of the involved systems. Even in the case in which this could happen, it is likely that the result would be a very unstable, short-living state, erased immediately by decoherence, therefore increasing the difficulty to detect it. In other words, decoherence helps us solve such an apparent paradox. Nevertheless, quantum effects at least at the mesoscopic level (that of big atoms and molecules) have been observed, and this is a promising field of investigation.[a]

---

[a] (Monroe *et al.*, 1996), (Brune *et al.*, 1996), (Leibfried *et al.*, 2005). We mention that the 2012 Noble prize in physics has been awarded to the principal investigators of the first two teams precisely for this kind of studies.

We have already considered the complementarity principle in Chapter 5. It is worth mentioning that in the original formulation[a] the complementary *wave-like* and *particle-like* behaviors were still understood as dependent on the macroscopic properties of the apparatus that we use to perform complementary experiments. This is commonly known as the Copenhagen interpretation, initially devised by Bohr and Heisenberg in the years 1924–27. We have already considered that the macroscopic bodies should be rather conceived as emergent from the microscopic quantum world. It is right that a macroscopic apparatus is necessary for assigning a specific property to a quantum system (that is, a certain value that we assign to an observable). However, as we have stressed, this is also what spontaneously happens in nature although in this case there is nobody reading, interpreting, and eventually storing the result (and so, properly speaking there is an outcome of the interaction but no property assignation).

Bohr's approach is quite understandable in a time in which quantum mechanics was still in its infancy. Indeed, on the one hand it was not possible at that time to perform experiments using apparatuses that are of the same size as the measured quantum system, as it is currently the case for the many novel experiments on mesoscopic systems. Moreover, there was no possibility to explore intermediate situations between corpuscular and undulatory behaviors as it is currently the case [see Section 5.5]. On the other hand, quantum theory was still conceptually dependent on classical physics and therefore quantum phenomena and quantities acquired a physical meaning only in the light of the classical theory (what remains today still true but only in the sense of the correspondence principle [see Principle 7.1]). Therefore, it is quite normal that this kind of assumptions or simplifications were made in the early development of quantum theory. Even today when we learn something new, as it happened when we began to learn algebra in our childhood, we need first to translate the new notions into the old ones. Only after a certain training we are able to deal with the new concepts in such a way that they become the new reference system in whose context we can also reinterpret the old notions.

---

[a](Bohr, 1928).

9.9 Summary

In this chapter we have

- Explained that there is no unitary transformation that can account for obtaining a single component out of a superposition state.
- Remarked the distinction between a unitary, reversible, linear and smooth dynamics ruled by the Schrödinger equation and a (local) random, irreversible, non-linear and abrupt transition when dealing with measurement.
- Reviewed the two main historically established subjectivist and objectivist positions on measurement and found reasons of dissatisfaction for both.
- Introduced the so-called many-worlds interpretation of quantum mechanics and the associated concept of relative states.
- Dealt with the problem of basis ambiguity and introduced the concept of premeasurement.
- Discussed decoherence as a solution to the measurement problem.
- Introduced the main notions of entropy and information.
- Considered quantum systems as information processors and shown that any irreversible transformation can be embedded in a larger reversible transformation.
- Distinguished between local irreversible processes, in which order and disorder are displaced, and global reversible dynamics.
- Introduced the Schrödinger's cat paradox and its resolution.
- Commented on the Copenhagen interpretation of quantum mechanics.

Chapter 10

Non-Locality and Non-Separability

In the previous chapter we have introduced the distinction between locality and non-locality. In this chapter we shall explore non-locality, one of the most amazing aspects of quantum mechanics. We shall discover that it consists in non-local correlations without transmission of signals. Along the way we will formulate and prove three important theorems as well as introduce the phenomenon of entanglement swapping that reveal unequivocally quantum non-locality.

10.1 EPR Paper

In a historical paper published in 1935, Einstein, Podolsky, and Rosen (EPR) asked whether quantum mechanics could be considered a complete description of microscopic systems.[a] The reason for this investigation is that quantum mechanics showed an irreducible probabilistic character [Sections 5.3–5.5 and 6.8–6.9]. Every physical description of reality up to that time used probabilities only as a statistical treatment for dealing with problems of high

---

[a](Einstein *et al.*, 1935).

Quantum Mechanics for Thinkers
Gennaro Auletta and Shang-Yung Wang
Copyright © 2014 Pan Stanford Publishing Pte. Ltd.
ISBN 978-981-4411-71-4 (Hardcover), 978-981-4411-72-1 (eBook)
www.panstanford.com

complexity, as those occurring in statistical mechanics, where the typical number of particles in a gas is of the order of 6.022×10^{23}, and hence in practice their motion is too complex to be treated with the traditional deterministic methods [see Section 1.2]. Therefore, it was well possible that quantum mechanics displayed intrinsic limitations in dealing with elusive entities which could be eventually overcome by another, more refined theory or at least approach. Indeed, the EPR paper represents a sort of general check of the theory and in this sense its study still represents an unavoidable step for truly understanding quantum mechanics.

Having this purpose, EPR formulated some general principles that any physical theory should satisfy. When judging a physical theory, one should inquire about both the correctness and completeness of that theory.[a] The *correctness* of a theory consists in the degree of agreement between the conclusions of the theory and experimental observation, i.e., the objective reality, while the notion of *completeness* can be best encapsulated in the following definition:

Definition 10.1 (Completeness) A theory is complete if every element of objective reality has a counterpart in it.

The aim of the EPR paper is to show the *incompleteness* of quantum mechanics in the sense of its inability to give a satisfactory explanation of entities which are considered fundamental. In a word, it is a "disproof" and not a positive proof. Indeed, theories can be disproved by experience and (even thought) experiments. This type of epistemology is the so-called *falsificationism*.[b] Given these premises, the core of the argument is the *separability principle*, which can be expressed as follows [see also Sections 1.1 and 7.9].

Principle 10.1 (Separability) *Two dynamically independent systems cannot influence each other.*

The separability principle consists in the assumption that any form of interdependence between physical systems is of dynamical and causal type, ultimately relying on local transmission of effects. Therefore, it is important to carefully distinguish the problem

---

[a](Auletta, 2000, Chapter 31), (Auletta *et al.*, 2009, Section 16.1).
[b](Peirce, 1866), (Peirce, 1877), (Peirce, 1878), (Popper, 1934).

of locality (i.e., the requirement of bounds in the transmission of signals and physical effects) from that of separability, which concerns only the impossibility of correlations between systems in the case in which there are *no dynamical and causal connections*. Part of the EPR argument is that in the absence of physical interactions, the systems are also separated.

Making use of the separability principle, EPR stated a *sufficient condition* for the reality of physical observables, which can be formulated as follows.

Principle 10.2 (Criterion of Physical Reality) *If, without in any way disturbing a system, we can predict with certainty the value of a physical quantity, then, independently of our measurement procedure, there exists an element of the physical reality corresponding to this physical quantity.*

The phrase "without in any way disturbing a system" means that the systems are considered dynamically independent. It may be noted that this principle is of absolute generality (since it may be applied to any scientific domain) and should therefore rather be considered a philosophical principle. The crucial point is that EPR have shown that the latter in conjunction with physical assumptions can lead to consequences that can be tested. Indeed, assuming separability and the sufficient condition of reality, EPR went on to argue that quantum mechanics is not complete. In logical terms, according to the EPR argument, the following statement holds for quantum mechanics:[a]

$$(\text{Suff. cond. of reality} \ \wedge \ \text{Separability}) \Rightarrow \neg\text{Completeness}, \quad (10.1)$$

where $\wedge$ is the logical conjunction, $\neg$ is the logical negation, and the arrow $\Rightarrow$ is the symbol for logical implication [see Section 9.7]. The implication $p \Rightarrow q$, where p and q are arbitrary statements, may be defined by the statement "p is false OR q is true." In other words, the implication is false only in the case that p is true AND q is false [see Table 10.1].

Before entering into details, it is very important to understand the abstract logic form of the argument. According to EPR, the

---

[a]In logic, a statement is an assertion that can be determined to be true or false.

Table 10.1 Truth table for logical implication

Input		Output
p	q	$p \Rightarrow q$
1	1	1
1	0	0
0	1	1
0	0	1

incompleteness of quantum mechanics would be a consequence of both separability and the sufficient condition of reality. Since the EPR argument has the logical structure of an implication, in order to invalidate the argument it suffices to show that *at least one* of the two assumptions is false. In fact, if one of the two assumptions is false then their joint assertion (i.e., Suff. cond. of reality ∧ Separability) is false as well [see Table 9.2], i.e., the antecedent of the implication is false. However, if the antecedent of an implication is false, its consequent (i.e., ¬Completeness) may be indifferently true or false, being, under this condition, the implication always true (see the last two lines of Table 10.1). It this case, the argument would prove neither the incompleteness, nor the completeness of quantum mechanics, and be finally inconclusive. As we shall see below, Schrödinger argued against the principle of separability, while Bohr tried to reject the sufficient condition of reality.

Now, the crucial issue for EPR was the following: how to find evidence showing that the implication (10.1) is true? To this purpose, the argument of EPR is structured as follows. From (i) Definition 10.1, (ii) Principle 10.2, and (iii) the fact that, according to quantum mechanics, two non-commuting observables cannot simultaneously have definite values [see Section 6.8], it follows that the following two statements are incompatible:

(i) The statement r that the quantum mechanical description of reality given by the wave function is not complete.

(ii) The statement s that when the operators describing two physical quantities do not commute, the two quantities cannot have simultaneous reality.

In formal terms, we have the statement

$$r \rtimes s, \tag{10.2}$$

where the symbol $\rtimes$ means a XOR [see Section 9.7]. The meaning of the statement (10.2) is the following: if it is possible to show (though some kind of experiment) that two non-commuting observables have in fact simultaneous reality, then we can logically conclude that quantum mechanics cannot be a complete description of reality. In other words, assuming the Sufficient Condition of Reality and by hypothetically adding the uncertainly principle (which EPR in fact did not acknowledge), we can derive the statement (10.2), that is, $r \rtimes s$. Now, if we perform an experiment and, by further assuming the Separability Principle (Principle 10.1), find in fact that $\neg s$ (i.e., we succeed in showing that the uncertainty principle is not valid), then we have in this way deduced r, that is, the incompleteness of quantum mechanics, which is in turn the consequent of the implication (10.1). Therefore, having derived this result from the two premises (Suff. cond. of reality and Separability) of the latter, we had also proved the truth of this implication.

Given the structure of the argument, EPR must therefore provide an example of wave function for which two non-commuting observables can have simultaneous reality, and, as mentioned, it is here that separability comes into play. Let us, for this purpose, consider a one-dimensional system S made of two subsystems S_1 and S_2, say two particles 1 and 2, which interact during the time interval between t_1 and t_2, after which they no longer interact. Let us write their respective momentum observables in the position representation as [see Eq. (6.116)]

$$\hat{p}_x^{(1)} = -i\hbar \frac{\partial}{\partial x_1} \quad \text{and} \quad \hat{p}_x^{(2)} = -i\hbar \frac{\partial}{\partial x_2}, \tag{10.3}$$

where x_1 and x_2 are the position variables used to describe particles 1 and 2, respectively. Recall that the eigenfunction of $\hat{p}_x^{(1)}$ with eigenvalue p_1 is given by [see Eq. (6.123)]

$$\varphi_{p_1}(x_1) = \frac{1}{\sqrt{2\pi\hbar}} e^{\frac{i}{\hbar} p_1 x_1}, \tag{10.4}$$

while the eigenfunction of $\hat{p}_x^{(2)}$ with eigenvalue p_2 is

$$\psi_{p_2}(x_2) = \frac{1}{\sqrt{2\pi\hbar}} e^{\frac{i}{\hbar} p_2 (x_2 - x_0)}, \tag{10.5}$$

where x_0 is some fixed position (constant). It is noted that the presence of constant x_0 in the exponent of the wave function above corresponds to a global phase factor $e^{-\frac{i}{\hbar}p_2 x_0}$. Let us suppose that the composite system is described by the wave function

$$\Psi(x_1, x_2) = \frac{1}{2\pi\hbar} \int_{-\infty}^{\infty} e^{\frac{i}{\hbar}p(x_1 - x_2 + x_0)} dp, \qquad (10.6)$$

which can be written in terms of the momentum eigenfunctions of particles 1 and 2 as

$$\Psi(x_1, x_2) = \int_{-\infty}^{\infty} \psi_{-p}(x_2)\varphi_p(x_1) dp. \qquad (10.7)$$

It is clear that $\Psi(x_1, x_2)$ is an entangled state and that the momenta of particles 1 and 2 are entangled [see Section 7.9]. Indeed, we recall that the concept of entanglement was introduced by Schrödinger *after* publication of the EPR paper and precisely as a reply to their proposal (in other words, EPR were unaware of how revolutionary was the formalism that they introduced). We now proceed as follows.

(a) We locally measure the momentum of particle 1 and the measurement result is some eigenvalue of $\hat{p}_x^{(1)}$, say p'.

(b) After the measurement the state $\Psi(x_1, x_2)$ given by Eq. (10.7) reduces to the state $\psi_{-p'}(x_2)\varphi_{p'}(x_1)$ [see Sections 9.1 and 9.2].

(c) It is evident that particle 2 must be in the state $\psi_{-p'}(x_2)$ and therefore its momentum must be $-p'$. This result can be predicted with absolute certainty.

(d) However, we were able to formulate such a prediction without disturbing particle 2 (by the assumption of separability).

(e) Thus, as a consequence of (c) and (d), and of the sufficient condition of reality, $\hat{p}_x^{(2)}$ is an element of reality.

Note that steps (a)–(c) are purely quantum mechanical. Only steps (d) and (e) are connected to the specific EPR argument. However, if we had chosen to consider the respective position observables $\hat{x}^{(1)}$ and $\hat{x}^{(2)}$ of particles 1 and 2, then we would have written the state $\Psi(x_1, x_2)$ of the composite system in term of the position eigenfunctions of particle 1 and 2 [see Section 9.4]. Let us denote the latter by $\varphi_x(x_1)$ and $\psi_x(x_2)$, respectively. Then we have [see Eq. (6.54)]

$$\varphi_x(x_1) = \delta(x - x_1), \qquad \psi_{x+x_0}(x_2) = \delta(x - x_2 + x_0), \qquad (10.8)$$

where x_0 is some fixed position (constant). Using Eq. (6.50), we can rewrite $\Psi(x_1, x_2)$ in terms of $\varphi_x(x_1)$ and $\psi_x(x_2)$ as

$$\Psi(x_1, x_2) = \int_{-\infty}^{\infty} \psi_{x+x_0}(x_2)\varphi_x(x_1)dx. \tag{10.9}$$

Again, it is clear that $\Psi(x_1, x_2)$ is an entangled state and that the positions of particles 1 and 2 are now entangled. Let us repeat the previous procedure for the position measurement.

(a') We locally measure the position of particle 1 and the measurement result is some eigenvalue of $\hat{x}^{(1)}$, say x'.

(b') After the measurement the state $\Psi(x_1, x_2)$ given by Eq. (10.9) reduces to the state $\psi_{x'+x_0}(x_2)\varphi_{x'}(x_1)$.

(c') It is evident that particle 2 must be in the state $\psi_{x'+x_0}(x_2)$ and therefore its position must be $x' + x_0$. This result can be predicted with absolute certainty.

(d') However, we were able to formulate such a prediction without disturbing particle 2 (by the assumption of separability).

(e') Thus, as a consequence of (c') and (d'), and of the sufficient condition of reality, $\hat{x}^{(2)}$ is an element of reality.

We note that conclusions (e) and (e') seem incompatible with each other on the basis of the fact that the position and momentum observables of particle 2 do not commute [see Section 6.8]. Going back to the statements r and s [see Eq. (10.2)], EPR have in this way shown that assuming r (the quantum mechanical description of reality is not complete) is false, then s (two physical quantities described by non-commuting operators cannot have simultaneous reality) is false as well since both $\hat{p}_x^{(2)}$ and $\hat{x}^{(2)}$ have therefore simultaneous reality. Then, the previous assumption must be rejected, and r must be true. Therefore, according to the EPR argument, quantum mechanics cannot be considered a complete theory and the description of reality as given by the wave function is not complete.

Problem 10.1 Have you understood the EPR argument? If not, your task is to read Section 9.7 again to understand the meaning of logical negation, logical conjunction, logical disjunction, and logical exclusive disjunction. Then read this section again to understand the notion of completeness, the sufficient condition of reality, and the meaning of logical implication. Once you have finished, read it again and verify if you understand the EPR argument fully. If not, start again the procedure until you have absolute clarity about the EPR argument.

10.2 Bohr's and Schrödinger's Criticism of EPR

As we have said, given the abstract logical structure (10.1) of the EPR argument, if we desire to reject the conclusion that quantum mechanics is incomplete, it is necessary to show the failure of the sufficient condition of physical reality in a quantum framework or the inconsistency of separability with quantum mechanics. In fact, they are the only non-quantum mechanical assumptions in steps (a)–(e) and (a')–(e'). EPR themselves have anticipated the former objection. In the end of the paper, they replied as follows.

> "One could object to this conclusion on the grounds that our criterion of reality is not sufficiently restrictive. Indeed, one would not arrive at our conclusion if one insisted that two or more physical quantities can be regarded as simultaneous element of reality *only when they can be simultaneously measured or predicted*. On this point of view, since either one or the other, but not both simultaneously, of the quantities P $[\hat{p}_x]$ and Q $[\hat{x}]$ can be predicted, they are not simultaneously real. This makes the reality of P $[\hat{p}_x]$ and Q $[\hat{x}]$ depend upon the process of measurement carried out on the first system, which does not disturb the second system in any way. No reasonable definition of reality could be expected to permit this."

In the same year 1935, Bohr rejected the sufficient condition of reality precisely along those lines. Bohr criticized the EPR argument by pointing out that, even if the EPR thought experiment excludes any direct physical interaction of the system with the measuring apparatus, the measurement process has an essential influence on the conditions on which the very definition of the physical observables in question rests.[a] And these conditions

---

[a](Bohr, 1935a), (Bohr, 1935b), see also (Jammer, 1974, pp. 195–197).

must be considered an inherent element of any phenomenon to which the term "physical reality" can be unambiguously applied. Bohr acknowledged that it is possible to determine experimental arrangements such that the measurement of the position or of the momentum of one particle automatically determines the position or the momentum of the other. However, such experimental arrangements for measuring momentum and position are incompatible with each other. Therefore, the central point of Bohr's criticism is that it is not possible to assign a reality to observables of quantum systems (and therefore certainly not properties to the involved systems) *independently* of the experimental context in which not only we actually interact with them but in which we *could* interact with them [see Sections 9.4 and 9.8].

In such a context, we wish to point out that the experimental procedures through which observables are determined are physical operations. Moreover, an observable precisely describes a possible operation that could be performed on a system either artificially in our laboratories or spontaneously in nature when certain kinds of interactions occur. In this way, an observable describes what would be the behavior of a system in certain conditions. Most of the properties and parameters that we assign even to classical and macroscopic systems show such a *dispositional* kind of reality.[a] Nevertheless, such dispositions are for us very real, otherwise the reality of most physical systems would dangerously evaporate. Then, nothing prevents us to consider observables as elements of reality independently of the fact that they may not commute. In this way, an important instance of the EPR argument is that to determine the way in which a system will behave in certain conditions *is* an ascription of reality. This point of view can be framed in a wider ontological context, in which state, observables, and properties show an increasing degree of determination. As a matter of fact, we are always allowed to assume the reality of the state of a quantum system, with all provisos about the word "reality" we shall deal with, provided that it is possible, at least in principle, to prepare the system in that state [see Sections 4.8, 5.6, and 6.9]. Moreover, it is also allowed to speak of observables as elements of reality,

---

[a](Hempel, 1953, Section 6).

provided that we are at least able to show a (possible) context of measurement, that is, a premeasurement (a suitable coupling) [see Section 9.5]. However, this also implies that to a certain extent we are accepting Bohr's instance too. The crucial point is that the possible context in which a certain observable is considered to be real is not an epistemic one as Bohr assumed, but both an ontological and epistemic context, and without an ontological context (that is, without some objective conditions, for instance of experimental kind) we could certainly make no reality ascription. As we have stressed, to be relative to a certain context does not imply to be dependent on a subjective consideration of the problem [see Section 9.7]. Therefore, Bohr's objection, although stimulating for refining the concept of reality, is not sufficient to demolish the EPR argument.

Another kind of problem is constituted by the fact that a certain observable being an element of reality does not imply that one of its possible properties (that is, one of its eigenvalues) is also a reality. In order to do this inference, we need an event, for instance a detection. In other words, we need to consider an *actual* detection event and not only a *possible objective* context, that is, a premeasurent. This is the essence of the other objection to EPR, a contribution of Schrödinger, which shows the true weakness of the EPR argument by removing an ambiguity that is present in their argumentation (it is also not completely clear whether Bohr's argument rejects reality ascription to observables or properties or, even more plausibly, to both). In a series of articles, Schrödinger answered to EPR by introducing the important concept of entanglement in quantum mechanics [see Section 7.9].[a] It is entanglement that (in the absence of events) prevents, in general, to attribute properties to a system or its subsystems. But, provided that there is an event, it is still entanglement that allows attribution of properties in a way that is classically unknown but also forbids other property assignments. In other words, quantum mechanical systems can show non-local correlations that make them non-separable even in the absence of any dynamical interaction (hence in disagreement with the separability principle). These correlations, once that we

---

[a](Schrödinger, 1935a), (Schrödinger, 1935b).

have performed a measurement on one particle, in turn prevent the possibility to assign properties to the entangled particle without taking into account the obtained result; but once those results are taken into account also allow new kinds of prediction. In finding entanglement as a resolution to a possible conflict between a thought experiment and quantum mechanical laws, Schrödinger was performing an abduction [see Section 4.1].[a] However, it is also important to note that Schrödinger considered entanglement to be far away from our common perception of reality. We recall indeed that his answer to EPR is developed in the same series of papers where he proposed for the first time the *Gedankenexperiment* (i.e., thought experiment) of Schrödinger's cat as a possible bizarre consequence of entanglement [see Section 9.8]. It is a sign of the highest quality in research and speculation when a scholar is forced to a conclusion that seems bizarre to him/her and finally accepts it at least as a hypothesis.

10.3 EPR–Bohm Experiment

The reformulation of the original EPR thought experiment proposed by Bohm in the 1950s deals with discrete observables like spins, instead of continuous ones such as position and momentum.[b] This step was originally understood as a further simplification of the EPR argument. It is an important step not only because it displays with high clarity that the non-local features of quantum theory are a consequence of entanglement, but also because it is experimentally realizable.

Let us consider two particles of spin $\frac{1}{2}$ that are in a state in which the total spin is zero, that is, they are in a singlet state [see Section 8.7]. They can be produced, for instance, by radioactive decay of a single particle of spin 0. After a time t_0 the two particles begin to separate and at time t_1 they no longer interact with each other [see Fig. 10.1]. On the hypothesis that they are not disturbed, the law of angular momentum conservation guarantees that they

---

[a] (Peirce CP, 2.96).
[b] (Bohm, 1951, pp. 614–623).

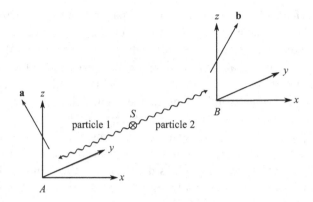

Figure 10.1 Schematic overview of the EPR–Bohm experiment. Two spin $\frac{1}{2}$ particles are produced in a spin singlet state from a common source S (e.g., by decay of a spin 0 particle). After the time the two particles no longer interact, the spin of particle 1 in a direction **a** and the spin of particle 2 in another direction **b** are measured with two apparatuses A and B, respectively. The Euclidean vectors **a** and **b** are taken to be unit vectors as they represent here spatial directions.

remain in a singlet state. Considering the projection of the spin along the z direction, the singlet state $|\Psi_0\rangle$ can be written in the form [see Eq. (8.140)]

$$|\Psi_0\rangle = \frac{1}{\sqrt{2}}(|\uparrow_z\rangle_1 \otimes |\downarrow_z\rangle_2 - |\downarrow_z\rangle_1 \otimes |\uparrow_z\rangle_2), \qquad (10.10)$$

where the subscripts 1 and 2 refer to the particles. As we have remarked, the singlet state $|\Psi_0\rangle$ is an entangled state. This implies that, if a measurement of the spin component in the z direction of particle 1 leads to the result $+\frac{1}{2}$ (in units of $\hbar$), the spin component of particle 2 in the same direction must give the value $-\frac{1}{2}$, and vice versa. It is straightforward to show that

$$(\hat{\sigma}_{1z} \otimes \hat{\sigma}_{2z})|\Psi_0\rangle = \frac{1}{\sqrt{2}}(\hat{\sigma}_{1z}|\uparrow_z\rangle_1 \otimes \hat{\sigma}_{2z}|\downarrow_z\rangle_2 - \hat{\sigma}_{1z}|\downarrow_z\rangle_1 \otimes \hat{\sigma}_{2z}|\uparrow_z\rangle_2)$$

$$= -\frac{1}{\sqrt{2}}(|\uparrow_z\rangle_1 \otimes |\downarrow_z\rangle_2 - |\downarrow_z\rangle_1 \otimes |\uparrow_z\rangle_2)$$

$$= -|\Psi_0\rangle, \qquad (10.11)$$

where $\hat{\sigma}_{1z}$ and $\hat{\sigma}_{2z}$ are the z components of the spin observable of particles 1 and 2 [see Eqs. (8.111)], respectively, and use has been made of the actions of $\hat{\sigma}_z$ on $|\uparrow_z\rangle$ and $|\downarrow_z\rangle$ that are tabulated in Table 8.1. The observable $\hat{\sigma}_{1z} \otimes \hat{\sigma}_{2z}$ represents the measurement of the spin component of both particles in the z direction. For the sake of presentational simplicity, here and in the next section we shall measure the spin of the two particles in units of $\frac{\hbar}{2}$. The above expression means that $|\Psi_0\rangle$ is an eigenstate of $\hat{\sigma}_{1z} \otimes \hat{\sigma}_{2z}$ with the eigenvalue given by -1. Indeed, it can be shown that this result holds true for the spin component of both particles in an arbitrary direction, namely, we have

$$(\hat{\boldsymbol{\sigma}}_1 \cdot \mathbf{a}) \otimes (\hat{\boldsymbol{\sigma}}_2 \cdot \mathbf{a})|\Psi_0\rangle = -|\Psi_0\rangle, \tag{10.12}$$

where $\mathbf{a}$ is an arbitrary direction (i.e., an arbitrary unit vector). In other words, the state $|\Psi_0\rangle$ is rotationally invariant. This is because it is an eigenstate of the magnitude squared of the total spin operator $(\hat{\boldsymbol{\sigma}}_1 + \hat{\boldsymbol{\sigma}}_2)^2$ with eigenvalue 0, as it is clear in Section 8.7. In this way, we have also confirmed that entanglement is a property of the state that is independent of the basis used [see Section 7.9]. In order to see this rotational invariance, it suffices to show that $|\Psi_0\rangle$ turns out to be also an eigenstate of $\hat{\sigma}_{1x} \otimes \hat{\sigma}_{2x}$ and $\hat{\sigma}_{1y} \otimes \hat{\sigma}_{2y}$. This can be achieved either by explicit calculations or by expressing $|\Psi_0\rangle$ in terms of the respective eigenstates of $\hat{\sigma}_x$ and $\hat{\sigma}_y$. We shall demonstrate in turn both approaches for the x direction and leave those for the y direction as an exercise for the reader [see Problems 10.2 and 10.3].

First, by applying the observable $\hat{\sigma}_{1x} \otimes \hat{\sigma}_{2x}$ to the state $|\Psi_0$ and using the actions of $\hat{\sigma}_x$ on $|\uparrow_z\rangle$ and $|\downarrow_z\rangle$ tabulated in Table 8.1, we obtain

$$(\hat{\sigma}_{1x} \otimes \hat{\sigma}_{2x})|\Psi_0\rangle = \frac{1}{\sqrt{2}}(\hat{\sigma}_{1x}|\uparrow_z\rangle_1 \otimes \hat{\sigma}_{2x}|\downarrow_z\rangle_2 - \hat{\sigma}_{1x}|\downarrow_z\rangle_1 \otimes \hat{\sigma}_{2x}|\uparrow_z\rangle_2)$$

$$= \frac{1}{\sqrt{2}}(|\downarrow_z\rangle_1 \otimes |\uparrow_z\rangle_2 - |\uparrow_z\rangle_1 \otimes |\downarrow_z\rangle_2)$$

$$= -|\Psi_0\rangle, \tag{10.13}$$

which means precisely $|\Psi_0\rangle$ is an eigenstate of the observable $\sigma_{1x} \otimes \sigma_{2x}$ with eigenvalue -1. Next, making use of the inverse of Eq. (8.114a),

$$|\uparrow_z\rangle = \frac{1}{\sqrt{2}}(|\uparrow_x\rangle + |\downarrow_x\rangle), \quad |\downarrow_z\rangle = \frac{1}{\sqrt{2}}(|\uparrow_x\rangle - |\downarrow_x\rangle), \tag{10.14}$$

we can rewrite $|\Psi_0\rangle$ in terms of $|\uparrow_x\rangle$ and $|\downarrow_x\rangle$ as

$$|\Psi_0\rangle = \frac{1}{\sqrt{2}}(|\uparrow_z\rangle_1 \otimes |\downarrow_z\rangle_2 - |\downarrow_z\rangle_1 \otimes |\uparrow_z\rangle_2)$$

$$= \frac{1}{2\sqrt{2}}[(|\uparrow_x\rangle_1 + |\downarrow_x\rangle_1) \otimes (|\uparrow_x\rangle_2 - |\downarrow_x\rangle_2)$$

$$- (|\uparrow_x\rangle_1 - |\downarrow_x\rangle_1) \otimes (|\uparrow_x\rangle_2 + |\downarrow_x\rangle_2)]$$

$$= -\frac{1}{\sqrt{2}}(|\uparrow_x\rangle_1 \otimes |\downarrow_x\rangle_2 - |\downarrow_x\rangle_1 \otimes |\uparrow_x\rangle_2), \qquad (10.15)$$

which, apart from the global phase factor -1, is a singlet state but with the projection of the spin along the x direction. Therefore, if the spin component in the x direction of particle 1 is measured first, followed by a measurement of the spin component of particle 2 in the same direction the next, then the two results must have the opposite values.

The spin singlet state $|\Psi_0\rangle$ is perhaps the simplest discrete version of the EPR state $\Psi(x_1, x_2)$, in that the form of the former given by Eqs. (10.10) and (10.15) correspond respectively the form of the latter given by Eqs. (10.7) and (10.9), and that the role of position and momentum is played by different spin components. Nonetheless, the advantage of the spin singlet state over the EPR state is that as we have mentioned the spin singlet state is an ideal candidate for the possible realization of experimental tests.

Problem 10.2 Using the actions of $\hat{\sigma}_y$ on $|\uparrow_z\rangle$ and $|\downarrow_z\rangle$ tabulated in Table 8.1, show that the singlet state $|\Psi_0\rangle$ given by Eq. (10.10) is an eigenstate of the observable $\sigma_{1y} \otimes \sigma_{2y}$ with eigenvalue -1.

Problem 10.3 Express the singlet state $|\Psi_0\rangle$ given by Eq. (10.10) in terms of the spin eigenstates $|\uparrow_y\rangle$ and $|\downarrow_y\rangle$ in the y direction.

10.4 Bell Theorem

The EPR paper raised a very important problem. If quantum mechanics should turn out to be incomplete, this would imply the existence of unknown variables able to determine the results that quantum theory describes in probabilistic terms. In other words, these variables would provide a full deterministic account in a classical sense of what quantum mechanics is meant to phenomenologically describe [see Chapter 1]. Such variables have been referred to as the *hidden variables* (or HVs for short), since it was assumed that the technological standard of that time was not sufficient for describing them. So, it is quite understandable that the main results after Bohm has proposed his model were of formal kind. In the 1960s Bell was able to prove the following striking result:[a] No deterministic local hidden variable theory can make predictions compatible with quantum mechanics.[b] By "local hidden variable theory" we mean a theory that does not make use of superluminal forms of communication. This achievement was of particular importance because it moved the discussion from a qualitative level to a strict quantitative (and therefore experimentally testable) ground. The incompatibility between a local hidden variable theory and quantum mechanics (that we have still to prove) raises the further difficult question as to what type of locality is violated by quantum mechanics. This discussion must be postponed for the moment, and here we will use the phrase "quantum non-locality" as a generic term that could cover two very different possibilities, i.e., violation of the separability principle [see Principle 10.1] and violation of Einstein's locality dictated by special relativity.

The *Gedankenexperiment* proposed by Bell was a further refinement of the EPR–Bohm experiment.[c] Bell assumed the existence of a hidden variable λ_{HV} such that it provides a full deterministic description of the measurement results obtained in the EPR–Bohm experiment. In specific, let us now imagine to perform a joint measurement in which the spin of particle 1 in a chosen direction **a** (i.e., the observable $\hat{\sigma}_1 \cdot \mathbf{a}$) is measured with apparatus A and the spin of particle 2 in another chosen direction **b** (i.e., the observable $\hat{\sigma}_2 \cdot \mathbf{b}$) is measured with apparatus B [see Fig. 10.1 and Section 8.6]. Then, given λ_{HV}, the results of this joint measurement are uniquely described by the quantity $M_{\mathbf{ab}}(\lambda_{HV})$, which is a deterministic function of the hidden variable λ_{HV}. It is noted that the Euclidean

---

[a] (Bell, 1964).
[b] (Gröblacher *et al.*, 2007).
[c] (Auletta, 2000, Section 35.1), (Auletta *et al.*, 2009, Section 16.4).

vectors **a** and **b** are taken to be unit vectors as they represent here spatial directions. Since the joint measurement is carried out after the time the two particles no longer interact, then the separability principle denies that there can be any form of interdependence between the two particles. This assumption can be mathematically formulated in a more rigorous way as the separability condition

$$M_{ab}(\lambda_{HV}) = A_a(\lambda_{HV}) B_b(\lambda_{HV}), \qquad (10.16)$$

which can be interpreted as a factorization rule. The function $A_a(\lambda_{HV})$ describes the result of measuring with apparatus A the spin of particle 1 in the direction **a**, regardless of the measurement result of particle 2, and the function $B_b(\lambda_{HV})$ the result of measuring with apparatus B the spin of particle 2 in the direction **b**, regardless of the measurement result of particle 1. Moreover, like $M_{ab}(\lambda_{HV})$ the functions $A_a(\lambda_{HV})$ and $B_b(\lambda_{HV})$ also are deterministic functions of the hidden variable λ_{HV}. Indeed, the separability condition (10.16) expresses the fact that the measurement results for the two "separated" particles are mutually independent.

Since $A_a(\lambda_{HV})$ and $B_b(\lambda_{HV})$ are functions describing the possible measurement results of spin components in certain directions, which in units of $\frac{\hbar}{2}$ can be either $+1$ (representing spin up) or -1 (representing spin down), then we have

$$A_a(\lambda) = \pm 1, \quad B_b(\lambda) = \pm 1, \qquad (10.17)$$

where for the sake of notational simplicity, here and henceforth, we have dropped the subscript HV from the hidden variable λ. While the nature of the hidden variable λ is unknown, it is conceivable and reasonable to assume that (i) the value of λ is real, (ii) all the possible values of λ form a (continuous) set, which we shall denoted by Λ, and (iii) a probability distribution $\wp(\lambda)$ can be assigned to λ such that the following normalization condition holds

$$\int_\Lambda \wp(\lambda)d\lambda = 1. \qquad (10.18)$$

Since we do not know the exact values of the hidden variable λ, in order to get rid of the λ dependence we integrate the measurement results over all the possible values of $\lambda \in \Lambda$, weighted by its probability distribution $\wp(\lambda)$. The resultant quantity can then be interpreted as the "expectation value" of the measurement in a local hidden variable theory. With the separability condition (10.16) and the above assumptions for the hidden variable λ, the expectation value of the product of the spin component $\hat{\sigma}_1 \cdot \mathbf{a}$ of particle 1 and the spin component $\hat{\sigma}_2 \cdot \mathbf{b}$ of particle 2 is given by

$$\langle(\hat{\sigma}_1 \cdot \mathbf{a})(\hat{\sigma}_2 \cdot \mathbf{b})\rangle = \int_\Lambda \wp(\lambda) A_\mathbf{a}(\lambda) B_\mathbf{b}(\lambda) d\lambda. \qquad (10.19)$$

Evidently, the above expectation value is the hidden-variable counterpart of the quantum expectation value

$$\langle(\hat{\sigma}_1 \cdot \mathbf{a})(\hat{\sigma}_2 \cdot \mathbf{b})\rangle_{\Psi_0} = \langle\Psi_0|(\hat{\sigma}_1 \cdot \mathbf{a})(\hat{\sigma}_2 \cdot \mathbf{b})|\Psi_0\rangle, \qquad (10.20)$$

where $|\Psi_0\rangle$ is the singlet state given by Eq. (10.10). From Eqs. (10.17) and (10.19), it follows that

$$\begin{aligned}
|\langle(\hat{\sigma}_1 \cdot \mathbf{a})(\hat{\sigma}_2 \cdot \mathbf{b})\rangle| &= \left|\int_\Lambda \wp(\lambda) A_\mathbf{a}(\lambda) B_\mathbf{b}(\lambda) d\lambda\right| \\
&\leq \int_\Lambda \wp(\lambda)|A_\mathbf{a}(\lambda)||B_\mathbf{b}(\lambda)|d\lambda \\
&= \int_\Lambda \wp(\lambda) d\lambda \\
&= 1, \qquad (10.21)
\end{aligned}$$

where use has been made of the triangle inequality for integrals (6.23). The above inequality implies that

$$-1 \leq \langle\mathbf{a}, \mathbf{b}\rangle \leq 1, \qquad (10.22)$$

where for the sake of notational simplicity we have used the shorthand notation $\langle\mathbf{a}, \mathbf{b}\rangle$ to denote $\langle(\hat{\sigma}_1 \cdot \mathbf{a})(\hat{\sigma}_2 \cdot \mathbf{b})\rangle$.

Our aim is to compare the prediction of a local hidden variable theory $\langle\mathbf{a}, \mathbf{b}\rangle$, as defined by Eq. (10.19), with the quantum mechanical prediction $\langle\mathbf{a}, \mathbf{b}\rangle_{\Psi_0}$, as defined by Eq. (10.20). The above quantum mechanical result can be computed as follows. We first write the dot product between the observable and the Euclidean vector in terms of the corresponding Cartesian components as

$$\hat{\sigma}_1 \cdot \mathbf{a} = a_x \hat{\sigma}_{1x} + a_y \hat{\sigma}_{1y} + a_z \hat{\sigma}_{1z}, \qquad (10.23a)$$
$$\hat{\sigma}_2 \cdot \mathbf{b} = b_x \hat{\sigma}_{2x} + b_y \hat{\sigma}_{2y} + b_z \hat{\sigma}_{2z}. \qquad (10.23b)$$

Thus the expectation value of the product $(\hat{\sigma}_1 \cdot \mathbf{a})(\hat{\sigma}_2 \cdot \mathbf{b})$ on the singlet state $|\Psi_0\rangle$ gives 9 terms, each of which can then be computed straightforwardly. Using the fact that $|\Psi_0\rangle$ is an eigenstate of the observables $\hat{\sigma}_{1x}\hat{\sigma}_{2x}$, $\hat{\sigma}_{1y}\hat{\sigma}_{2y}$, and $\hat{\sigma}_{1z}\hat{\sigma}_{2z}$ with eigenvalue -1 [see Section 10.3 and Problem 10.2], we find that the three direct terms (i.e., terms resulting from the product between the same components) are given by

$$\langle\Psi_0|a_x b_x \hat{\sigma}_{1x}\hat{\sigma}_{2x}|\Psi_0\rangle = -a_x b_x, \tag{10.24a}$$

$$\langle\Psi_0|a_y b_y \hat{\sigma}_{1y}\hat{\sigma}_{2y}|\Psi_0\rangle = -a_y b_y, \tag{10.24b}$$

$$\langle\Psi_0|a_z b_z \hat{\sigma}_{1z}\hat{\sigma}_{2z}|\Psi_0\rangle = -a_z b_z. \tag{10.24c}$$

The remaining six cross terms (i.e., terms resulting from the product between different components) are instead all zero. For instance, using the actions of Pauli matrices on $|\uparrow_z\rangle$ and $|\downarrow_z\rangle$ tabulated in Table 8.1, we find

$$
\begin{aligned}
\langle\Psi_0|\hat{\sigma}_{1x}\hat{\sigma}_{2y}|\Psi_0\rangle &= \langle\Psi_0|\hat{\sigma}_{1x}\hat{\sigma}_{2y}\left(|\uparrow_z\rangle_1|\downarrow_z\rangle_2 - |\downarrow_z\rangle_1|\uparrow_z\rangle_2\right)\\
&= \langle\Psi_0|\left[(-i)|\downarrow_z\rangle_1|\uparrow_z\rangle_2 - i|\uparrow_z\rangle_1|\downarrow_z\rangle_2\right]\\
&= -i(\langle\uparrow_z|_1\langle\downarrow_z|_2 - \langle\downarrow_z|_1\langle\uparrow_z|_2)\\
&\quad\left(|\uparrow_z\rangle_1|\downarrow_z\rangle_2 + |\downarrow_z\rangle_1|\uparrow_z\rangle_2\right)\\
&= 0, \tag{10.25a}
\end{aligned}
$$

$$
\begin{aligned}
\langle\Psi_0|\hat{\sigma}_{1x}\hat{\sigma}_{2z}|\Psi_0\rangle &= \langle\Psi_0|\hat{\sigma}_{1x}\hat{\sigma}_{2z}\left(|\uparrow_z\rangle_1|\downarrow_z\rangle_2 - |\downarrow_z\rangle_1|\uparrow_z\rangle_2\right)\\
&= \langle\Psi_0|\left(-|\downarrow_z\rangle_1|\downarrow_z\rangle_2 - |\uparrow_z\rangle_1|\uparrow_z\rangle_2\right)\\
&= -(\langle\uparrow_z|_1\langle\downarrow_z|_2 - \langle\downarrow_z|_1\langle\uparrow_z|_2)\\
&\quad\left(|\uparrow_z\rangle_1|\uparrow_z\rangle_2 + |\downarrow_z\rangle_1|\downarrow_z\rangle_2\right)\\
&= 0, \tag{10.25b}
\end{aligned}
$$

etc. [see Problem 10.4], where we recall that the scalar product of the direct product states is defined by Eq. (7.125). Collecting the above results, we finally conclude that

$$\langle\mathbf{a},\mathbf{b}\rangle_{\Psi_0} = -(a_x b_x + a_y b_y + a_z b_z) = -\mathbf{a}\cdot\mathbf{b}. \tag{10.26}$$

When the two directions $\mathbf{a}$ and $\mathbf{b}$ are parallel, we have (recall that $\mathbf{a}$ and $\mathbf{b}$ are taken to be unit vectors)

$$\langle\mathbf{a},\mathbf{a}\rangle_{\Psi_0} = -1, \tag{10.27}$$

as it should be since there is a perfect *anticorrelation* (spin up versus spin down, and vice versa) between the results of the two measurements.

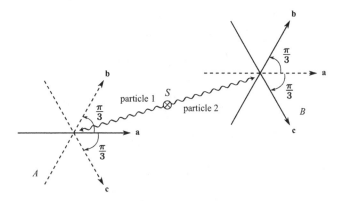

Figure 10.2 A scheme depiction of the experimental setup used in the counterexample proposed for proving the Bell theorem, in which there are three coplanar directions **a**, **b**, and **c**, with both **b** and **c** making an angle of $\frac{\pi}{3}$ with **a** (but in the opposite sense) and the angle between **b** and **c** begin $\frac{2\pi}{3}$.

Since the quantum mechanical result for perfect anticorrelation is an experimental fact, any local hidden variable theory must also satisfy this requirement. On the other hand, the hidden-variable prediction $\langle \mathbf{a}, \mathbf{a} \rangle = -1$ holds if and only if we have

$$A_{\mathbf{a}}(\lambda) = -B_{\mathbf{a}}(\lambda), \tag{10.28}$$

for all directions **a** and all values of $\lambda \in \Lambda$. In this case, $\langle \mathbf{a}, \mathbf{b} \rangle$ also reaches its minimum value [see Eq. (10.22)]. With the above requirement, we can drop any reference to apparatus B and rewrite $\langle \mathbf{a}, \mathbf{b} \rangle$ as [see Eq. (10.19)]

$$\langle \mathbf{a}, \mathbf{b} \rangle = - \int_{\Lambda} \wp(\lambda) A_{\mathbf{a}}(\lambda) A_{\mathbf{b}}(\lambda) d\lambda. \tag{10.29}$$

Let us now consider two alternative directions, say **b** and **c**, for the spin measurement of particle 2 [see Fig. 10.2]. Then, we have

$$\langle \mathbf{a}, \mathbf{b} \rangle - \langle \mathbf{a}, \mathbf{c} \rangle = - \int_{\Lambda} \wp(\lambda) [A_{\mathbf{a}}(\lambda) A_{\mathbf{b}}(\lambda) - A_{\mathbf{a}}(\lambda) A_{\mathbf{c}}(\lambda)] d\lambda$$

$$= \int_{\Lambda} \wp(\lambda) A_{\mathbf{a}}(\lambda) A_{\mathbf{b}}(\lambda) [A_{\mathbf{b}}(\lambda) A_{\mathbf{c}}(\lambda) - 1] d\lambda, \tag{10.30}$$

where use had been made of Eq. (10.29) and the property [see Eq. (10.17)]

$$A_{\mathbf{a}}(\lambda) A_{\mathbf{b}}(\lambda) A_{\mathbf{b}}(\lambda) A_{\mathbf{c}}(\lambda) = A_{\mathbf{a}}(\lambda) A_{\mathbf{c}}(\lambda). \tag{10.31}$$

Starting from Eq. (10.30) as well as using the triangle inequality for integrals (6.23) and the property (10.17), we have

$$|\langle \mathbf{a}, \mathbf{b} \rangle - \langle \mathbf{a}, \mathbf{c} \rangle| = \left| \int_\Lambda \wp(\lambda) A_\mathbf{a}(\lambda) A_\mathbf{b}(\lambda) [A_\mathbf{b}(\lambda) A_\mathbf{c}(\lambda) - 1] d\lambda \right|$$

$$\leq \int_\Lambda \wp(\lambda) |A_\mathbf{a}(\lambda)| |A_\mathbf{b}(\lambda)| |A_\mathbf{b}(\lambda) A_\mathbf{c}(\lambda) - 1| d\lambda$$

$$= \int_\Lambda \wp(\lambda) [1 - A_\mathbf{b}(\lambda) A_\mathbf{c}(\lambda)] d\lambda$$

$$= 1 - \int_\Lambda \wp(\lambda) A_\mathbf{b}(\lambda) A_\mathbf{c}(\lambda) d\lambda. \tag{10.32}$$

Therefore, we finally obtain the sought-after result

$$|\langle \mathbf{a}, \mathbf{b} \rangle - \langle \mathbf{a}, \mathbf{c} \rangle| - \langle \mathbf{b}, \mathbf{c} \rangle \leq 1, \tag{10.33}$$

where $\langle \mathbf{b}, \mathbf{c} \rangle$ is the expectation value of the product of the spin component $\hat{\sigma}_1 \cdot \mathbf{b}$ of particle 1 and the spin component $\hat{\sigma}_2 \cdot \mathbf{c}$ of particle 2, i.e.,

$$\langle \mathbf{b}, \mathbf{c} \rangle = - \int_\Lambda \wp(\lambda) A_\mathbf{b}(\lambda) A_\mathbf{c}(\lambda) d\lambda. \tag{10.34}$$

It is noted that the inequality (10.33) is valid for arbitrary directions $\mathbf{a}$, $\mathbf{b}$, and $\mathbf{c}$. This property allows us to rewrite it in a more symmetric form as

$$|\langle \mathbf{a}, \mathbf{b} \rangle + \langle \mathbf{a}, \mathbf{c} \rangle| + \langle \mathbf{b}, \mathbf{c} \rangle \leq 1. \tag{10.35}$$

The above result is obtained by setting $\mathbf{c}$ to $-\mathbf{c}$ in the original inequality, which corresponds to the situation that we had chosen in the first place the two alternative directions $\mathbf{b}$ and $-\mathbf{c}$ for the spin measurement of particle 2. The inequality (10.35) is the first of a family of inequalities, collectively known as the *Bell inequalities*. Its importance lies in the fact that it sets precise quantitative bounds on the prediction of *any* deterministic local hidden variable theory.

A generalization of the Bell inequality can be derived if we consider spin measurement in alternative directions for each particles. Let $\mathbf{a}$ and $\mathbf{a}'$ denote two alternative directions for particle 1, while $\mathbf{b}$ and $\mathbf{b}'$ two alternative directions for for particle 2. Since the spin component of each particle is measured in two alternative directions, there are four possible joint measurement pairs for the two particles. They are represented by the four operators $(\hat{\sigma}_1 \cdot \mathbf{a})(\hat{\sigma}_2 \cdot \mathbf{b})$,

$(\hat{\sigma}_1 \cdot \mathbf{a})(\hat{\sigma}_2 \cdot \mathbf{b}')$, $(\hat{\sigma}_1 \cdot \mathbf{a}')(\hat{\sigma}_2 \cdot \mathbf{b})$, and $(\hat{\sigma}_1 \cdot \mathbf{a}')(\hat{\sigma}_2 \cdot \mathbf{b}')$, in terms of which we can then define the Bell operator[a] by

$$\begin{aligned}
\hat{B} &= (\hat{\sigma}_1 \cdot \mathbf{a})(\hat{\sigma}_2 \cdot \mathbf{b} + \hat{\sigma}_2 \cdot \mathbf{b}') + (\hat{\sigma}_1 \cdot \mathbf{a}')(\hat{\sigma}_2 \cdot \mathbf{b} - \hat{\sigma}_2 \cdot \mathbf{b}') \\
&= (\hat{\sigma}_1 \cdot \mathbf{a})(\hat{\sigma}_2 \cdot \mathbf{b}) + (\hat{\sigma}_1 \cdot \mathbf{a})(\hat{\sigma}_2 \cdot \mathbf{b}') + (\hat{\sigma}_1 \cdot \mathbf{a}')(\hat{\sigma}_2 \cdot \mathbf{b}) \\
&\quad - (\hat{\sigma}_1 \cdot \mathbf{a}')(\hat{\sigma}_2 \cdot \mathbf{b}').
\end{aligned} \tag{10.36}$$

Although relatively complicated, the Bell operator is not in principle different from the basic expression of any quantum observable (here for a composite system). According to the second line of Eq. (10.36), the expectation value of $\hat{B}$ in a local hidden variable theory can be written as [see also Eqs. (10.29) and (10.19)]

$$B = \langle \mathbf{a}, \mathbf{b} \rangle + \langle \mathbf{a}, \mathbf{b}' \rangle + \langle \mathbf{a}', \mathbf{b} \rangle - \langle \mathbf{a}', \mathbf{b}' \rangle, \tag{10.37}$$

where we recall the shorthand notation $\langle \mathbf{a}, \mathbf{b} \rangle = \langle (\hat{\sigma}_1 \cdot \mathbf{a})(\hat{\sigma}_2 \cdot \mathbf{b}) \rangle$, etc. The so-called CHSH inequality (short for Clauser, Horne, Shimony, and Holt) is given by[b]

$$|B| = \left| \langle \mathbf{a}, \mathbf{b} \rangle + \langle \mathbf{a}, \mathbf{b}' \rangle + \langle \mathbf{a}', \mathbf{b} \rangle - \langle \mathbf{a}', \mathbf{b}' \rangle \right| \le 2, \tag{10.38}$$

which is a generalized reformulation of the Bell inequality (10.35).

We are now in a position to formulate the Bell theorem in terms of the Bell inequality.

Theorem 10.1 (Bell) A deterministic local hidden variable theory, which acknowledges the separability principle, must satisfy an inequality of the type given by (10.35) or (10.38). The predictions of quantum mechanics on the contrary violate such an inequality.

The first part of the Bell theorem has been already proved since we have obtained the Bell inequality by assuming the separability condition (10.16). In order to prove the second part, it suffices to show the contradiction of the Bell inequality (10.35) or the CHSH inequality (10.38) with quantum mechanics by means of counterexamples.[c] We shall first consider a specific counterexample of the Bell inequality. Let us take the unit vectors $\mathbf{a}$, $\mathbf{b}$, and $\mathbf{c}$ to be coplanar, with both $\mathbf{b}$ and $\mathbf{c}$ making an angle of $\frac{\pi}{3}$ with $\mathbf{a}$ (but in the opposite sense) and the angle between $\mathbf{b}$ and $\mathbf{c}$ begin $\frac{2\pi}{3}$ [see

---

[a](Braunstein *et al.*, 1992).
[b](Clauser *et al.*, 1969).
[c](Clauser/Shimony, 1978, pp. 1888–1890).

Fig. 10.2]. Using Eq. (10.26), we find that the quantum expectation values in this case are given by

$$\langle \mathbf{a}, \mathbf{b} \rangle_{\Psi_0} = -\mathbf{a} \cdot \mathbf{b} = -\cos\frac{\pi}{3} = -\frac{1}{2}, \tag{10.39a}$$

$$\langle \mathbf{a}, \mathbf{c} \rangle_{\Psi_0} = -\mathbf{a} \cdot \mathbf{c} = -\cos\frac{\pi}{3} = -\frac{1}{2}, \tag{10.39b}$$

$$\langle \mathbf{b}, \mathbf{c} \rangle_{\Psi_0} = -\mathbf{b} \cdot \mathbf{c} = -\cos\frac{2\pi}{3} = \frac{1}{2}, \tag{10.39c}$$

where use has been made of Eq. (3.16). From the above results, we obtain

$$\left| \langle \mathbf{a}, \mathbf{b} \rangle_{\Psi_0} + \langle \mathbf{a}, \mathbf{c} \rangle_{\Psi_0} \right| + \langle \mathbf{b}, \mathbf{c} \rangle_{\Psi_0} = \frac{3}{2} > 1, \tag{10.40}$$

which evidently violates the Bell inequality (10.35). We then consider a specific counterexample of the CHSH inequality. We choose the following directions of spin measurement for particles 1 and 2:

$$\mathbf{a} = \mathbf{e}_x, \quad \mathbf{a}' = \mathbf{e}_z, \quad \mathbf{b} = -\frac{1}{\sqrt{2}}(\mathbf{e}_z + \mathbf{e}_x), \quad \mathbf{b}' = \frac{1}{\sqrt{2}}(\mathbf{e}_z - \mathbf{e}_x), \tag{10.41}$$

where $\mathbf{e}_x$, $\mathbf{e}_y$, and $\mathbf{e}_z$ are unit vectors in the x, y, and z directions, respectively. Using again Eq. (10.26), we find that the quantum expectation values in this case are given by [see Box 3.1]

$$\langle \mathbf{a}, \mathbf{b} \rangle_{\Psi_0} = -\mathbf{a} \cdot \mathbf{b} = \frac{1}{\sqrt{2}}, \quad \langle \mathbf{a}, \mathbf{b}' \rangle_{\Psi_0} = -\mathbf{a} \cdot \mathbf{b}' = \frac{1}{\sqrt{2}}, \tag{10.42a}$$

$$\langle \mathbf{a}', \mathbf{b} \rangle_{\Psi_0} = -\mathbf{a}' \cdot \mathbf{b} = \frac{1}{\sqrt{2}}, \quad \langle \mathbf{a}', \mathbf{b}' \rangle_{\Psi_0} = -\mathbf{a}' \cdot \mathbf{b}' = -\frac{1}{\sqrt{2}}, \tag{10.42b}$$

where use has been made of Eqs. (3.16) and (3.18). Consequently, we have

$$\left| \langle \hat{B} \rangle_{\Psi_0} \right| = \left| \langle \mathbf{a}, \mathbf{b} \rangle_{\Psi_0} + \langle \mathbf{a}, \mathbf{b}' \rangle_{\Psi_0} + \langle \mathbf{a}', \mathbf{b} \rangle_{\Psi_0} - \langle \mathbf{a}', \mathbf{b}' \rangle_{\Psi_0} \right| = 2\sqrt{2} \geq 2, \tag{10.43}$$

which is obviously in contradiction to the CHSH inequality (10.38). The construction of other counterexamples of the Bell and CHSH inequalities is left as an exercise for the reader [see Problem 10.5].

There have been several experiments confirming that indeed quantum mechanics violates the Bell inequality, thus providing experimental verification of the Bell theorem. The first series of experiments began in the 1970s and have made use essentially

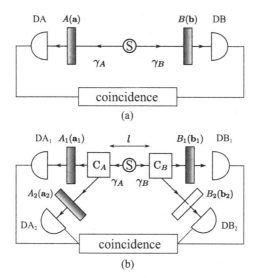

Figure 10.3 (a) Friedman–Clauser experiment. The correlated photons γ_A, γ_B coming from the source S impinge upon the linear polarizers A and B oriented in directions **a** and **b**, respectively. The rate of joint detection by the photomultipliers is monitored for various combinations of orientations. (b) Experiment proposed by Aspect. The optical commutator C_A directs the photon γ_A either towards polarizer A_1 with orientation $\mathbf{a}_1$ or to polarizer A_2 with orientation $\mathbf{a}_2$. Similarly for C_B to direct the photon γ_B to polarizer B_1 with orientation $\mathbf{b}_1$ or polarizer B_2 with orientation $\mathbf{b}_2$. The two commutators work independently (the time intervals between two commutations are taken to be stochastic). The four joint detection rates are monitored and the orientations $\mathbf{a}_1$, $\mathbf{a}_2$, $\mathbf{b}_1$, and $\mathbf{b}_2$ are not changed for the whole experiment. The distance l is the separation between the switches. Adapted from (Auletta *et al.*, 2009, p. 600).

of particles decay, according to Bohm's original model [see Section 10.3]. Prototypical is the experiment performed by Friedman and Clauser[a] [see Fig. 10.3(a)]. This kind of experiment was later on shown to present several flaws.[b] For this reason, in the 1980s a new generation of experiments was performed starting from the historical result of Aspect and collaborators[c] [see Fig. 10.3(b)].

---

[a](Freedman/Clauser, 1972).
[b](Auletta, 2000, Sections 35.3–35.4), (Auletta *et al.*, 2009, Section 16.5). See also Selleri (1988), Santos (1991).
[c](Aspect *et al.*, 1982).

Problem 10.4 Verify that the cross terms of the expectation value $\langle \mathbf{a}, \mathbf{b} \rangle_{\psi_0}$ given by Eq. (10.26) all vanish.

Problem 10.5 Try to construct another counterexample to the Bell inequality (10.35).

10.5 Entanglement Swapping

In this section we try to make a significant step towards the understanding of the concept of entanglement in its generality, namely, if it is possible for systems that have never directly interacted before to get entangled. This novel phenomenon reveals clearly that entanglement is not a dynamical interdependence between systems and, in particular, that entanglement is not associated with causal relations. Such genuine quantum mechanical characteristics clashes with the assumptions made in the EPR argument, as far as EPR postulated that there are no connections between systems if their interdependence is not of dynamical and therefore of causal type [see Section 10.1]. The milestone step towards the generalization of the entanglement concept was the *entanglement swapping* experiment performed by Zeilinger and collaborators in 1993.[a] Consider two pairs of entangled photons emitted by two independent sources as shown in Fig. 10.4. The states of the two photon pairs are given by

$$|\psi\rangle_{12} = \frac{1}{\sqrt{2}} \left(|h\rangle_1 |v\rangle_2 - |v\rangle_1 |h\rangle_2 \right), \tag{10.44a}$$

$$|\psi\rangle_{34} = \frac{1}{\sqrt{2}} \left(|h\rangle_3 |v\rangle_4 - |v\rangle_3 |h\rangle_4 \right), \tag{10.44b}$$

where the subscripts 12 and 34 denote the photon pairs composed of photons 1 and 2, and photons 3 and 4, respectively. Evidently, the state of the composite four-photon system $|\Psi\rangle$ is factorized, i.e., we have

$$|\Psi\rangle = |\psi\rangle_{12} \otimes |\psi\rangle_{34}. \tag{10.45}$$

In other words, while the two photons in each photon pair are entangled, there is *no* entanglement between either of the two photons in different photon pairs.

---

[a](Zukowski *et al.*, 1993).

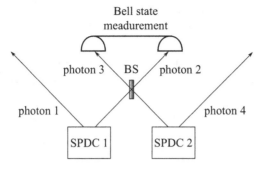

Figure 10.4 Schematic setup of entanglement swapping. Two pumped spontaneous parametric down-conversion sources SPDC 1 and SPDC 2 each emit a photon pair (1–2 and 3–4, respectively). Note that photons 2 and 3 merge through a beam splitter BS before being measured by a Bell state measurement device. Spontaneous parametric down-conversion consists in pumping a non-linear crystal with ultraviolet (high-energy) light. An ultraviolet photon may eventually decay into a couple (of less energetic) photons, obeying the laws of conservation of energy and momentum. If the orientation of the nonlinear optical crystal is chosen appropriately, two such possible decay processes become indistinguishable and lead to quantum correlations between the daughter photons. Adapted from (Auletta *et al.*, 2009, p. 611).

If we now perform a particular type of joint measurement on photons 2 and 3, we are able to project photons 1 and 4 onto an entangled state that depends on the result of the measurement on photons 2 and 3. To be specific, let us use the so-called Bell states for photons 2 and 3, which in the basis $\{|h\rangle, |v\rangle\}$ are given by[a]

$$|\Psi^+\rangle_{23} = \frac{1}{\sqrt{2}}(|h\rangle_2|v\rangle_3 + |v\rangle_2|h\rangle_3), \qquad (10.46a)$$

$$|\Psi^-\rangle_{23} = \frac{1}{\sqrt{2}}(|h\rangle_2|v\rangle_3 - |v\rangle_2|h\rangle_3), \qquad (10.46b)$$

$$|\Phi^+\rangle_{23} = \frac{1}{\sqrt{2}}(|h\rangle_2|h\rangle_3 + |v\rangle_2|v\rangle_3), \qquad (10.46c)$$

$$|\Phi^-\rangle_{23} = \frac{1}{\sqrt{2}}(|h\rangle_2|h\rangle_3 - |v\rangle_2|v\rangle_3). \qquad (10.46d)$$

Although we are considering here photon polarization states, with a suitable reformulation it may be shown that the Bell states are

---

[a](Braunstein *et al.*, 1992).

eigenstates of the Bell operator $\hat{B}$ given by (10.36). Indeed, the Bell states are four maximally entangled states of photons 2 and 3 and also constitute a complete orthonormal basis of the Hilbert space of photons 2 and 3. In order to find the effect of the joint measurement that projects $|\Psi\rangle$ onto one of the Bell states given by Eqs. (10.46), we rewrite the state $|\Psi\rangle$ given by Eq. (10.45) in terms of the latter as

$$|\Psi\rangle = \frac{1}{2}\big(|\Psi^+\rangle_{14}|\Psi^+\rangle_{23} - |\Psi^-\rangle_{14}|\Psi^-\rangle_{23} - |\Phi^+\rangle_{14}|\Phi^+\rangle_{23}$$
$$+ |\Phi^-\rangle_{14}|\Phi^-\rangle_{23}\big), \tag{10.47}$$

where the states for photons 1 and 4 have been expressed now in terms of their own Bell states $|\Psi^+\rangle_{14}$, $|\Psi^-\rangle_{14}$, $|\Phi^+\rangle_{14}$, and $|\Phi^-\rangle_{14}$. Note also that in the above expansion the components for photons 1 and 4 are paired with components of the same type for photons 2 and 3 (i.e., each "plus"-type state for photons 1 and 4 is paired with the corresponding "plus"-type state for photons 2 and 3, and the same for each "minus"-type state). It follows from Eq. (10.47) that the projection of the state $|\Psi\rangle$ onto a Bell state of photons 2 and 3, also projects photons 1 and 4 onto an entangled state precisely of the same form. In other words, by performing a joint measurement on photons 2 and 3 that entangles the two initial unentangled photons, we are able to entangle two initially unentangled photons 1 and 4. It is noted that photons 1 and 4 never met and therefore are not involved at all in this dynamical interaction. Therefore we are finally justified in saying that entanglement represents a form of interdependence in which no dynamic interaction is necessary. We remark that entanglement swapping of photon is not the only phenomenon of genuine quantum interference of systems from different sources. Indeed, it has been shown that electrons also show a similar behavior.[a]

Problem 10.6 Show that the state $|\Psi\rangle$ given by Eq. (10.47) is indeed the same as that given by Eq. (10.45). It is a cumbersome but very instructive calculation.

---

[a](Neder *et al.*, 2007).

10.6 Eberhard Theorem

Until now we have postponed the general issue of which kind of non-locality is displayed by quantum mechanics. Although entanglement does not imply exchange of signals, we cannot exclude that this will happen in other situations. Since 1978 Eberhard has focused on the conceptual distinction between separability and locality.[a] Let us consider the problem in all its generality. In other words, we do not need to restrict ourselves to the specific spin model previously introduced. To this purpose, we consider a composite system S consists of two subsystems S_1 and S_2, which are set to be measured with apparatuses A and B, respectively. Let $\hat{O}_1$ and $\hat{O}_2$ be the respective observables of systems S_1 and S_2. Now, given that the respective settings of apparatuses A and B are $\mathbf{a}$ and $\mathbf{b}$, suppose that the probability that a joint measurement of $\hat{O}_1$ on S_1 and $\hat{O}_2$ on S_2 yield the respective results o_a and o_b is given by $\wp(o_a, o_b|\mathbf{a}, \mathbf{b})$. The latter is the kind of probability called *conditional probability* [see Box 10.1].

Let $\wp(o_a|\mathbf{a})$ denote the probability that a measurement of $\hat{O}_1$ on S_1 yield the outcome o_a given the setting of apparatus A is $\mathbf{a}$, regardless of the setting of apparatus B and the measurement outcome of $\hat{O}_2$ on S_2. From the addition rule (10.55), the conditional probability $\wp(o_a|\mathbf{a})$ is obtained by summing the joint probabilities $\wp(o_a, o_b|\mathbf{a}, \mathbf{b})$ over all the possible outcomes o_b of $\hat{O}_2$ on S_2, i.e.,

$$\wp(o_a|\mathbf{a}) = \sum_{o_b} \wp(o_a, o_b|\mathbf{a}, \mathbf{b}). \qquad (10.48)$$

A similar probability $\wp(o_b|\mathbf{b})$ can be defined for the measurement outcome o_b of $\hat{O}_2$ on S_2 when the setting of apparatus B is $\mathbf{b}$, regardless of the setting of apparatus A and the measurement outcome of $\hat{O}_1$ on S_1. Likewise, the probability $\wp(o_b|\mathbf{b})$ is given by

$$\wp(o_b|\mathbf{b}) = \sum_{o_a} \wp(o_a, o_b|\mathbf{a}, \mathbf{b}). \qquad (10.49)$$

The careful reader may have noticed that *neither* the dependences of the setting $\mathbf{b}$ on both sides of Eq. (10.48) *nor* the dependences

---

[a](Eberhard, 1978), (Auletta, 2000, Section 36.5).

Box 10.1 Conditional probability

Conditional probability deals with the probability of an event while we have the information that another event has already occurred. To be specific, the conditional probability $\wp(A|B)$ that an event A occurs given the known occurrence of an event B is defined by

$$\wp(A|B) = \frac{\wp(A, B)}{\wp(B)}, \tag{10.50}$$

where $\wp(A, B)$ is the joint probability that the events A and B both occurred and $\wp(B) \neq 0$. Similarly, the conditional probability of B given A is

$$\wp(B|A) = \frac{\wp(A, B)}{\wp(A)}, \tag{10.51}$$

where $\wp(A) \neq 0$. Rearranging and combining these two equations, we find

$$\wp(A|B)\wp(B) = \wp(B|A)\wp(A), \tag{10.52}$$

Dividing both sides by $\wp(B)$, we obtain

$$\wp(A|B) = \frac{\wp(B|A)\wp(A)}{\wp(B)}, \tag{10.53}$$

which is known as the *Bayes theorem*. If $\wp(A, B) = \wp(A)\wp(B)$, then the events A and B are said to be statistically independent (or uncorrelated). It is noted that statistical independence of A and B implies $\wp(A|B) = \wp(A)$ and $\wp(B|A) = \wp(B)$. Since conditional probability is probability, it also satisfies the basic probability rules [see Eq. (2.2)]:

$$0 \leq \wp(A|B) \leq 1 \text{ for all } A \subset \Omega, \tag{10.54a}$$

$$\wp(B|B) = 1, \tag{10.54b}$$

$$\wp(A_1 \cup A_2|B) = \wp(A_1|B) + \wp(A_2|B) \text{ if } A_1 \cap A_2 = \emptyset, \tag{10.54c}$$

$$\wp(A_1 \cup A_2|B) = \wp(A_1|B) + \wp(A_2|B) - \wp(A_1, A_2|B), \tag{10.54d}$$

provided that $\wp(B) \neq 0$. A generalization of the addition rule (10.54c) is that for a set of mutually exclusive events A_j whose union is the event A, we have

$$\wp(A|B) = \sum_j \wp(A_j|B), \tag{10.55}$$

where B is an arbitrary event with $\wp(B) \neq 0$. Moreover, if the set of mutually exclusive events A_j exhausts the sample space Ω, then we have

$$\wp(B) = \sum_j \wp(B, A_j), \tag{10.56}$$

which in turn can be expressed in terms of the conditional probability $\wp(B|A_j)$ as

$$\wp(B) = \sum_j \wp(B|A_j)\wp(A_j). \tag{10.57}$$

The above expression is of great use when we discuss information acquisition in Section 12.1.

of the setting **a** on both sides of Eq. (10.49) are consistent. Indeed, in writing the above two equations we have implicitly taken into account the locality requirement. By "locality requirement" we mean that the conditional probability $\wp(o_a|\mathbf{a})$ has to be independent of the setting **b** of apparatus B and similarly the conditional probability $\wp(o_b|\mathbf{b})$ has to be independent of the setting **a** of apparatus A. In other words, the conditional probabilities $\wp(o_a|\mathbf{a})$ and $\wp(o_b|\mathbf{b})$ must depend only on the local setting.

According to Eberhard, if the locality requirement were violated we would have a *non-local causal interdependence* between the two subsystems. This is because, by changing the setting of one apparatus, we would be able to have influence on the measurement outcomes of the other apparatus, and hence, if we performed experiments on subsystems that are space-like separated (i.e., not connected by a light signal), we would be able to transmit a message at superluminal or even infinite speed. Actually, as we shall see, the violation of the above requirement does not necessarily imply a non-local causal interconnection because there could still be some form of interdependence between possible apparatus *settings* of the kind that quantum mechanics acknowledges for possible measurement outcomes, as we have seen to be the case for entanglement. To exclude also the latter possibility sets a stronger bound than to exclude a violation of Einstein's locality because it would represent also a violation of the separability between apparatus settings. Eberhard himself interpreted the theorem in the weaker form (no causal interconnection between apparatus settings) that implies the stronger formulation (no violation of the separability between apparatus settings). Here, we shall formulate the theorem in the most general terms as denying any kind of non-local correlations between apparatus settings and shall back on this issues later on.

Theorem 10.2 (Eberhard) Quantum mechanics does not allow non-local correlations between apparatus settings.

In order to prove the above theorem, we consider the situation in which we first perform a measurement of $\hat{O}_1$ on S_1, followed by a subsequent measurement of $\hat{O}_2$ on S_2. Let us denote by $|o_a, \mathbf{a}\rangle$ the state of subsystem S_1 when the setting of apparatus A is $\mathbf{a}$ and the measurement outcome of $\hat{O}_1$ is o_a, and by $|o_b, \mathbf{b}\rangle$ the state of subsystem S_2 when the setting of apparatus B is $\mathbf{b}$ and the measurement outcome of $\hat{O}_2$ is o_b. Then the conditional probability of obtaining the outcome o_a when the setting of apparatus A is $\mathbf{a}$, can be expressed in terms of the density matrix of the composite system $\hat{\rho}$ as [see Section 9.2]

$$\wp(o_a|\mathbf{a}) = \mathrm{Tr}\left(\hat{P}_{o_a,\mathbf{a}}\,\hat{\rho}\right), \tag{10.58}$$

where $\hat{P}_{o_a,\mathbf{a}}$ is the projector on the state $|o_a, \mathbf{a}\rangle$ of S_1, i.e.,

$$\hat{P}_{o_a,\mathbf{a}} = |o_a, \mathbf{a}\rangle\langle o_a, \mathbf{a}|. \tag{10.59}$$

After the measurement of $\hat{O}_1$ on S_1 when the setting of apparatus A is $\mathbf{a}$ and the outcome is o_a, the density matrix $\hat{\rho}$ of the composite system reduces to [see Eq. (9.10)]

$$\hat{\rho}' = \frac{\hat{P}_{o_a,\mathbf{a}}\,\hat{\rho}\,\hat{P}_{o_a,\mathbf{a}}}{\wp(o_a|\mathbf{a})}. \tag{10.60}$$

If we perform a subsequent measurement of $\hat{O}_2$ on S_2, then the conditional probability of obtaining the outcome o_b when the setting of apparatus B is $\mathbf{b}$, is given by

$$\wp(o_b|o_a, \mathbf{a}, \mathbf{b}) = \mathrm{Tr}\left(\hat{P}_{o_b, \mathbf{b}}\, \hat{\rho}'\right)$$
$$= \frac{\mathrm{Tr}\left(\hat{P}_{o_b, \mathbf{b}}\hat{P}_{o_a, \mathbf{a}}\, \hat{\rho}\, \hat{P}_{o_a, \mathbf{a}}\right)}{\wp(o_a|\mathbf{a})}, \tag{10.61}$$

where $\hat{P}_{o_b, \mathbf{b}}$ is the projector on the state $|o_b, \mathbf{b}\rangle$ of S_2, i.e.,

$$\hat{P}_{o_b, \mathbf{b}} = |o_b, \mathbf{b}\rangle\langle o_b, \mathbf{b}|. \tag{10.62}$$

Therefore, the joint probability of obtaining the outcomes o_a and o_b, given the respective settings $\mathbf{a}$ and $\mathbf{b}$ of apparatus A and B, is given by [see Eq. (10.50)]

$$\wp(o_a, o_b|\mathbf{a}, \mathbf{b}) = \wp(o_a|\mathbf{a})\, \wp(o_b|o_a, \mathbf{a}, \mathbf{b})$$
$$= \wp(o_a|\mathbf{a})\frac{\mathrm{Tr}\left(\hat{P}_{o_b, \mathbf{b}}\hat{P}_{o_a, \mathbf{a}}\hat{\rho}\hat{P}_{o_a, \mathbf{a}}\right)}{\wp(o_a|\mathbf{a})}$$
$$= \mathrm{Tr}\left(\hat{P}_{o_b, \mathbf{b}}\hat{P}_{o_a, \mathbf{a}}\hat{\rho}\hat{P}_{o_a, \mathbf{a}}\right) \tag{10.63}$$

By using the cyclic property of the trace (7.115c), the fact that the projectors $\hat{P}_{o_a, \mathbf{a}}$ and $\hat{P}_{o_b, \mathbf{b}}$ commute because they pertain to different subsystems, and the general property $\hat{P}_{o_a, \mathbf{a}}^2 = \hat{P}_{o_a, \mathbf{a}}$ [see Eq. (3.62)], we may further simplify the above expression to

$$\wp(o_a, o_b|\mathbf{a}, \mathbf{b}) = \mathrm{Tr}\left(\hat{P}_{o_a, \mathbf{a}}\hat{P}_{o_b, \mathbf{b}}\hat{P}_{o_a, \mathbf{a}}\, \hat{\rho}\right)$$
$$= \mathrm{Tr}\left(\hat{P}_{o_b, \mathbf{b}}\hat{P}_{o_a, \mathbf{a}}\hat{P}_{o_a, \mathbf{a}}\, \hat{\rho}\right)$$
$$= \mathrm{Tr}\left(\hat{P}_{o_b, \mathbf{b}}\hat{P}_{o_a, \mathbf{a}}\, \hat{\rho}\right). \tag{10.64}$$

From Eq. (10.49), we sum the above result over all possible outcomes of o_a to obtain the probability $\wp(o_b|\mathbf{b})$ as

$$\wp(o_b|\mathbf{b}) = \sum_{o_a} \wp(o_a, o_b|\mathbf{a}, \mathbf{b})$$
$$= \sum_{o_a} \mathrm{Tr}\left(\hat{P}_{o_b, \mathbf{b}}\hat{P}_{o_a, \mathbf{a}}\, \hat{\rho}\right)$$
$$= \mathrm{Tr}\left[\hat{P}_{o_b, \mathbf{b}}\left(\sum_{o_a}\hat{P}_{o_a, \mathbf{a}}\right)\hat{\rho}\right]$$
$$= \mathrm{Tr}\left(\hat{P}_{o_b, \mathbf{b}}\, \hat{\rho}\right), \tag{10.65}$$

where use has been made of the property that $\sum_{o_a} \hat{P}_{o_a,\mathbf{a}} = \hat{I}$ for any complete set of orthogonal projectors [see Section 3.7]. We note that the result for $\wp(o_b|\mathbf{b})$ given by the last equality in Eq. (10.65) is indeed independent of the setting $\mathbf{a}$ of apparatus A. This is in agreement with locality requirement. The same conclusion applies to $\wp(o_a|\mathbf{a})$ if we had started the proof by considering a measurement of $\hat{O}_2$ on S_2, followed by a measurement of $\hat{O}_1$ on S_1 [see Problem 10.7].

The Eberhard theorem is of particular importance in our understanding of the conceptual aspects of quantum mechanics. In fact, it proves that although the probability distributions of possible measurement outcomes of two entangled systems are not independent, as predicted by quantum mechanics, there is no correlation between possible apparatus settings. This excludes the possibility to influence the probability distributions of the outcomes of measurement on one system by changing the setting of the apparatus for measurement on another system, which in turn would imply the possibility of exchanging superluminal signals. Although in the following we shall often use the term *non-locality* as a shorthand way to refer to all quantum correlations associated with entanglement, as a fulfillment of the examination developed in the previous sections we assume as proved that the kind of non-local interdependence expressed as quantum features is not of dynamical type but of pure relations or constraints of a new kind. Moreover, as anticipated, we take for proved that quantum mechanical correlations in fact only concern the possible outcomes but never settings, although this will be also the object of a later examination. A philosophical remark may be opportune here. Already the great philosopher Peirce has distinguished between static relations, that is, correlations, and dynamical ones, that is, interactions.[a]

Problem 10.7 Prove the Eberhard theorem by starting from the conditional probability $\wp(o_a|o_b, \mathbf{b}, \mathbf{a})$.

---

[a](Peirce CP, 1.293; 1.303–1.332; 3.472–3.473), see also (Auletta, 2006a).

10.7 Kochen–Specker Theorem

We wish to briefly present in this section an important general result represented by a theorem proved by Kochen and Specker in 1967,[a] which shows the incompatibility between quantum mechanics and hidden variable theories without any separability assumption. We may recall that the EPR argument can be circumvented by either denying the separability assumption (and we have followed this first development until now) or by rejecting the principle of physical reality [see Sections 10.1 and 10.2]. The latter is the subject of the present section.[b] Let $\{a_1, a_2, \ldots, a_n\}$ be a set of n atomic statements, while b_{ij} and c_{ijk} be compound statements of the form[c]

$$b_{jk} = \neg(a_j \wedge a_k), \quad c_{ijk} = a_i \vee a_j \vee a_k. \tag{10.66}$$

By De Morgan's law (9.49), we have $b_{ij} = \neg a_j \vee \neg a_k$, which means that classically the disjunction of $\neg a_j$ and $\neg a_k$ can be true if at least one of the two statements a_j and a_k is false. On the contrary, in quantum mechanics the statement b_{ij} can be true even if both a_j and a_k are true. Indeed, we known that in an interferometer experiment we can affirm that the photon is in a state such that it can be found in one of the two paths, down or up, but in most cases we cannot affirm that it is indeed in either the down or the up path [see Chapter 2]. Formally, $d \vee u$ can be true even when neither u or d is. Similarly, in quantum mechanics the statement c_{ijk} can be true even if a_i, a_j, and a_k are all false.

We now consider a statement d composed of ten atomic statements a_j (with $j = 0, 1, \ldots, 9$)

$$d = b_{01} \wedge b_{02} \wedge b_{08} \wedge b_{13} \wedge b_{15} \wedge b_{24} \wedge b_{26} \wedge b_{35} \wedge b_{37} \wedge b_{46}$$
$$\wedge \, b_{47} \wedge b_{56} \wedge b_{78} \wedge b_{79} \wedge b_{89} \wedge c_{135} \wedge c_{346} \wedge c_{789}. \tag{10.67}$$

The graphical representation of the statement d is depicted in Fig. 10.5, in which the vertices are the atomic statements. The two vertices a_j and a_k are connected by a straight line if and only if b_{jk} is

---

[a](Kochen/Specker, 1967).
[b]We follow here a simplified exposition given by Pitowsky in (Pitowsky, 1989, pp. 109–117).
[c]A statement which cannot be broken down into other simpler statements is called an atomic statement.

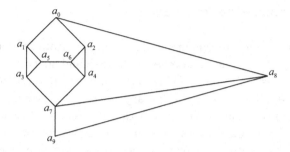

Figure 10.5 The 10-point Kochen–Specker graph, i.e., the graphical representation of the statement d given by Eq. (10.67).

in d, while the vertices a_i, a_j, and a_k constitute a triangle if and only if c_{ijk} is in d. Let $f(a)$ denotes the truth value of the statement a. If $f(d) = 1$, then no pair of atomic statements connected by a straight line can have truth value 1. This is because the statement d is false if one of its constituent statements b_{jk} is false. Formally, we have the statement

$$[f(a_j) = f(a_k) = 1 \Rightarrow f(b_{jk}) = 0] \Rightarrow f(d) = 0. \qquad (10.68)$$

Indeed, on the one hand, the disjunction b_{jk} is true only if at least one of the two statements $\neg a_j$ and $\neg a_k$ is true (or only if a_j or a_k is false). On the other hand, if b_{jk} is false, then the compound statement d is also false as the latter is composed of conjunctions between b_{jk} and c_{ijk}, which implies that neither b_{jk} nor c_{ijk} in d can be false [see Table 9.3]. Now we shall prove the following lemma:

Lemma 10.1 (Pitowsky) If both $f(d) = 1$ and $f(a_0) = 1$, then $f(a_9) = 1$.

A quick proof by *reductio ad absurdum* can be obtained by inspecting the graph of d depicted in Fig. 10.5. Since a_0 is true, we have that all the statements connected to it through a straight line are false, that is, a_1, a_2, and a_8 are all false. But if we assume for the sake of the argument that a_9 is also false, this then implies that a_7 must be true since $c_{789} = a_7 \vee a_8 \vee a_9$ must hold true if d is true. However, if a_7 is true, then a_3 and a_4 must be false because both are connected with a_7 by a straight line. However, since both c_{135} and c_{246} must be true, this implies that both a_5 and a_6 are true. But this is a contradiction since the a_5 and a_6 are connected by a straight line.

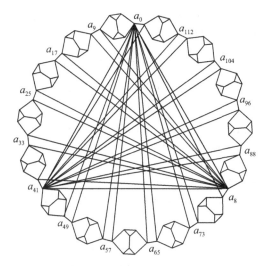

Figure 10.6 The 117-point Kochen–Specker graph, i.e., the graphical representation of the statement s' given by Eq. (10.69), is constructed by using 15 copies of the 10-point Kochen–Specker graph depicted in Fig. 10.5 (consider especially the subgraph containing a_0, a_8, and a_9). The vertices are numbered consecutively in a counterclockwise direction from a_0 to a_{119} except for a_8 and a_{41}. Moreover, vertices a_0 and a_{40} are identified with each other, and so are a_8 and a_{119}, and a_{41} and a_{80}.

With the above lemma, we now proceed to the formulation and proof of the Kochen–Specker theorem. Consider a compound statement s' constructed as follows by using 15 copies of the d graph considered above [see Fig. 10.6]. There are three groups of five interlocking copies of the d graph such that in each group every a_9 of one copy of d is identified with every a_0 of the next copy and all copies of a_8 are identified with each other. Then, the three groups are cyclically interlocked together by identifying a_9 of the last copy of d and a_8 in one group with a_0 of the first copy of d in the next group and that in the next next group, respectively. Finally, a_0, a_8, and a_{41} constitute a triangle. Let us now denote by $d_{0,8,9}$ the statement given by Eq. (10.67), where the subscripts 0, 8, and 9 are the vertices a_0, a_j, and a_k to be identified with those of the next copy in the above construction, and by $d'_{0,8,9}$ the same statement but with b_{08} removed. To avoid confusion, here we have used commas to separate numerical subscripts. Then the compound statement s' can be expressed in terms of d and d' as

$$s' = d_{0,8,9} \wedge d'_{9,8,17} \wedge d'_{17,8,25} \wedge d'_{25,8,33} \wedge d'_{33,8,41} \wedge d_{41,0,49} \wedge d'_{49,0,57}$$
$$\wedge\, d'_{57,0,65} \wedge d'_{65,0,73} \wedge d'_{73,0,8} \wedge d_{8,41,88} \wedge d'_{88,41,96} \wedge d'_{96,41,104}$$
$$\wedge\, d'_{104,41,112} \wedge d'_{112,41,0} \wedge c_{0,8,41}. \tag{10.69}$$

Moreover, from the 15 copies of the d graph used in the process of constructing s', we subtract those vertices that were identified with each other and end up with 117 different vertices, which correspond to the 117 constituting atomic statements of s'. The reason why we have denoted such a statement as s' is that it can be considered opposite to the statement s involved in the EPR argument [see Section 10.1].

We are now in a position to formulate the Kochen–Specker theorem in terms of the compound statement that we have just constructed.

Theorem 10.3 (Kochen–Specker) The statement s' is classically a logical falsity, but there are cases in which it is true in quantum mechanics.

The first part of the Kochen–Specker theorem, that is, the statement s' is classically a logical falsity, can be proved by *reductio ad absurdum*. Suppose that s' is true, then both $c_{0,8,41} = a_0 \vee a_8 \vee a_{41}$ and $d_{0,8,9}$ must hold true because they are subgraphs of the graph of s' [see Eq. (10.69) and Fig. 10.6]. The logical truth of $c_{0,8,41}$ implies that at least one of a_0, a_8, and a_{41} is true. We assume for the sake of the argument that a_0 is true. Together with the logical truth of $d_{0,8,9}$, by Lemma 10.1, this implies that a_9 is also true. The same argument can be repeated to show that a_{17}, a_{25}, a_{33}, and a_{41} are all true. But a_0 and a_{41} are connected by a straight line, which implies $b_{1,41}$ is false and hence s' is also false. This is a contradiction, therefore we conclude that the statement s' is false.[a]

In order to prove the second part, it suffices to provide a quantum mechanical counterexample to the logical falsity of the statement s'.

---

[a] Recently, an experiment on a single system with two degrees of freedom has been reported in (Hasegawa *et al.*, 2006), which demonstrates the validity of a Kochen–Specker-like argument and shows that quantum mechanics is a contextual theory.

To this purpose, let us consider a single-particle spin 1 system and the measurement of the squared spin components $\hat{S}_x^2$, $\hat{S}_y^2$, and $\hat{S}_z^2$. It can be shown that the latter operators are mutually commuting (while $\hat{S}_x$, $\hat{S}_y$, and $\hat{S}_z$ themselves are not). In the common eigenbasis of $\hat{\mathbf{S}}^2$ and $\hat{S}_z$, i.e., the $\{|1, 1\rangle, |1, 0\rangle, |1, -1\rangle\}$ basis, the operators $\hat{S}_x$, $\hat{S}_y$, and $\hat{S}_z$ are respectively represented by the three 3×3 matrices given by Eq. (8.45). Thus, the corresponding squared spin components are given by (in units of $\hbar^2$)

$$\hat{S}_x^2 = \frac{1}{2} \begin{bmatrix} 0 & 1 & 0 \\ 1 & 0 & 1 \\ 0 & 1 & 0 \end{bmatrix} \begin{bmatrix} 0 & 1 & 0 \\ 1 & 0 & 1 \\ 0 & 1 & 0 \end{bmatrix} = \frac{1}{2} \begin{bmatrix} 1 & 0 & 1 \\ 0 & 2 & 0 \\ 1 & 0 & 1 \end{bmatrix}, \tag{10.70a}$$

$$\hat{S}_y^2 = \frac{1}{2} \begin{bmatrix} 0 & -i & 0 \\ i & 0 & -i \\ 0 & i & 0 \end{bmatrix} \begin{bmatrix} 0 & -i & 0 \\ i & 0 & -i \\ 0 & i & 0 \end{bmatrix} = \frac{1}{2} \begin{bmatrix} 1 & 0 & -1 \\ 0 & 2 & 0 \\ -1 & 0 & 1 \end{bmatrix}, \tag{10.70b}$$

$$\hat{S}_z^2 = \begin{bmatrix} 1 & 0 & 0 \\ 0 & 0 & 0 \\ 0 & 0 & -1 \end{bmatrix} \begin{bmatrix} 1 & 0 & 0 \\ 0 & 0 & 0 \\ 0 & 0 & -1 \end{bmatrix} = \begin{bmatrix} 1 & 0 & 0 \\ 0 & 0 & 0 \\ 0 & 0 & 1 \end{bmatrix}. \tag{10.70c}$$

With the above expressions, it is now straightforward to verify that $\hat{S}_x^2$, $\hat{S}_y^2$, and $\hat{S}_z^2$ are mutually commuting. Indeed, we have

$$[\hat{S}_x^2, \hat{S}_y^2] = \hat{S}_x^2 \hat{S}_y^2 - \hat{S}_y^2 \hat{S}_x^2$$

$$= \frac{1}{4} \begin{bmatrix} 0 & 0 & 0 \\ 0 & 4 & 0 \\ 0 & 0 & 0 \end{bmatrix} - \frac{1}{4} \begin{bmatrix} 0 & 0 & 0 \\ 0 & 4 & 0 \\ 0 & 0 & 0 \end{bmatrix} = 0, \tag{10.71a}$$

$$[\hat{S}_x^2, \hat{S}_z^2] = \hat{S}_x^2 \hat{S}_z^2 - \hat{S}_z^2 \hat{S}_x^2$$

$$= \frac{1}{2} \begin{bmatrix} 1 & 0 & 1 \\ 0 & 0 & 0 \\ 1 & 0 & 1 \end{bmatrix} - \frac{1}{2} \begin{bmatrix} 1 & 0 & 1 \\ 0 & 0 & 0 \\ 1 & 0 & 1 \end{bmatrix} = 0, \tag{10.71b}$$

$$[\hat{S}_y^2, \hat{S}_z^2] = \hat{S}_y^2 \hat{S}_z^2 - \hat{S}_z^2 \hat{S}_y^2$$

$$= \frac{1}{2} \begin{bmatrix} 1 & 0 & -1 \\ 0 & 0 & 0 \\ -1 & 0 & 1 \end{bmatrix} - \frac{1}{2} \begin{bmatrix} 1 & 0 & -1 \\ 0 & 0 & 0 \\ -1 & 0 & 1 \end{bmatrix} = 0. \tag{10.71c}$$

It is important to note that while the above calculation is carried out using the squared spin components in the x, y, and z directions, the same conclusion holds for the squared spin components in three arbitrary mutually orthogonal directions. This is because commutation relations are invariant under a unitary transformation [see Eq. (7.41)], and the unitary transformation here is simply a rotation about an arbitrary direction [see Section 8.6]. Therefore, for a spin 1 particle it is always possible (at least in principle) to simultaneously measure the squared spin components in three arbitrary mutually orthogonal directions. Moreover, for a spin 1 particle we have (again, in units of $\hbar^2$)

$$\hat{\mathbf{S}}^2 = \hat{S}_x^2 + \hat{S}_y^2 + \hat{S}_z^2 = 2\hat{I}, \tag{10.72}$$

which can be easily verified by an explicit calculation using the matrices given by Eq. (10.70). This in turn implies that the sum of the eigenvalues of the three operators $\hat{S}_x^2$, $\hat{S}_y^2$, and $\hat{S}_z^2$ is equal to 2. Since the eigenvalues of the operators $\hat{S}_x^2$, $\hat{S}_y^2$, and $\hat{S}_z^2$ are either 0 or 1, we conclude that the eigenvalue of one, and only one, of these operators is 0.

Now, let us assign a unit Euclidean vector $\mathbf{a}_j$ to each of the 117 vertices a_j in the graph of s', such that we can build from these unit vectors triplets of mutually orthogonal vectors. Indeed, each of these triplets can be obtained from a rotation of the basis vectors $\mathbf{e}_x$, $\mathbf{e}_y$, and $\mathbf{e}_z$ about some direction. The rule for our assignment of vectors is that when the vertices a_i, a_j, and a_k constitute a triangle, the corresponding vectors $\mathbf{a}_i$, $\mathbf{a}_j$, and $\mathbf{a}_k$ are one of these triplets of mutually orthogonal vectors. Moreover, since a triangle also corresponds to three pairs of vertices, each connected by a straight line, our above assignment implies that if a_j and a_k are connected by a straight line, then two orthogonal vectors $\mathbf{a}_j$ and $\mathbf{a}_k$ must be assigned to them. Consider all the 117 atomic statements a_j meaning "The squared spin component of the particle in the direction $\mathbf{a}_j$ is zero." Then, since from our above analysis in each triplet of mutually orthogonal directions the squared spin component of the particle is zero only in one direction, we find that all of the statements $b_{jk} = \neg(a_j \wedge a_k)$ as well as all of the statements $c_{ijk} = a_i \vee a_j \vee a_k$ hold true in quantum mechanics (at least) for the single-particle spin 1 system under consideration. This in turn means that the compound statement s' hold true in quantum mechanics despite the fact that it is classically a logical falsity.

In conclusion, the Kochen–Specker theorem provides a powerful argument against the possibility of interpreting quantum mechanics in terms of hidden variables. The fact that in quantum mechanics there can be value assignments that are not classically meaningful, means acknowledging that we cannot have a classical value assignment for all eigenvalues of quantum observables. This is to a certain extent not far away form the spirit of Bohr's argument [see Section 10.2], in which it is argued that such a value assignment should be done in the framework of a specific context that turns out to be experimental. Only that, we have proved that the true issue here, in the absence of any detection events, is ascription of reality to *properties* but not to observables. We have mentioned that Bohr's argument was still not completely clear about this point. In fact, this is what made his criticism of EPR weak and not sufficient to disprove EPR's principle of physical reality (one of the two main assumptions of EPR). If clarified in such a way, both Schrödinger's and Bohr's arguments, although in different ways, deal with properties.

10.8 Summary

In this chapter, we have

- Analyzed the path-breaking paper of Einstein, Podolsky, and Rosen (EPR), in which the authors aimed at showing the incompleteness of quantum mechanics.
- Considered Bohr's and Schrödinger's replies to the EPR argument.
- Presented Bohm's model for testing the EPR thought experiment using a spin singlet state.
- Proved the Bell theorem, which shows that local hidden variable theories do not satisfy quantum predictions and are in contradiction to experimental results.
- Introduced the amazing concept of entanglement swapping, which shows that entanglement does not need a direct dynamical interaction.
- Proved the Eberhard theorem, which shows that quantum systems display correlations between possible outcomes violating

separability but not Einstein's locality principle. More strongly, they are not interdependences between apparatus settings.

- Proved the Kochen–Specker theorem, which shows that it is not possible to have a classical value assignment for all eigenvalues of quantum observables.

Chapter 11

Quantum Information

We have already hinted at the relevance of information for measurement. In the last chapter we have considered the relevance of non-local correlations. It is time to deal again with measurement and non-locality from the point of view of information. Indeed, in this chapter, we shall learn how to treat quantum systems as information processors along the lines of quantum reversibility presented in Section 9.7. After providing a brief discussion on the nature of information, we introduce the concepts of information accessibility and partial information. We shall present a basic analysis of some commonly used quantum gates in quantum computation. Quantum teleportation and the quantum key distribution problem are then analyzed. Finally, the important concept of mutual information and its relation to quantum non-separability are discussed.

11.1 Nature of Information

Let us consider two entangled spin $\frac{1}{2}$ particles in a singlet state. If we measure their spin, we shall find out that, if one particle is spin up in a certain direction, the other particle is necessarily in spin down state in the same direction, and vice versa. That is,

Quantum Mechanics for Thinkers
Gennaro Auletta and Shang-Yung Wang
Copyright © 2014 Pan Stanford Publishing Pte. Ltd.
ISBN 978-981-4411-71-4 (Hardcover), 978-981-4411-72-1 (eBook)
www.panstanford.com

we shall find out either down–up or up–down, but never up–up or down–down. A statistics in which we would obtain all the four possible outcomes up–up, down–down, up–down, and down–up with equiprobability would be completely random, as can be seen by comparing, for instance, the entangled photon polarization state given by Eq. (7.131) with the product state given by Eq. (7.128). A random statistics is precisely what we should expect if the world consisted only of random events. If, on the contrary, we found that only two of the four possible outcomes occur, this would be really amazing and we will be forced to search for some explanation. In particular, we should infer that there is a correlation between the possible measurement outcomes. We note that to get a subset of possibilities in a completely random set represents an increase in order. Indeed, if we get spin up, we are able to predict that our partner will get spin down, and vice versa. Therefore, the key point we are trying to make here is that the existence of quantum correlations or, at a more general level, constraints of a new kind, is precisely the piece of the world that allows us to make predictions, to formulate theories and laws, and to have an ontological import of these theories and laws. Hence, the irreducibility of the randomness of quantum events does not prevent us from attributing to the quantum world an aspect of regularity that, although somehow opposite to the irreducibility of randomness, may well be articulated together with the latter [see Sections 7.2 and 7.3].

The natural question arises about the kind of reality quantum correlations consist in. We have already considered them in terms of the features that characterize quantum states and have also assigned to them a certain ontological import [see Sections 4.8 and 5.6]. Now, we need to deepen this examination. The term *feature* has a more physical flavor and stresses the particularity of *quantum* correlations. However, the concept of *correlation* has also a certain advantage, not only because it also applies to the classical case [see Section 7.9]. What we would like to stress here is that we live in a world that is physical but correlations are by definition formal, and this may well be the main reason why EPR postulated separability [see Section 10.1]: they wold avoid any non-physical factor [see also Section 9.3]. The key point is now that correlations and physical quantities (i.e., quantum observables, to which EPR would ascribe reality [see Section 10.2]) are tightly connected. It suffices to consider that correlations are mediated by, or instantiated in, physical interactions involving exchanges of dynamical quantities like momentum and energy. Now, how is it possible to put together something formal with something else that is not? We need a sort of quantity that is both formal and notwithstanding linked with the material dimension. We have seen that, given two entangled spin $\frac{1}{2}$ particles in a singlet state, the statistics of the measurement outcome is more ordered (with two out of four possible outcomes) than when there is no entanglement between the particles (with four out of four possible outcomes). There is a language for dealing in the most general way with such kind of problems, namely, the language of information. As a matter of fact, information is a dimensionless quantity that is basic for the understanding of many physical systems, among them the quantum ones.

First, let us consider an example drawn from the microscopic world. DNA (deoxyribonucleic acid) is a good example of the encoding of information.[a] As first discovered by Watson and Crick in 1953,[b] the structure of DNA consists of two long polymers of simple units called nucleotides, with backbones made of sugars and phosphate groups joined by ester bonds. It is quite interesting to observe that, in this case, a sharp separation between chemical bonds and information combinatorics is necessary. As we have said, information is a formal quantity. For instance, the sequence of words on a written page cannot depend on the chemistry of the page. Also in the case of DNA, the four nucleobases (or bases for short) A, G, C, and T are not involved in the chemical bonds constituting the phosphate sugar backbone along a single strand, that is, the linkage of a nucleotide with another is not constrained by the chemical details of the bases. This is the reason why the bases may instantiate truly informational connections along the same strand. In fact, this ensures that any combination is in principle possible and, therefore, that DNA is able to store information. Therefore, we have some kind of formal entity that, although grafted onto the physics or chemistry of life, it is also somehow independent.

---

[a](Auletta, 2011a, Chapter 7).
[b](Watson/Crick, 1953).

Information, together with its measure, the Shannon entropy, is precisely concerned with the issue of singling out a subset of elements (e.g., the message or the information we like to acquire) from a larger set (e.g., the set of all possible messages or at least the set of elementary units out of which all possible messages can be composed, like the alphabet). The Shannon entropy measures the unlikeness to perform such an extrapolation, which is expressed by the probability that the latter does not occur or is not chosen [see Section 9.6]. Obviously, if there is no (syntactical) order among the units and the sequence of the latter is random, to acquire or to guess the right message will be much more difficult than the case in which there is some rule (for instance, in some cases, we can understand that an encrypted message represents an English rather than an Italian sentence because of the bigger or smaller frequency of some letters). This fully justifies our previous treatment of entanglement, as we shall see now.

In conclusion, we stress that what is interesting about information is that it is a quantity "sitting" somehow between pure mathematics and physical reality. It must necessarily be instantiated in some physical media but, as we have seen, both the information content and its syntax are not dependent on the specific physical characteristics of the medium (although the latter must satisfy some general requirements to carry information). Instead, the information carried by a physical medium resides in the structure, the order, the configuration, the pattern, or the combination of discrete physical elements. Let us now consider these issues in details.

11.2 Information Accessibility

Let us consider a very useful geometrical representation of pure states of a two-state quantum system, the so-called Bloch sphere (also known as the Poincaré sphere). Recall that according of the superposition principle [see Principle 2.1], given an orthonormal basis represented by the vectors $|0\rangle$ and $|1\rangle$, the state vector of a two-state system can be written as

$$|\psi\rangle = c_0|0\rangle + c_1|1\rangle, \tag{11.1}$$

where c_0 and c_1 are complex coefficients satisfying the normalization condition $|c_0|^2 + |c_1|^2 = 1$. Since each complex number can be specified by a non-negative modulus and a phase, for two complex coefficients whose square moduli summed to 1 we need one modulus and two phases to uniquely specify a state vector like $|\psi\rangle$ (as one of the modulus can be expressed in terms of the other). However, thanks to the fact that state vectors that differ by a global phase factor (i.e., an overall complex factor with unit modulus) represent the same physical state vector [see Section 5.2], it turns out that we only need one modulus (varying between 0 and 1) and one relative phase (ranging from 0 to 2π) to uniquely specify a *physical* state vector. The Bloch sphere representation provides a convenient geometric parameterization of the single modulus and the relative phase respectively in terms of the polar and azimuthal angles of a point on a unit sphere [see also Figs. 1.1 and 8.6]. Indeed, in the Bloch sphere representation the state vector $|\psi\rangle$ may be written as (up to a global phase factor)[a]

$$|\psi\rangle = \cos\frac{\theta}{2}|0\rangle + e^{i\phi}\sin\frac{\theta}{2}|1\rangle, \qquad (11.2)$$

which is a superposition of the basis states $|0\rangle$ and $|1\rangle$, represented by the north and south poles of the sphere, respectively [see Fig. 11.1]. In the above expression, the parameter θ is the polar angle (with $0 \leq \theta \leq \pi$), which covers the meridians and represents the relative contributions of the components $|0\rangle$ and $|1\rangle$ to the superposition [see Section 5.5], while the parameter ϕ is the azimuthal angle (with $0 \leq \phi < 2\pi$), which covers the parallels of the sphere and represents the relative phase between the components $|0\rangle$ and $|1\rangle$ of the superposition [see Section 2.6]. We recall that a relative phase is the distance (modulo the wavelength) between a peak in one component of the superposition and the corresponding peak in the other component [see Fig. 2.1].

The state $|\psi\rangle$ represents a new information entity called the *quantum bit* (or *qubit* for short). In other words, quantum superposition spread what is classically a one-bit state (either 0 or 1, geometrically represented here by the two poles) first to an infinite number of possible combinations (thanks to the parameter θ that accounts for their relative contribution) and then to a

---

[a](Auletta, 2005).

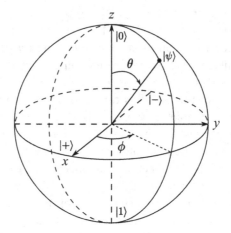

Figure 11.1 The state vector of a two-state quantum system, for instance $|\psi\rangle$, can be represented by a point on a unit sphere, called the Bloch sphere. The parameters ϕ (the angle between the x axis and the projection of $|\psi\rangle$ on the equatorial plane) and θ (the angle between $|\psi\rangle$ and the z axis) are sufficient to specify its location. The states $|+\rangle$ and $|-\rangle$ lie on the equatorial plane and represent two symmetric superpositions of $|0\rangle$ and $|1\rangle$, located at the north and south poles, respectively. (For the sake of convenience, we have chosen the diameter connecting them to be along the x axis as well as the diameter connecting $|0\rangle$ and $|1\rangle$ along the z axis.) In other words, orthogonal states that were previously represented by orthogonal vectors in a two-dimensional Hilbert space are here represented by antipodal points on the Bloch sphere.

further dimension of infinite possibilities of interference (thanks to the parameter ϕ that accounts for the relative phase). To a first approximation, we are authorized to assume that the two orthogonal states $|0\rangle$ and $|1\rangle$ represent information since they indeed represent the measurement outcomes of some observable with two distinct eigenvalues. We also stress that these two states are not different at all from the superposition of them given by Eq. (11.2). In fact, if one chooses to measure another observable, whose eigenstates are represented by, say,

$$|+\rangle = \frac{1}{\sqrt{2}}\left(|0\rangle + |1\rangle\right), \quad |-\rangle = \frac{1}{\sqrt{2}}\left(|0\rangle - |1\rangle\right), \quad (11.3)$$

where $|+\rangle$ and $|-\rangle$ are orthogonal to each other and hence they also constitute an orthonormal basis, the possible outcomes would be $|+\rangle$ and $|-\rangle$ [see Sections 8.6 and 9.4]. This is due the fact that superposition is a relative concept [see Section 4.5]. Here, for the sake of simplicity, we have chosen the states $|+\rangle$ and $|-\rangle$ as symmetric superpositions of the states $|0\rangle$ and $|1\rangle$ such that in the Bloch sphere they are located on the equator with equal distance from the poles [see Fig. 11.1]. Since the states $|+\rangle$ and $|-\rangle$ are possible measurement outcomes (if we decided to measure the observable of which they are eigenstates), they also will represent information in spite of the fact that they are superpositions of the states $|0\rangle$ and $|1\rangle$.

To demonstrate the above exposition with an specific example, let us consider measuring the spin component of a spin $\frac{1}{2}$ particles. If we choose to measure the spin component in the z direction, the possible outcomes would be $|\uparrow_z\rangle$ and $|\downarrow_z\rangle$, which constitute the usual orthonormal basis for the spin $\frac{1}{2}$ space. Since there is nothing special about the z direction, we may choose to measure the spin component of the particle in other direction, say in the x direction. Then the possible outcomes would be $|\uparrow_x\rangle$ and $|\downarrow_x\rangle$, which also constitute an orthonormal basis for the spin $\frac{1}{2}$ space. If, for some reason of convenience, we decide to denote the states $|\uparrow_z\rangle$ and $|\downarrow_z\rangle$ by $|0\rangle$ and $|1\rangle$, then according to Eq. (8.114a) the states $|\uparrow_x\rangle$ and $|\downarrow_x\rangle$ will be written exactly as the superposition states $|+\rangle$ and $|-\rangle$ given by Eq. (11.3). We therefore conclude that information represented by a qubit is the information contained in the superposition of two arbitrary orthogonal states, and is independent of the information acquisition procedures (measurement is indeed a particular form of information acquisition). There are several requisites for the encoding of information, and these are satisfied by quantum systems:

(i) There exist (at least) two orthogonal states that represent two mutually exclusive measurement outcomes of a quantum system. According to the superposition principle [see Principle 2.1], these two orthogonal states constitute an orthonormal basis and can be linearly combined into an arbitrary superposition state. Note that while not all combination rules are linear, linearity however is necessary for quantum information encoding. The resultant orthonormal basis is referred to as the computational basis, and the two computational basis states are conventionally denoted by $|0\rangle$ and $|1\rangle$.

(ii) By varying the coefficients of the superposition (i.e., the parameters θ and ϕ in the Bloch sphere representation), we can obtain an infinite number of possible superposition states. In other words, it is in principle possible to represent an arbitrary qubit by means of a superposition of the two computational basis states $|0\rangle$ and $|1\rangle$. Therefore, the coefficients of the superposition are the rules according to which we combine the two computational basis states $|0\rangle$ and $|1\rangle$, and the latter can be understood as a (binary) code, as in quantum computation.

(iii) It is always possible to choose another set of orthogonal states, say $|+\rangle$ and $|-\rangle$ as in the above example, as the computational basis states. In other words, different computational bases are different codes used to encode quantum information. For a given state, there are specific rules that allow the translation of expressions from one computational basis to another. These translation rules are precisely unitary transformations for the corresponding changes of basis [see Section 4.6]. Indeed, this is a necessary requirement for having information.

Having established these basic elements, we can now deepen our investigation. At a formal level or *a priori* we have said that there are no appreciable differences between an initial state in superposition and a measurement outcome: indeed, we have remarked that a quantum state represents information since it is a possible measurement outcome, and have added that such information is independent of the information acquisition procedures (i.e., local measurements). However, this is only one side of the coin. From the point of view of this information acquisition or *a posteriori* there is a remarkable difference. First, let us stress which kind of information is represented by a qubit and consider to this purpose the state $|\psi\rangle$ of a system given by Eq. (11.2). What is really amazing here is that the amount of information contained in a qubit is infinite, since, as mentioned, the parameters ϕ and θ allow the state $|\psi\rangle$ to be located *anywhere* on the Bloch sphere (which obviously has an infinite number of points). However, what is crucial is that such an infinite amount of information contained in the state $|\psi\rangle$ is not only inaccessible to an observer at a given moment, but also inaccessible to any observer at any time. As a a matter of fact, every time we try to measure the system we need two steps after the initial preparation. We are first obliged to *choose* a specific observable [see Section 6.9]. Suppose that we choose to measure an observable whose eigenstates are $|0\rangle$ and $|1\rangle$. Then, we will detect the system and obtain either $|0\rangle$ or $|1\rangle$ as the measurement outcome. In other words, the state of the system after the measurement will be reduced with a certain probability to either $|0\rangle$ or $|1\rangle$. However, since both $|0\rangle$ and $|1\rangle$ represent a single point on the Bloch sphere (the two poles, here), then, when we obtain one or the other as the outcome of a measurement, we have acquired *at most one bit* of information, a much smaller amount of information than what is initially contained in a qubit. This is the essence of the *Holevo theorem*,[a] which can be formulated as follows.

Theorem 11.1 (Holevo) The information that can be acquired from one qubit is at most one bit.

In other words, when measuring a system we dump into the environment precisely the non-local features that contribute to the infinite amount of information contained in the state $|\psi\rangle$, that is, that spread the information on the whole Bloch sphere [see Section 9.5]. In fact, this could be another way to consider the reduction of the wave packet that occurs during a measurement [see Section 9.3]. This allows us to reconcile the *a priori* (encoding) and the *a posteriori* (information acquiring) understanding of information. Every time the appropriate conditions for the dynamic interaction between two open systems are satisfied (e.g., whenever we have a measurement or a measurement-like interaction process), we are able to acquire a classical bit of information in the form of a measurement outcome. The fact that we are free to measure either the observable whose eigenstates are $|0\rangle$ and $|1\rangle$ or the observable whose eigenstates are $|+\rangle$ and $|-\rangle$ shows clearly that what is inaccessible is not a superposition of $|0\rangle$ and $|1\rangle$ *as such*, but the whole amount of information contained in an *arbitrary unknown state* $|\psi\rangle$ (independently of whether we decide to measured it or

---

[a](Holevo, 1998).

not), which besides the eigenstates also comprehends all possible interdependences between them, i.e., features. This definitively establishes that it is not only what is actually communicated and received that can be called information.

We note however that there could be a plausible argument against the validity of the above analysis *if* it is possible to create identical copies of an arbitrary unknown quantum state. Indeed, by doing so an observer would be able to make a large (possibly infinite) number of identical copies of the state $|\psi\rangle$. Then, by measuring different observables on those identical copies of the state $|\psi\rangle$, the observer could in principle measure all the possible observables on the *same* state $|\psi\rangle$ over and over again. Since from each measurement outcome the observer would acquire one bit of information, in this way the infinite amount of information contained in the state $|\psi\rangle$ could be retrieved and therefore be accessible to the observer. As a matter of fact, such a possibility is ruled out by the celebrated *no-cloning theorem*, first proved independently by Wootters and Zurek,[a] and by Dieks[b] in 1982.

Theorem 11.2 (No-Cloning) It is not possible to create identical copies of an arbitrary unknown quantum state.

We shall prove the no-cloning theorem by *reductio ad absurdum*. For the sake of simplicity, let us restrict ourselves to the two-state quantum system that is under current consideration. The generalization to many-state quantum systems is straightforward. Suppose that cloning of an arbitrary unknown state $|\psi\rangle$ is possible, then this would mean preparing a second qubit initially in some "blank state" $|B\rangle$ and copying the original state $|\psi\rangle$ on to it. This in turn implies the existence of a unitary transformation $\hat{U}_{\text{clone}}$ such that

$$|\psi\rangle \otimes |B\rangle \xrightarrow{\hat{U}_{\text{clone}}} |\psi\rangle \otimes |\psi\rangle. \tag{11.4}$$

For this to be a true cloning transformation it is necessary for the transformation to hold for arbitrary qubit states. In particular, the cloning transformation on the computational basis states $|0\rangle$ and $|1\rangle$

---

[a](Wootters/Zurek, 1982).
[b](Dieks, 1982).

is given by

$$|0\rangle \otimes |B\rangle \xrightarrow{\hat{U}_{\text{clone}}} |0\rangle \otimes |0\rangle, \quad |1\rangle \otimes |B\rangle \xrightarrow{\hat{U}_{\text{clone}}} |1\rangle \otimes |1\rangle. \quad (11.5)$$

Then, due to linearity of the unitary transformation, the cloning transformation on the superposition state $|\psi\rangle = c_0|0\rangle + c_1|1\rangle$ of $|0\rangle$ and $|1\rangle$ is given by

$$|\psi\rangle \otimes |B\rangle = (c_0|0\rangle + c_1|1\rangle) \otimes |B\rangle \xrightarrow{\hat{U}_{\text{clone}}} c_0|0\rangle \otimes |0\rangle + c_1|1\rangle \otimes |1\rangle, \quad (11.6)$$

where c_0 and c_1 are arbitrary complex amplitudes such that the normalization condition $|c_0|^2 + |c_1|^2 = 1$ holds. It is obvious that the transformed state is not equal to the expected state $|\psi\rangle \otimes |\psi\rangle$, except for the case in which one of the two amplitudes c_0 and c_1 is identically zero. Therefore, we conclude that there does not exist a unitary transformation that can clone an arbitrary unknown quantum state.[a]

In classical physics, instead, it is assumed that the state of a physical system is an observable, which means that it is possible to extract all the information contained in an arbitrary physical state [see Section 7.8]. This is however not supported by empirical facts. Indeed, let us consider the case in which we desire to measure exactly the circumference of a ring (e.g., an ordinary wedding ring). Let us avoid any complications that derive from the matter structure and consider a pure ideal case in which matter would be totally uniform (continuous) and static. In this case, we would very soon run into the difficulty that the circumference as well as probably the radius of the ring would be a real number. Now, we cannot (even in principle) exhaust the infinite sequence of decimals that constitute a real number. This means that we cannot measure the circumference of the ring with infinite precision [see also Section 6.3]. In other words, we cannot acquire the whole information contained in a system because of the finite resolution of measurement (i.e., the impossibility to reduce to zero the measurement error).

We may summarize what have been said in an *information accessibility principle*, which can be formulated as follows.

Principle 11.1 (Information Accessibility) *The whole information contained in a system can only be partially accessed.*

---

[a](D'Ariano/Yuen, 1996).

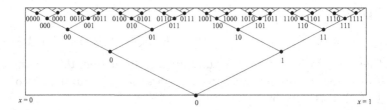

Figure 11.2 The continuous unit interval [0, 1] can be viewed as a limiting case of digitization in which the number of digits approaches infinity. Adapted from (Auletta, 2011a, p. 38).

From a quantum mechanical perspective, the reason is that we have non-local features that cannot be classically acquired as information [see also Section 6.9]. Classically, this is due to the finite resolution of measurement. However, in both cases the ultimate source of that impossibility seems to be the discrete nature of encoding [see Fig. 11.2] and therefore also of *any* information acquisition.

Problem 11.1 Show that each pair of antipodal points on the Bloch sphere corresponds to mutually orthogonal state vectors.

11.3 Potential Information

Information as such is only a formal quantity. It needs to be activated or acquired. As mentioned, and as we shall see below, information acquisition is a dynamical process in which the entropic flux between the involved interacting systems is also relevant. Therefore, the difference between a qubit and a bit is related to the measurement procedure within a local context [see Section 9.7], that is, related to a given environment. A bit can be understood as a qubit that has been made active or has been accessed.[a] Both can be interpreted as the minimum information entities. The only difference is that a qubit is a bit once it is obtained, that is, if someone had chosen or will choose to perform a possible operation (or if some objective conditions had been or will be spontaneously produced) through which this state has been or could be obtained. In

---

[a](Auletta, 2006b).

other words, it is not the *form* of the superposition [see Eq. (11.3)] that distinguishes $|0\rangle$ from $|+\rangle$, but the *fact* that we have obtained $|0\rangle$ instead of $|+\rangle$ as the actual information, given (i) the local choice of measuring the observable of which $|0\rangle$ is an eigenstate, and (ii) the selection of $|0\rangle$ in certain environmental conditions. Therefore, London and Bauer's point of view that the final selection act is relevant to the measurement problem is ultimately not ungrounded [see Sections 9.3 and 9.5], provided that we free it from any subjective form.

There is also a further and important difference between a prepared state and a measurement outcome. As we have said, the states $|0\rangle$ and $|1\rangle$ as such (as well as the states $|+\rangle$ and $|-\rangle$ as such) can be taken to be the *possible* but not actual measurement results. In this sense, they represent *potential* information, i.e., information that has not yet been acquired (and possibly never will be).[a] In other words, the potential information is the relation between possible events, so that we can take this to be the definition of information as such.[b] We note that one however often speaks of potential information in the sense of information that can be later received. Here, we mean potentiality in a more radical sense as information that is to be but has not yet been acquired. For this reason, the concept of the potential information allows us to deal with both the classical and the quantum case because in both cases the whole information contained in a system can only be partially acquired or accessed [see Principle 11.1].

Let us give an evidence supporting this view. When two qubits are entangled, they constitute a further new information entity called the *entangled bit* (or *ebit* for short). The interdependence displayed by the two entangled qubits (or an ebit) is not mediated by any detectable physical signal, and therefore there is no causal connection either, as shown in the previous chapter. It is non-local in nature and is an immediate consequence of features. The reason is again to be found in quantum information. Since information expresses the relation between possible events, it is

---

[a](Von Weizsäcker, 1972), (Auletta, 2011a, Chapter 2), see also (Küppers, 1990, pp. 36–38).
[b]A. Zeilinger, private communication.

also independent of space and time, and entanglement is a natural consequence. An ebit is therefore a kind of quantum channel (especially when qubits are maximally entangled), i.e., a typical quantum mechanical information sharing between systems. This is very relevant since an ebit allows things to be done that cannot be performed classically, like transmitting the information contained in an unknown quantum state in quantum teleportation [see Section 11.5] or encrypting a text against eavesdropping in quantum cryptography [see Section 11.6]. Thus, entanglement can be interpreted as a potential *resource* to transfer additional quantum information in the future at no further cost.[a] This justifies the notion of potential information. Bell seemed to share the viewpoint expressed here when he introduced the concept *beable* for expressing the capability of quantum systems to become actual or to be.[b]

Therefore, the *actual* information is only the information which has been actually acquired. While following the previous examination let us call *encoded information* the information contained in any combination of mutually orthogonal states of a physical system according to a certain set of prescribed rules. Each set of mutually orthogonal states is called the alphabet, that is, a collection of symbols (or characters), while the set of rules is known as the code system. It is noted that encoded information is always potential, both classically and quantum mechanically. A good measure of the classical information content of encoded information is represented by the Kolmogorov measure of complexity, namely, the measure of the computational resources needed to specify the string of characters instantiating encoded information.[c] This measure has also been generalized to quantum mechanics.[d] Therefore, both classically and quantum mechanically we may posit an information activation principle as follows.

Principle 11.2 (Information Activation) *Any encoded information is as such potential, and may be activated only by actual (external) conditions.*

---

[a] (Horodecki *et al.*, 2005).
[b] (Bell, 1973, 1976).
[c] (Kolmogorov, 1963).
[d] (Berthiaume *et al.*, 2000).

Moreover, quantum systems can be understood as *information processors*, in the sense that they evolve in time by changing their states according to a unitary transformation and hence can be thought of as transforming an initial encoded information into some other encoded information. We note that this information processing is reversible, provided that there is no information selection and therefore no measurements involved. This aspect is crucial and even the most basic one when dealing with quantum mechanics, and we have already seen its relevance to the measurement problem when a larger system that is reversible is considered [see Section 9.7]. This is also the backbone of the fast growing field of quantum computing.[a] For these reasons, it is conceivable that all quantum systems can be considered in terms of information.[b]

Now, the issue is how can we make understandable that information is processed in a reversible way and what is the relation between this reversibility and the irreversible information acquisition represented by measurement. A natural question, indeed, is whether it is possible to have information processing without any energy expenditure. We have already said that it is possible to process information in a reversible way. However, this issue demands some additional considerations. Landauer showed that *throwing bits away* (i.e., selecting them), not processing them, requires an expenditure of energy.[c] This is the Landauer's principle. In this way, as anticipated in Section 9.6, we have in fact a connection between information selection, informational entropy, and thermodynamic entropy. Indeed, the cost of information erasure has been precisely quantified. The erasure of a classical bit of information will cost dissipating a minimum possible amount of energy $k_B T \ln 2 \approx 0.6931 k_B T$ into the environment.[d] Successively, Bennett explained how a computer could be designed that would not discard information and thus virtually dissipate no energy.[e] Bennett

---

[a] (Nielsen/Chuang, 2011).
[b] (Wheeler, 1990), (Chiribella *et al.*, 2011), (D'Ariano, 2012).
[c] (Landauer, 1961), (Landauer, 1996), (Landauer, 1991).
[d] (Plenio/Vitelli, 2001).
[e] (Bennett, 1973), (Bennett, 1982), (Bennett/Landauer, 1985).

showed that each step in the computation can be carried out in a way that allows not only the output to be deduced from the input but also the input to be deduced from the output. In other words, the computer can also run backwards. Such a computer, after having processed information in the ordinary way, could put itself into the reverse mode so that each step is undone. No information is erased here and accordingly no energy is dissipated (as opposed to a measurement process). This is precisely the way in which a quantum system behaves when not interacting with other systems or when a sufficiently large system is considered [see Section 9.6]. On the contrary, during a measurement process part of the information contained in a quantum system is selected and dumped into the environment. Because and only because of this selection, there is energy expenditure. This energy expenditure together with the increase in entropy makes the measurement process for quantum systems locally irreversible.

11.4 Quantum Computation

In this and in the following sections, we shall first explore the non-local and reversible aspects of the quantum information (those that in fact make the difference with respect to the classical case). The first idea of quantum computation capable of exploiting the potentialities represented by the qubits was developed in the 1980s.[a] It mainly relies on previous ideas that a Turing machine can be simulated by unitary transformations of a quantum system. A *Turing machine*[b] is a a recursive device that manipulates symbols on a strip of tape according to a table of rules [see Box 11.1]. In the classical case, a Turing machine is an irreversible device, while, if we desire to make use of the specificity of quantum information, we need unitary transformations, which are reversible. As we have already shown at a general level, an irreversible device can be embedded in a larger device that is reversible [see Section 9.7].

---

[a] (Feynman, 1982), (Feynman, 1986), (Deutsch, 1985).
[b] (Turing, 1936), (Turing, 1937).

Box 11.1 Turing machine

A Turing machine is a theoretical computing machine invented by Turing during 1936–37 to serve as an idealized model for mathematical calculation. A Turing machine is composed of four parts [see Fig. 11.3]:

(i) A *tape* that is divided into cells, one next to the other. Each cell can have a symbol from a finite alphabet.

(ii) A *head* that can read and write symbols on the tape and move along the tape left or right one cell at a time.

(iii) A *register* that stores the state of the Turing machine, one of finitely many.

(iv) A *table* of a finite collection of instructions that tell the machine how operations are performed.

The action of the Turing machine is made up of discrete steps, and each step is determined by two initial conditions: the current state of the machine and the symbol currently scanned by the head. Given these two conditions, the machine performs the following operations in sequence:

(i) Assume the same or a new state of the machine.

(ii) Write a symbol into the scanned cell.

(iii) Move the head left or right along the tape or stop.

The Turing machine is the simplest form of a computer. It plays an important role in the development of modern computer science as the world knows it.

A classical or quantum computer consists of *wires* and *gates* [see Fig. 9.4]. The wires transmit information, whereas the gates perform transformations of bits or qubits to process information. A wire can be represented by a gate that does nothing, i.e., the identity gate, whose function is precisely to further transmit information without changing it. According to the number of the input bits, quantum gates can be classified as either one-qubit or multi-qubit gates. The number of qubits in the input and output of a quantum gate have to be equal. Quantum gates are represented by unitary matrices, and

tape

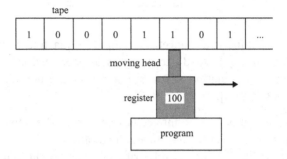

Figure 11.3 Scheme of a Turing machine. A moving head goes (here from the left to the right) along a tape that is divided into cells (here, storing binary encoded information) with the task to read and eventually to modify (write) the content of each cell. A register stores the state of the Turing machine. Finally the program is the appropriate set of instructions.

a quantum gate which acts on k qubits is represented by a $2^k \times 2^k$ unitary matrix. An elementary one-qubit gate is the phase shift gate, which on the computational basis states $|0\rangle$ and $|1\rangle$ acts as

$$\hat{U}_\phi |0\rangle = |0\rangle, \quad \hat{U}_\phi |1\rangle = e^{i\phi} |1\rangle. \tag{11.7}$$

As usual writing the basis states $|0\rangle$ and $|1\rangle$ in component form as

$$|0\rangle = \begin{pmatrix} 1 \\ 0 \end{pmatrix} \quad \text{and} \quad |1\rangle = \begin{pmatrix} 0 \\ 1 \end{pmatrix}, \tag{11.8}$$

respectively, then we can express $\hat{U}_\phi$ in matrix form as

$$\hat{U}_\phi = \begin{bmatrix} 1 & 0 \\ 0 & e^{i\phi} \end{bmatrix}, \tag{11.9}$$

which is precisely the unitary operator representing the phase shifter PS on the upper path in the Mach–Zehnder interferometer that produces a phase shift ϕ to the state $|u\rangle$ [see Section 2.6 and Problem 4.10]. Some common examples of the phase shift gates are the $\frac{\pi}{8}$ gate for $\phi = \frac{\pi}{4}$,[a] the phase gate for $\phi = \frac{\pi}{2}$, and the Pauli-Z gate

---

[a]The curious reader might wonder why $\hat{U}_{\frac{\pi}{4}}$ is called the $\frac{\pi}{8}$ gate when the phase shift is actually $\frac{\pi}{4}$. The reason for this is simply because the gate has been historically written as

$$\hat{U}_{\frac{\pi}{4}} = e^{i\pi/8} \begin{bmatrix} e^{-i\pi/8} & 0 \\ 0 & e^{i\pi/8} \end{bmatrix},$$

which, up to an irrelevant global phase, shifts the phase of $|0\rangle$ and $|1\rangle$ by $\frac{\pi}{8}$ but in the opposite sense.

for $\phi = \pi$ (see below). Another elementary one-qubit gate is the NOT gate, whose action on the computational basis states $|0\rangle$ and $|1\rangle$ is to change $|0\rangle$ to $|1\rangle$ and $|1\rangle$ to $|0\rangle$. In other words, we have

$$\hat{U}_{NOT}|0\rangle = |1\rangle, \quad \hat{U}_{NOT}|1\rangle = |0\rangle. \qquad (11.10)$$

In matrix form $\hat{U}_{NOT}$ is given by

$$\hat{U}_{NOT} = \begin{bmatrix} 0 & 1 \\ 1 & 0 \end{bmatrix}. \qquad (11.11)$$

It is noted that $\hat{U}_{NOT}$ for a single qubit is exactly identical to the Pauli spin matrix $\hat{\sigma}_x$ [see Eq. (8.111)]. For this reason, the NOT gate is also called the Pauli-X gate (or X gate for short). It represents a rotation of the Bloch sphere about the x axis by π radians. Similarly, the Pauli matrices $\hat{\sigma}_y$ and $\hat{\sigma}_z$ also represent two one-qubit gates, called respectively the Pauli-Y gate (or Y gate for short) and Pauli-Z gate (or Z gate for short). They represent respectively rotations of the Bloch sphere about the y and z axes by π radians.

A more involved (and also useful) one-qubit gates is the so-called Hadamard gate, whose action on the computational basis states $|0\rangle$ and $|1\rangle$ is given by[a]

$$\hat{U}_H|0\rangle = \frac{1}{\sqrt{2}}(|0\rangle + |1\rangle) = |+\rangle, \qquad (11.12a)$$

$$\hat{U}_H|1\rangle = \frac{1}{\sqrt{2}}(|0\rangle - |1\rangle) = |-\rangle. \qquad (11.12b)$$

It follows that $\hat{U}_H$ can be expressed in matrix form as

$$\hat{U}_H = \frac{1}{\sqrt{2}} \begin{bmatrix} 1 & 1 \\ 1 & -1 \end{bmatrix}. \qquad (11.13)$$

For obvious reason, the operator $\hat{U}_H$ is also called the Hadamard operator. Comparing the above equation with Eq. (4.35), we find that the Hadamard gate represents the action of a 50–50 beam splitter. Also note that, as can been seen from Eq. (11.3) and the Bloch sphere depicted in Fig. 11.1, the Hadamard gate transforms the state $|0\rangle$ (or $|1\rangle$) to a state "halfway" between the state and its "negation" (i.e., orthogonal) state $|1\rangle$ (or $|0\rangle$). However, two consecutive applications of the Hadamard gate are not equal to the NOT gate but to the

---

[a](Cerf *et al.*, 1998).

identity gate, due to the fact that the Hadamard gate is the inverse of itself. As for the Pauli matrices, this happens because $\hat{U}_H$ is both Hermitian and unitary. When we have an initial two-qubit state $|00\rangle = |0\rangle_1 \otimes |0\rangle_2$, the Hadamard transform takes the form

$$\hat{U}_H \otimes \hat{U}_H |00\rangle = \hat{U}_H |0\rangle_1 \otimes \hat{U}_H |0\rangle_2$$

$$= \frac{1}{2}(|0\rangle_1 + |1\rangle_1)(|0\rangle_2 + |1\rangle_2)$$

$$= \frac{1}{2}(|00\rangle + |01\rangle + |10\rangle + |11\rangle), \qquad (11.14)$$

which is the symmetric superposition of the conventional binary encoding of the four (2^2) integer 0, 1, 2, and 3. In the case of three qubits, we have

$$\hat{U}_H^{\otimes 3} |000\rangle = \frac{1}{2\sqrt{2}}(|0\rangle_1 + |1\rangle_1)(|0\rangle_2 + |1\rangle_2)(|0\rangle_3 + |1\rangle_3)$$

$$= \frac{1}{2\sqrt{2}}(|000\rangle + |001\rangle + |010\rangle + |011\rangle + |100\rangle$$

$$+ |101\rangle + |110\rangle + |111\rangle), \qquad (11.15)$$

where we have used the shorthand notation $\hat{U}_H^{\otimes 3} = \hat{U}_H \otimes \hat{U}_H \otimes \hat{U}_H$. The above expression again is the symmetric superposition of the conventional binary encoding of the eight (2^3) integers 0, 1, 2, ..., 7. These results can be generalized straightforwardly to the case of n qubits, in which we obtain the symmetric superposition of the binary encoding of the 2^n integers 0, 1, 2, ..., 2^{n-1}.

Since the actions of a 50–50 beam splitter and a phase shifter are represented respectively by the Hadamard gate and the phase shift gates, the Mach–Zehnder interferometer considered in Chapter 2 can be viewed as a quantum device composed of wires and quantum gates. In other words, the series of beam splitter 1, phase shifter, and beam splitter 2 in Fig. 2.3 can be thought of as a succession of a Hadamard gate, a phase shift gate, and a second Hadamard gate as depicted in Fig. 11.4. Suppose that the input state of the qubit is given by $|0\rangle$, then we have the transformation

$$|0\rangle \xrightarrow{\hat{U}_H} \frac{1}{\sqrt{2}}(|0\rangle + |1\rangle)$$

$$\xrightarrow{\hat{U}_\phi} \frac{1}{\sqrt{2}}(|0\rangle + e^{i\phi}|1\rangle)$$

$$\xrightarrow{\hat{U}_H} \frac{1}{2}\left[(1 + e^{i\phi})|0\rangle + (1 - e^{i\phi})|1\rangle\right], \qquad (11.16)$$

$$|i\rangle \quad\boxed{\text{H}}\quad\boxed{\phi}\quad\boxed{\text{H}}\quad|f\rangle$$

Figure 11.4 Graphical representation of the quantum device equivalent to a Mach–Zehnder interferometer. The wires and the quantum gates are represented by lines and the squares, respectively. The diagram are to be read from *left to right*, with the line on the left representing the input state $|i\rangle$ of the qubit and the line on the right the output state $|f\rangle$ as transformed by a Hadamard gate, a phase shift gate, and a second Hadamard gate.

which, when compared with Eq. (2.29), is exactly the state of the photon after it leaves the beam splitter BS2 but before it is set to be detected at D1 or D2.

One of the most important two-qubit gate is the controlled-NOT (CNOT) gate. It leaves the control qubit unchanged and performs a transformation on a target qubit depending on the state of the control qubit. If the control qubit is in the state $|0\rangle$ then the target qubit is unchanged, but if it is in the state $|1\rangle$ then the quantum NOT gate is applied to the target qubit. Thus, we have

$$\hat{U}_{\text{CNOT}}|00\rangle = |00\rangle, \quad \hat{U}_{\text{CNOT}}|01\rangle = |01\rangle, \tag{11.17a}$$

$$\hat{U}_{\text{CNOT}}|10\rangle = |11\rangle, \quad \hat{U}_{\text{CNOT}}|11\rangle = |10\rangle, \tag{11.17b}$$

where the first qubit is the control qubit and the second the target qubit. In the computational basis of the two-qubit Hilbert space

$$|00\rangle = \begin{pmatrix} 1 \\ 0 \\ 0 \\ 0 \end{pmatrix}, \quad |01\rangle = \begin{pmatrix} 0 \\ 1 \\ 0 \\ 0 \end{pmatrix}, \quad |10\rangle = \begin{pmatrix} 0 \\ 0 \\ 1 \\ 0 \end{pmatrix}, \quad |11\rangle = \begin{pmatrix} 0 \\ 0 \\ 0 \\ 1 \end{pmatrix}, \tag{11.18}$$

the CNOT gate can be expressed in matrix form as

$$\hat{U}_{\text{CNOT}} = \begin{bmatrix} 1 & 0 & 0 & 0 \\ 0 & 1 & 0 & 0 \\ 0 & 0 & 0 & 1 \\ 0 & 0 & 1 & 0 \end{bmatrix}. \tag{11.19}$$

From the above expression, it is obvious that $\hat{U}_{\text{CNOT}}$ is both Hermitian and unitary. The CNOT gate can be thought of as the quantum analogue of the classical XOR gate introduced in Section 9.7 [see Table 9.4]. In fact, it is easy to show that when the CNOT gate acts on

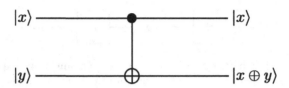

Figure 11.5 Graphical representation of the CNOT gate. The top line represents the control qubit, the bottom line the target qubit. The states $|x\rangle$ and $|y\rangle$ represent the computational basis states $|0\rangle$ and $|1\rangle$, and $\oplus$ denotes modulo-2 addition.

the computational basis states, the output state of the target qubit may be expressed in terms of modulo-2 addition (i.e., XOR) of the input states of the control and target qubits as [see Fig. 11.5 and Eqs. (11.17)]

$$|x\rangle|y\rangle \xrightarrow{\hat{U}_{\text{CNOT}}} |x\rangle|x \oplus y\rangle, \tag{11.20}$$

where $|x\rangle$ and $|y\rangle$ are the computational basis states of the control and target qubits, respectively, and $\oplus$ denotes modulo-2 addition, an operation in modular arithmetic [see Box 11.2]. Moreover, the CNOT gate is sometimes called the measurement gate, because, if the target qubit is prepared in the state $|0\rangle$ then its output state is always the

---

Box 11.2 Modular arithmetic

The idea of modular arithmetic can be understood as addressed to solve the shift (or "wrap around") in time of a 12–hour clock. If now the clock indicates the 9 position (it does not matter whether am or pm) and we ask which position it will occupy after 7 hours, the answer is $9 + 7 - 12 = 4$ and not $9 + 7 = 16$, as the ordinary addition would require.

Given two integers a and b, then a modulo b means the remainder of the division of a by b. For instance, in the clock example above, 16 divided by 12 gives the quotient 1 and the remainder 4, so that 16 modulo 12 is equal to 4. In other words, we have that $16 - 4 = 12$ is an integer multiple of 12. In general, given integers a, b, and n, if $a = b + kn$ for some integer k then we say that a is congruent to b modulo n. This relation is usually expressed as $a \equiv b \bmod n$.

It is fairly easy to show that for any integers a, b, c, and $m \neq 0$, the following properties hold:

$$a \equiv a \bmod n, \tag{11.21a}$$

$$a \equiv b \bmod n \Rightarrow b \equiv a \bmod n, \tag{11.21b}$$

$$a \equiv b \bmod n, b \equiv c \bmod n \Rightarrow a \equiv c \bmod n. \tag{11.21c}$$

If $a \equiv b \bmod n$, then it can be shown that

$$ka \equiv kb \bmod n \quad \text{and} \quad a^n \equiv b^n \bmod n, \tag{11.22}$$

where k is an integer and n a non-negative integer. Let $a_1 \equiv b_1 \bmod n$ and $a_2 \equiv b_2 \bmod n$, then we have

$$a_1 + a_2 \equiv b_1 + b_2 \bmod n \quad \text{and} \quad a_1 a_2 \equiv b_1 b_2 \bmod n. \tag{11.23}$$

Moreover, modular arithmetic satisfies the following properties:

$$(a + b) \bmod n = [(a \bmod n) + (b \bmod n)] \bmod n, \tag{11.24a}$$

$$(a \times b) \bmod n = [(a \bmod n) \times (b \bmod n)] \bmod n. \tag{11.24b}$$

In the case of binary computation, the rules for modulo-2 addition are given by

$$0 \oplus 0 = 0, \quad 0 \oplus 1 = 1 \oplus 0 = 1, \quad 1 \oplus 1 = 0. \tag{11.25}$$

Arithmetic modulo 2 is sometimes referred to as Boolean arithmetic and it plays an important role in computer science.

same as the state of the control qubit [see Eqs. (11.17)]. Another important feature of the CNOT gate is its ability to entangle and disentangle states of a pair of qubits. Let us consider, for instance, a device composed of a Hadamard gate followed by a CNOT gate as depicted in Fig. 11.6. It is easy to see that the actions of this device on the unentangled product states $|0\rangle|0\rangle, |0\rangle|1\rangle, |1\rangle|0\rangle$, and $|1\rangle|1\rangle$ of the first (control) and second (target) qubits are respectively given by

$$|0\rangle|0\rangle \xrightarrow{\hat{U}_H \otimes \hat{I}} \frac{1}{\sqrt{2}}(|0\rangle|0\rangle + |1\rangle|0\rangle) \xrightarrow{\hat{U}_{CNOT}} \frac{1}{\sqrt{2}}(|0\rangle|0\rangle + |1\rangle|1\rangle),$$

$$\tag{11.26a}$$

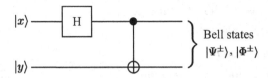

Figure 11.6 Graphical representation of the device for generating the Bell states. The states $|x\rangle$ and $|y\rangle$ represent the computational basis states $|0\rangle$ and $|1\rangle$. When running backwards, this device also serves as a Bell state measurement device.

$$|0\rangle|1\rangle \xrightarrow{\hat{U}_H \otimes \hat{I}} \frac{1}{\sqrt{2}}(|0\rangle|1\rangle + |1\rangle|1\rangle) \xrightarrow{\hat{U}_{CNOT}} \frac{1}{\sqrt{2}}(|0\rangle|1\rangle + |1\rangle|0\rangle),$$

$$\text{(11.26a)}$$

$$|1\rangle|0\rangle \xrightarrow{\hat{U}_H \otimes \hat{I}} \frac{1}{\sqrt{2}}(|0\rangle|0\rangle - |1\rangle|0\rangle) \xrightarrow{\hat{U}_{CNOT}} \frac{1}{\sqrt{2}}(|0\rangle|0\rangle - |1\rangle|1\rangle),$$

$$\text{(11.26b)}$$

$$|1\rangle|1\rangle \xrightarrow{\hat{U}_H \otimes \hat{I}} \frac{1}{\sqrt{2}}(|0\rangle|1\rangle - |1\rangle|1\rangle) \xrightarrow{\hat{U}_{CNOT}} \frac{1}{\sqrt{2}}(|0\rangle|1\rangle - |1\rangle|0\rangle).$$

$$\text{(11.26c)}$$

The output states are recognized respectively as the four Bell states $|\Phi^+\rangle$, $|\Psi^+\rangle$, $|\Phi^-\rangle$, and $|\Psi^-\rangle$ given by Eqs. (10.46) [see also Eqs. (11.43)] and therefore are maximally entangled states of the two qubits. Moreover, since both $\hat{U}_H$ and $\hat{U}_{CNOT}$ are the inverse of themselves (i.e., they are both Hermitian and unitary), the device can also run backwards so as to disentangle the Bell states into unentangled product computational basis states. In other words, when running backwards, this device also serves as a Bell state measurement device.

The CNOT gate belongs to a larger classes of possible quantum gates called the controlled-unitary (CU) gates, the graphical representation of which is depicted in Fig. 11.7. The action of these gates is as follows. If the control qubit is in the state $|0\rangle$ then the target qubit is left unchanged, but if it is in the state $|1\rangle$ then the unitary operator $\hat{U}$ is applied to the target qubit. The CNOT gate is clearly a simple example of the controlled-unitary gate, in which the unitary operator is the Pauli matrix $\hat{\sigma}_x$ (i.e., the Pauli-X gate). Other common examples of the controlled-unitary gates are

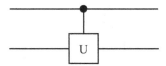

Figure 11.7 Graphical representation of the CU gate. The top line represents the control qubit, the bottom line the target qubit, and the U gate a unitary operator $\hat{U}$.

the controlled-Pauli-Z gate and controlled-phase gate. Moreover, any of the two-qubit gates can be also used in a controlled-unitary gate to make a controlled three-qubit gate. Indeed, the simplest three-qubit gate is the controlled-controlled-NOT (CCNOT) or Toffoli gate. Its action on three qubits is that if the first two (control) qubits are in the state $|1\rangle$ then it applies a NOT on the third (target) qubit, otherwise it does nothing. Therefore, we have

$$\hat{U}_{\text{Toffili}}|000\rangle = |000\rangle, \quad \hat{U}_{\text{Toffili}}|001\rangle = |001\rangle, \tag{11.27a}$$

$$\hat{U}_{\text{Toffili}}|010\rangle = |010\rangle, \quad \hat{U}_{\text{Toffili}}|011\rangle = |011\rangle, \tag{11.27b}$$

$$\hat{U}_{\text{Toffili}}|100\rangle = |100\rangle, \quad \hat{U}_{\text{Toffili}}|101\rangle = |101\rangle, \tag{11.27c}$$

$$\hat{U}_{\text{Toffili}}|110\rangle = |111\rangle, \quad \hat{U}_{\text{Toffili}}|111\rangle = |110\rangle, \tag{11.27d}$$

where the first two qubit are the control qubits and the third qubit is the target qubit. In the computational basis of the three-qubit Hilbert space

$$|000\rangle = \begin{pmatrix} 1 \\ 0 \\ 0 \\ 0 \\ 0 \\ 0 \\ 0 \\ 0 \end{pmatrix}, \quad |001\rangle = \begin{pmatrix} 0 \\ 1 \\ 0 \\ 0 \\ 0 \\ 0 \\ 0 \\ 0 \end{pmatrix}, \quad |010\rangle = \begin{pmatrix} 0 \\ 0 \\ 1 \\ 0 \\ 0 \\ 0 \\ 0 \\ 0 \end{pmatrix}, \quad |011\rangle = \begin{pmatrix} 0 \\ 0 \\ 0 \\ 1 \\ 0 \\ 0 \\ 0 \\ 0 \end{pmatrix},$$

$$\tag{11.28a}$$

$$|100\rangle = \begin{pmatrix} 0 \\ 0 \\ 0 \\ 0 \\ 1 \\ 0 \\ 0 \\ 0 \end{pmatrix}, \quad |101\rangle = \begin{pmatrix} 0 \\ 0 \\ 0 \\ 0 \\ 0 \\ 1 \\ 0 \\ 0 \end{pmatrix}, \quad |110\rangle = \begin{pmatrix} 0 \\ 0 \\ 0 \\ 0 \\ 0 \\ 0 \\ 1 \\ 0 \end{pmatrix}, \quad |111\rangle = \begin{pmatrix} 0 \\ 0 \\ 0 \\ 0 \\ 0 \\ 0 \\ 0 \\ 1 \end{pmatrix}.$$

$$(11.28b)$$

the Toffoli gate can be expressed in matrix form as

$$\hat{U}_{\text{Toffoli}} = \begin{bmatrix} 1 & 0 & 0 & 0 & 0 & 0 & 0 & 0 \\ 0 & 1 & 0 & 0 & 0 & 0 & 0 & 0 \\ 0 & 0 & 1 & 0 & 0 & 0 & 0 & 0 \\ 0 & 0 & 0 & 1 & 0 & 0 & 0 & 0 \\ 0 & 0 & 0 & 0 & 1 & 0 & 0 & 0 \\ 0 & 0 & 0 & 0 & 0 & 1 & 0 & 0 \\ 0 & 0 & 0 & 0 & 0 & 0 & 0 & 1 \\ 0 & 0 & 0 & 0 & 0 & 0 & 1 & 0 \end{bmatrix}, \qquad (11.29)$$

which leaves all basis states unchanged except for the last two, that is, only in the cases in which both control qubits are in the state $|1\rangle$, the target qubit is flipped. Moreover, as the name suggests, the Toffoli gate is the quantum analogue of the classical Toffoli gate introduced in Section 9.7 [see Table 9.6]. Indeed, when the quantum Toffoli gate acts on the computational basis states, the output state of the target qubit may be expressed in terms of multiplication (i.e., AND) and modulo-2 addition (i.e., XOR) as [see Fig. 11.8]

$$|x\rangle|y\rangle|z\rangle \xrightarrow{\hat{U}_{\text{Toffoli}}} |x\rangle|y\rangle|(x \cdot y) \oplus z\rangle, \qquad (11.30)$$

where $|x\rangle$, $|y\rangle$, and $|z\rangle$ are the computational basis states $|0\rangle$ and $|1\rangle$, and $\oplus$ denotes modulo-2 addition. The validity of the above expression can be verified straightforwardly by using Eqs. (11.27).

The most fascinating and innovative aspect of quantum computation is that it allows to solve certain problems in an efficient way that classical computation cannot. An *algorithm* is a method in which a set of well-defined instructions, applied to a certain initial state, can complete a required task or solve a given problem. Then the question

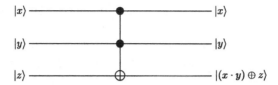

Figure 11.8 Graphical representation of the Toffoli gate, where $|x\rangle$, $|y\rangle$, and $|z\rangle$ are the computational basis states $|0\rangle$ and $|1\rangle$, and $\oplus$ denotes modulo-2 addition. The top two qubits represent the control qubits, the bottom qubit the target qubit.

arises naturally as to how to quantify the efficiency of an algorithm. It is common sense that given some problem the algorithm that takes a short time in solving the problem is more efficient then the one that takes a longer time. From a general viewpoint, let us consider an algorithm which takes an input of n digits and gives an answer after a certain amount of time T. Obviously, the time T is a function of the number of digits n. An algorithm is said to be of *polynomial time* if its running time increases no faster than a polynomial function of the size of the input for the algorithm, i.e.,

$$T(n) = O(n^k), \tag{11.31}$$

where k is some non-negative integer. In the above expression, the big O notation denotes the upper bound on the limiting behavior of a function when the argument of the function approaches infinity. On the other hand, an algorithm is said to be of *exponential time* if its running time increases no faster than an exponential function of the size of the input for the algorithm, i.e.,

$$T(n) = O(e^n). \tag{11.32}$$

From Eq. (6.93), it is evident that the exponential time is much longer than the polynomial time. Therefore, an algorithm of polynomial time is said to be easy or efficient, while an algorithm of exponential time is called hard or inefficient. Many problems can be solved by using efficient algorithms, and a notable example is calculating the greatest common divisor. However, there is a class of problems for which no efficient algorithms are known but for which the solutions, once found, can be verified as correct in polynomial time. Among these problems perhaps the most known example is integer factorization. The obvious way to factorize a very large integer n is to try dividing it by each of the primes less than $\sqrt{n}$ in order to determine which are its factors. Running on a computer, this is known to be a CPU intensive and time consuming task. However, once the integer has been factorized, it is relatively easy to verify on a computer that its factors when multiplied together give the integer.

Let us now discuss one of the first examples demonstrating the efficiency of quantum computation. As we have mentioned in Section 9.7, a Boolean function $f : \{0, 1\} \rightarrow \{0, 1\}$ is a function which maps the truth values $\{0, 1\}$ to the truth values $\{0, 1\}$ [see Eq. (9.48)]. A Boolean function f is called *constant* if it always returns the same value, i.e., either $f(0) = f(1) = 0$ or $f(0) = f(1) = 1$, and *balanced* if it returns 1 for half of the input domain and 0 for the other half. Suppose now that there is a device that can evaluate the function f and that it is allowed to run only once. The so-called *Deutsch problem* is to ask whether, under these conditions, it is possible to determine if the function f is constant or balanced.[a] It is easy to see that classically it is impossible to answer this question by running the device only once. However, this can be done quantum mechanically. In order to see this, let us consider first the quantum implementation of a Boolean function as a quantum gate. For a given Boolean function f, its quantum implementation is the gate whose action is defined by the following unitary transformation (which for the moment will be referred to as simply the Boolean gate) [see Fig. 11.9]

$$|x\rangle|y\rangle \xrightarrow{\hat{U}_f} |x\rangle|f(x) \oplus y\rangle, \tag{11.33}$$

where $|x\rangle$ and $|y\rangle$ are the computational basis states $|0\rangle$ and $|1\rangle$, and $\oplus$ denotes modulo-2 addition. The power of a quantum device lies largely in the fact that due to linearity we can input not just the computational basis states $|0\rangle$ and $|1\rangle$, but an arbitrary superposition of them. Therefore, in its generality, the transformation $\hat{U}_f$ can be implemented by

$$\sum_j c_j |x_j\rangle|y\rangle \xrightarrow{\hat{U}_f} \sum_j c_j |x_j\rangle|f(x_j) \oplus y\rangle, \tag{11.34}$$

---

[a](Deutsch, 1985).

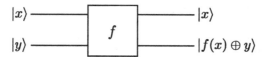

Figure 11.9 The implementation of a Boolean function f as a quantum gate, where $|x\rangle$ and $|y\rangle$ are the computational basis states $|0\rangle$ and $|1\rangle$, and $\oplus$ denotes modulo-2 addition.

where $|x_j\rangle$, $|y\rangle = |0\rangle$, $|1\rangle$ and c_j are coefficients. It is interesting to note that in the above expression the state created by the Boolean gate is an entangled state. Now consider the two-qubit quantum device shown in Fig. 11.10, in which the input state is set to be $|0\rangle|1\rangle$. The two qubits are first transformed separately by two Hadamard gates

$$|0\rangle|1\rangle \xrightarrow{\hat{U}_H \otimes \hat{U}_H} \hat{U}_H|0\rangle\hat{U}_H|1\rangle = \frac{1}{2}(|0\rangle + |1\rangle)(|0\rangle - |1\rangle). \quad (11.35)$$

For later convenience, the resultant state can be rewritten as

$$|\Psi\rangle = \frac{1}{2}[(|0\rangle + |1\rangle)|0\rangle - (|0\rangle + |1\rangle)|1\rangle]. \quad (11.36)$$

The state $|\Psi\rangle$ is then processed by the Boolean gate, leading to four possible results depending on the nature of the function f. If f is a

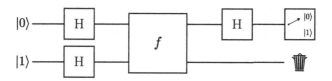

Figure 11.10 Device for solving the Deutsch problem of determining if a Boolean function f is constant or balanced in a single run. A square with a dial denotes a measurement of the qubit in the computational basis, while a trash bin means the qubit is discarded rather than measured. The top and bottom qubits are initially in the states $|0\rangle$ and $|1\rangle$, respectively. A measurement on the output state of the first qubit gives the requested answer.

constant function, then we obtain

$$\hat{U}_f(0, 0)|\Psi\rangle = \frac{1}{2}(|0\rangle|0 \oplus 0\rangle + |1\rangle|0 \oplus 0\rangle - |0\rangle|0 \oplus 1\rangle - |1\rangle|0 \oplus 1\rangle)$$

$$= \frac{1}{2}[(|0\rangle + |1\rangle)|0\rangle - (|0\rangle + |1\rangle)|1\rangle]$$

$$= \frac{1}{2}(|0\rangle + |1\rangle)(|0\rangle - |1\rangle)$$

$$= |+\rangle|-\rangle, \tag{11.37a}$$

$$\hat{U}_f(1, 1)|\Psi\rangle = \frac{1}{2}(|0\rangle|1 \oplus 0\rangle + |1\rangle|1 \oplus 0\rangle - |0\rangle|1 \oplus 1\rangle - |1\rangle|1 \oplus 1\rangle)$$

$$= \frac{1}{2}(|0\rangle|1\rangle + |1\rangle|1\rangle - |0\rangle|0\rangle - |1\rangle|0\rangle)$$

$$= -\frac{1}{2}(|0\rangle + |1\rangle)(|0\rangle - |1\rangle)$$

$$= -|+\rangle|-\rangle, \tag{11.37b}$$

where use has been made of Eq. (11.34) and the shorthand notation $\hat{U}_f(0, 0) = \hat{U}_{f(0)=f(1)=0}$ and $\hat{U}_f(1, 1) = \hat{U}_{f(0)=f(1)=1}$. Similarly, if on the other hand f is a balanced function then we have

$$\hat{U}_f(0, 1)|\Psi\rangle = \frac{1}{2}(|0\rangle|0 \oplus 0\rangle + |1\rangle|1 \oplus 0\rangle - |0\rangle|0 \oplus 1\rangle - |1\rangle|1 \oplus 1\rangle)$$

$$= \frac{1}{2}(|0\rangle|0\rangle + |1\rangle|1\rangle - |0\rangle|1\rangle - |1\rangle|0\rangle)$$

$$= \frac{1}{2}(|0\rangle - |1\rangle)(|0\rangle - |1\rangle)$$

$$= |-\rangle|-\rangle \tag{11.37c}$$

$$\hat{U}_f(1, 0)|\Psi\rangle = \frac{1}{2}(|0\rangle|1 \oplus 0\rangle + |1\rangle|0 \oplus 0\rangle - |0\rangle|1 \oplus 1\rangle - |1\rangle|0 \oplus 1\rangle)$$

$$= \frac{1}{2}(|0\rangle|1\rangle + |1\rangle|0\rangle - |0\rangle|0\rangle - |1\rangle|1\rangle)$$

$$= -\frac{1}{2}(|0\rangle - |1\rangle)(|0\rangle - |1\rangle)$$

$$= -|-\rangle|-\rangle, \tag{11.37d}$$

where $\hat{U}_f(0, 1) = \hat{U}_{f(0)=0, f(1)=1}$ and $\hat{U}_f(1, 0) = \hat{U}_{f(0)=1, f(1)=0}$. The first qubit of each of the above four results (two for the constant case and two for the balanced case) is further processed by the Hadamard

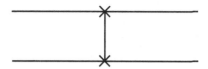

Figure 11.11 Graphical representation of the swap gate.

gate, leading to the final results

$$\pm|+\rangle|-\rangle \xrightarrow{\hat{U}_H \otimes \hat{I}} \pm|0\rangle|-\rangle \quad \text{if } f \text{ is constant,} \tag{11.38a}$$

$$\pm|-\rangle|-\rangle \xrightarrow{\hat{U}_H \otimes \hat{I}} \pm|1\rangle|-\rangle \quad \text{if } f \text{ is balanced.} \tag{11.38b}$$

This shows clearly that, apart from an irrelevant global phase factor (i.e., the overall sign of the output state), a measurement on the first qubit immediately gives the requested answer: the state $|0\rangle$ implies that the function is constant, while the state $|1\rangle$ implies that the function is balanced.

Problem 11.2 Perform a computation similar to that given by Eq. (11.16) for the case in which the input state is $|1\rangle$.

Problem 11.3 Express the controlled-Pauli-Z (CZ) gate in matrix form, compute its action on a two-qubit input and use the result to verify its unitarity.

Problem 11.4 Check that the transformations (11.26) and (11.27) are indeed reversible.

Problem 11.5 Check the results given by Eqs. (11.38). (*Hint*: Recall that the Hadamard gate is the inverse of itself.)

Problem 11.6 Another commonly used two-qubit gate is the swap gate, the graphical representation of which is depicted in Fig. 11.11. Its action is to interchange the two input qubits, i.e.,

$$\hat{U}_{\text{swap}}|\psi\rangle|\phi\rangle = |\phi\rangle|\psi\rangle, \tag{11.39}$$

where $|\psi\rangle$ and $|\phi\rangle$ are arbitrary qubit states. Express the swap gate in matrix form.

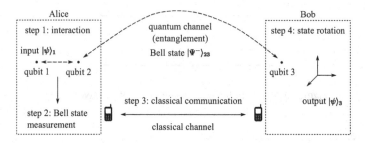

Figure 11.12 Schematic setup of quantum teleportation. In step 1, Alice let qubit 1 (the qubit to be teleported) interact with quit 2 which is entangled with qubit 3, previously given to Bob, in the Bell state $|\Psi^-\rangle_{23}$. In step 2 she reads the outcome of the Bell measurement performed on qubits 1 and 2. In step 3, she classically communicates this result to Bob. With this piece of classical information, in step 4 Bob rotates the state of qubit 3 and obtains as the output state $|\psi\rangle_3$ the state $|\psi\rangle_1$ of qubit 1.

11.5 Quantum Teleportation

In this section we describe one of the most striking examples demonstrating the power of quantum information processing: quantum teleportation. By this term, we mean a procedure that is able to transfer with certainty the state of an input quantum system onto the state of a distant output system of the same type. As we know, an instantaneous transfer of information is not possible [see Section 10.6]. In other words, due to the intrinsic randomness of quantum mechanics we cannot manipulate at a distance the measurement outcomes of a system, neither their probability distributions although that system may be entangled with another one that we are able to manipulate locally (obviously, by choosing certain measurement procedures but not by controlling the possible outcomes). However, it is possible to "teleport" some information by exploiting entanglement and without violating the Einstein locality.[a]

Suppose that a sender, conventionally called Alice, wishes to send to a receiver, conventionally called Bob, the information about a quantum system, say a qubit labeled 1, prepared in the state $|\psi\rangle_1$ *unknown* to her as well as to Bob. To this purpose, Alice allows

---

[a](Bennett/Wiesner, 1992), (Bennett *et al.*, 1993).

qubit 1, called the *ancilla*, to interact with qubit 2 that is entangled with qubit 3, previously given to Bob [see Fig. 11.12]. Now, Alice performs a special kind of measurement on the her composite two-qubit system 12, so that, by classically communicating with Bob the result of her measurement, Bob can reconstruct the same state of qubit 1 onto qubit 3. In order to discuss this procedure specifically, let the entangled qubits 2 and 3 be in the EPR state [see Eq. (10.10)]

$$|\Psi^-\rangle_{23} = \frac{1}{\sqrt{2}}(|0\rangle_2|1\rangle_3 - |1\rangle_2|0\rangle_3). \tag{11.40}$$

We note that to make connection with previous discussions presented in Sections 10.3 and 10.4, the computational basis states $|0\rangle$ and $|1\rangle$ may be thought of respectively as the spin up and spin down states $|\uparrow_z\rangle$ and $|\downarrow_z\rangle$ in the z direction. If qubit 1 is in the unknown state

$$|\psi\rangle_1 = \alpha|0\rangle_1 + \beta|1\rangle_1, \tag{11.41}$$

where α and β are complex amplitudes, then before the measurement the state of the whole composite three-qubit system 123 is given by the product state

$$\begin{aligned}|\Psi\rangle_{123} &= |\psi\rangle_1|\Psi^-\rangle_{23} \\ &= \frac{\alpha}{\sqrt{2}}(|0\rangle_1|0\rangle_2|1\rangle_3 - |0\rangle_1|1\rangle_2|0\rangle_3) \\ &\quad + \frac{\beta}{\sqrt{2}}(|1\rangle_1|0\rangle_2|1\rangle_3 - |1\rangle_1|1\rangle_2|0\rangle_3). \end{aligned} \tag{11.42}$$

Alice could chose to measure qubits 1 and 2 in the Bell basis composed of the four orthonormal Bell states for the two qubits [see Eqs. (10.46)]

$$|\Psi^+\rangle_{12} = \frac{1}{\sqrt{2}}(|0\rangle_1|1\rangle_2 + |1\rangle_1|0\rangle_2), \tag{11.43a}$$

$$|\Psi^-\rangle_{12} = \frac{1}{\sqrt{2}}(|0\rangle_1|1\rangle_2 - |1\rangle_1|0\rangle_2), \tag{11.43b}$$

$$|\Phi^+\rangle_{12} = \frac{1}{\sqrt{2}}(|0\rangle_1|0\rangle_2 + |1\rangle_1|1\rangle_2), \tag{11.43c}$$

$$|\Phi^-\rangle_{12} = \frac{1}{\sqrt{2}}(|0\rangle_1|0\rangle_2 - |1\rangle_1|1\rangle_2). \tag{11.43d}$$

To this end, the state $|\Psi\rangle_{123}$ is written in terms of the above Bell states as

$$|\Psi\rangle_{123} = -\frac{1}{2}\big[|\Psi^-\rangle_{12}(\alpha|0\rangle_3 + \beta|1\rangle_3) + |\Psi^+\rangle_{12}(\alpha|0\rangle_3 - \beta|1\rangle_3)$$
$$- |\Phi^-\rangle_{12}(\alpha|1\rangle_3 + \beta|0\rangle_3) - |\Phi^+\rangle_{12}(\alpha|1\rangle_3 - \beta|0\rangle_3)\big].$$
$$(11.44)$$

This expression can be further simplified to

$$|\Psi\rangle_{123} = -\frac{1}{2}\big[|\Psi^-\rangle_{12}|\psi\rangle_3 + |\Psi^+\rangle_{12}\,\hat{\sigma}_z|\psi\rangle_3 - |\Phi^-\rangle_{12}\,\hat{\sigma}_x|\psi\rangle_3$$
$$+ i|\Phi^+\rangle_{12}\,\hat{\sigma}_y|\psi\rangle_3\big], \qquad (11.45)$$

where $\hat{\sigma}_x$, $\hat{\sigma}_y$, and $\hat{\sigma}_z$ are the usual Pauli matrices, whose actions on the computational basis states are tabulated in Table 8.1. After Alice's measurement, the three-qubit system is projected into one of the four states superposed in $|\Psi\rangle_{123}$, depending on the measurement outcome. According to Eq. (11.45), this means that given Alice's measurement outcome, the state of Bob's qubit (up to an irrelevant global phase factor) is given by

$$|\Psi^-\rangle_{12} \longrightarrow |\psi\rangle_3, \qquad (11.46a)$$

$$|\Psi^+\rangle_{12} \longrightarrow \hat{\sigma}_z|\psi\rangle_3, \qquad (11.46b)$$

$$|\Phi^-\rangle_{12} \longrightarrow \hat{\sigma}_x|\psi\rangle_3, \qquad (11.46c)$$

$$|\Phi^+\rangle_{12} \longrightarrow \hat{\sigma}_y|\psi\rangle_3. \qquad (11.46d)$$

It is noted that three of the four possible states of qubit 3 (Bob's one) are simply related to the state $|\psi\rangle_1$ of qubit 1 (the one Alice wished to teleport) by a unitary transformation, namely, the Pauli matrix $\hat{\sigma}_x$, $\hat{\sigma}_y$, or $\hat{\sigma}_z$.

In the case of the first measurement outcome $|\Psi^-\rangle_{12}$, the state of qubit 3 is the same as that of qubit 1 except for an irrelevant phase factor, so that Bob needs to do nothing further to recover the state of qubit 1. In the three other cases, in order to convert the state of qubit 3 to the state of qubit 1, Bob must apply to qubit 3 one of the unitary operators $\hat{\sigma}_x$, $\hat{\sigma}_y$, and $\hat{\sigma}_z$ (or the Pauli-X, Pauli-Y, and Pauli-Z gates), which as we have said in the previous section represents respectively rotations of the Bloch sphere about the x, y, and z axis by π radians. The possible operations involved in the

Table 11.1 Possible events in a teleportation process. The fact that each of Alice's measurement result is uniquely mapped (through the ebit $|\Psi^-\rangle_{23}$) to the input state $|\psi\rangle_1$ of qubit 1 (which could represent a code) allows that Alice classical instructs Bob about the kind of operation to be performed so as to recover the state $|\psi\rangle_1$ of qubit 1 on qubit 3

Alice's measurement	State of Bob's qubit	Bob's operation
$|\Psi^-\rangle_{12}$	$|\psi\rangle_3$	$\hat{I}$
$|\Psi^+\rangle_{12}$	$\hat{\sigma}_z|\psi\rangle_3$	$\hat{\sigma}_z$
$|\Phi^-\rangle_{12}$	$\hat{\sigma}_x|\psi\rangle_3$	$\hat{\sigma}_x$
$|\Phi^+\rangle_{12}$	$\hat{\sigma}_y|\psi\rangle_3$	$\hat{\sigma}_y$

teleportation procedure are summarized in Table 11.1. What Bob has to do, obviously depends on the classical communication of Alice's measurement result, say via phone call or email. In other words, once Alice obtains a certain result she sends to Bob a simple instruction. For instance the instruction may reads "Do nothing," "Perform the rotation about the z axis," "Perform the rotation about the x axis," or "Perform the rotation about the y axis." For this reason, both Alice and Bob could be even computers with no understanding whatsoever of quantum mechanics and therefore no knowledge of the initial conditions and the equations describing the states of the three qubits. It suffices that two bits of classical information is sent by Alice (according to a previously established code, like the binary numbers 00, 01, 10, and 11) so that Bob may perform the required operation as per the prearranged rules. Thanks to the shared information, i.e., the entanglement between qubits 2 and 3, Bob is able to recover the encoded quantum information. In short, depending on Alice's measurement result, Bob can recover with certainty the original state $|\psi\rangle_1$ with the proper application of one of the above unitary transformations. Finally, we note that quantum teleportation has been experimentally realized, first with photons.[a] The setup of a typical quantum teleportation experiment is schematically shown in Fig. 11.13.

In conclusion, quantum teleportation is based on two channels: a *quantum channel* (the shared ebit composed of qubits 2 and

---

[a](Bouwmeester *et al.*, 1997), (Furusawa *et al.*, 1998).

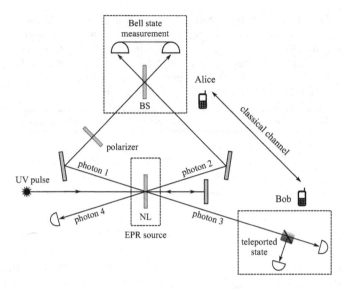

Figure 11.13 Experimental realization of quantum teleportation by Zeilinger and colleagues. A pulse of ultraviolet radiation passing through a non-linear crystal NL produces the EPR photon pair 2–3. After reflection during its second passage through NL, the radiation creates another pair, 1–4, of which the photon 1 is that to be teleported and photon 4 is a trigger indicating that the other photon is under way. Alice looks for coincidence counts of photons 1 and 2, after they pass through the beam splitter BS. Finally, Bob, after receiving two bits of classical information, may retrieve the input state of photon 1 through appropriate detection. Adapted from (Auletta *et al.*, 2009, p. 646).

3), by which Alice teleport to Bob the state $|\psi\rangle_1$ of qubit 1 [see Section 11.3], and a *classical channel*, by which Alice communicates to Bob the result of her measurement. After having received the classical information about Alice's measurement result, and hence without violating relativistic locality, Bob uses the entanglement between qubits 2 and 3 to recover exactly the state $|\psi\rangle_1$ of qubit 1 on qubit 3. However, the original state $|\psi\rangle_1$ of qubit 1 is destroyed as a result of the entanglement between qubits 1 and 2 created by Alice's measurement. We note that this is in accordance with the no-cloning theorem [see Theorem 11.2], which dictates the impossibility to clone arbitrary unknown quantum states.

Since an *ebit* is the amount of entanglement between a maximally entangled pair of qubits (for example two spin $\frac{1}{2}$ particles in a singlet state), by teleportation we transmit a qubit by means of a shared ebit and two bits of classical information. It is noted that an ebit is a weaker resource than a qubit. In fact, the transmission of a qubit can always be used to create one ebit, while the sharing of an ebit (or many ebits) does not suffice to transmit a qubit as we also need classical information to make an appropriate state rotation. This is another way to say that without additional classical and local resources quantum correlations do not allow transmission of information.

Problem 11.7 Work out the explicit calculation that leads to the expression given by Eq. (11.44).

Problem 11.8 Can quantum teleportation be implemented by using the Bell state $|\Psi^+\rangle_{23}$, $|\Phi^+\rangle_{23}$, or $|\Phi^-\rangle_{23}$ instead of $|\Psi^-\rangle_{23}$?

11.6 Quantum Cryptography

One of the most important aspects of information technology in modern society is how to establish a secure way to exchange secret messages. By secure we mean here that the message cannot be eavesdropped by extraneous agencies. Cryptography is precisely the technique that allows the sender, traditionally called Alice, to encode a message and the receiver, traditionally called Bob, to decode it by using a certain key with a procedure such that the eventual eavesdropper, traditionally called Eve, is not able to break the secrecy of the encoded message.

There are several possible schemes of cryptography. Two best known examples are represented by the public-key protocols and the private-key protocols. In the public-key protocols, the encryption key may be publicly announced by the receiver to the sender, but the decryption key is only known to the former. This ensures that only the receiver is able to decrypt the message. The public-key protocols rely on the enormous difficulty in inverting certain mathematical operations. For instance, multiplying two large prime numbers is relatively easy but, on the contrary, factorizing the result into its prime factors may take a very long time [see Section 11.4]. For this reason, public-key protocols are also called *asymmetric protocols*. Unfortunately, it has never been proved that the mathematical operations used in the public-key protocols are truly difficult (i.e., the time required to solve them grows exponentially with the size of the input). The only thing we know so far is that there are no classical algorithms that can solve them in a polynomial time. On the other hand, private-key protocols are called *symmetric protocols* since they use the same key for both encryption and decryption. They rely on secure communication in order to establish a shared secret key. Such a communication is often performed by a human courier that is however not completely reliable. Indeed, there does not exist a classical procedure that allows for a complete safe key distribution, since any key can in principle be intercepted without the knowledge of the interested parties. This problem is known as the *key distribution problem*.

Quantum information has brought a new light to the key distribution problem. As a matter of fact, through *quantum key distribution*, quantum mechanics allows two parties to establish a totally secure private key, by transmitting information in quantum superpositions or entangled states. Essentially, what is used here are some basic quantum principles:

(i) A non-interaction-free measurement will irreversibly *perturb* the state of the measured system [see Sections 5.2–5.3].

(ii) It is impossible to *distinguish* with a single measurement two non-orthogonal states [see Section 7.8].

(iii) It is impossible to *clone* an unknown quantum state [see Section 11.2].

Quantum cryptography was initiated by a pioneering study of Wiesner, proposed in the 1970's but only published later on in 1983.[a] Bennett and Brassard short thereafter in 1984 published a seminal paper in which they proposed the first protocol for quantum key distribution.[b] In Bennett and Brassard's protocol, now know universally as the BB84 protocol, Alice and Bob are

---

[a](Wiesner, 1983).
[b](Bennett/Brassard, 1984).

connected by a quantum communication channel which allows the transmission of quantum states. In addition, they may communicate via a public classical channel. Both these channels may possibly be not secure. In particular, if the quantum channel is represented by the transmission of photons (e.g., in an optical fiber or in free space), they can make use of two possible bases for photon polarization states[a]

$$\oplus = \{|\leftrightarrow\rangle, |\updownarrow\rangle\} \quad \text{and} \quad \otimes = \{|\nearrow\rangle, |\nwarrow\rangle\}, \tag{11.47}$$

which represent, respectively, the horizontal–vertical polarization basis and the 45°–135° polarization basis. First, Alice and Bob establish a one-to-one correspondence between each basis state of the two bases and the classical bits 0 and 1 that they desires to communicate with each other. They may agree to use, for instance, the following correspondence

$$0 \Longleftrightarrow |\leftrightarrow\rangle, |\nearrow\rangle, \quad \text{and} \quad 1 \Longleftrightarrow |\updownarrow\rangle, |\nwarrow\rangle. \tag{11.48}$$

Then, Alice sends to Bob a sequence of classical bits, encoded in photon polarization states using one or the other polarization basis chosen at random. Bob decodes the transmitted bits by measuring the polarization state of the transmitted photons, using the basis randomly chosen from one of the two bases. After the measurement, they publicly inform each other via a classical channel which basis they have chosen to use for each transmitted bit. They can immediately discard the bits for which the two chosen bases do not match (on average 50% of the transmitted bits). The bits for which Alice and Bob used the same basis constitute a shared secret key. An example of the transmission sequence is schematically shown in Table 11.2.

It is obvious that Alice and Bob should take into account the possible presence of eavesdroppers. In order to exclude any possibility of eavesdropping, Bob takes a subset of the shared key and publicly compares the bits with those sent by Alice. If they do not match, this means that Eve has intercepted and measured

---

[a]It is noted that the horizontal–vertical polarization states $|\leftrightarrow\rangle$ and $|\updownarrow\rangle$ may be put in correspondence to the spin states $|\uparrow_z\rangle$ and $|\downarrow_z\rangle$ of a spin $\frac{1}{2}$ particle, while the 45°–135° polarization states $|\nearrow\rangle$ and $|\nwarrow\rangle$ to the spin states $|\uparrow_x\rangle$ and $|\downarrow_x\rangle$ of a spin $\frac{1}{2}$ particle.

Table 11.2 An example of the transmission sequence in the BB84 protocol in the absence of eavesdroppers. At the end of the protocol, Alice and Bob have the shared key 011010

Alice's bits	0	1	1	1	0	0	1	0	0	1
Alice's basis	⊕	⊕	⊗	⊗	⊗	⊕	⊕	⊗	⊕	⊕
Alice's states	$\lvert\leftrightarrow\rangle$	$\lvert\updownarrow\rangle$	$\lvert\nwarrow\rangle$	$\lvert\nwarrow\rangle$	$\lvert\nearrow\rangle$	$\lvert\leftrightarrow\rangle$	$\lvert\updownarrow\rangle$	$\lvert\nearrow\rangle$	$\lvert\leftrightarrow\rangle$	$\lvert\updownarrow\rangle$
Bob's basis	⊕	⊗	⊗	⊗	⊗	⊗	⊕	⊕	⊕	⊗
Bob's states	$\lvert\leftrightarrow\rangle$	$\lvert\nearrow\rangle$	$\lvert\nwarrow\rangle$	$\lvert\nwarrow\rangle$	$\lvert\nearrow\rangle$	$\lvert\nwarrow\rangle$	$\lvert\updownarrow\rangle$	$\lvert\leftrightarrow\rangle$	$\lvert\leftrightarrow\rangle$	$\lvert\nearrow\rangle$
Shared key	0	–	1	1	0	–	1	–	0	–

Table 11.3 An example of the transmission sequence in the BB84 protocol in the presence of an eavesdropper. At the end of the protocol, Alice and Bob are left with the partially correlated bit strings 011010 and 010010, respectively. A comparison containing the third bit (1 for Alice, but 0 for Bob) will reveal the presence of an eavesdropper

Alice's bits	0	1	1	1	0	0	1	0	0	1
Alice's basis	⊕	⊕	⊗	⊗	⊗	⊕	⊕	⊗	⊕	⊕
Alice's states	$\lvert\leftrightarrow\rangle$	$\lvert\updownarrow\rangle$	$\lvert\nwarrow\rangle$	$\lvert\nwarrow\rangle$	$\lvert\nearrow\rangle$	$\lvert\leftrightarrow\rangle$	$\lvert\updownarrow\rangle$	$\lvert\nearrow\rangle$	$\lvert\leftrightarrow\rangle$	$\lvert\updownarrow\rangle$
Eve's basis	⊗	⊗	⊕	⊕	⊗	⊕	⊕	⊕	⊗	⊕
Eve's states	$\lvert\nearrow\rangle$	$\lvert\nearrow\rangle$	$\lvert\updownarrow\rangle$	$\lvert\leftrightarrow\rangle$	$\lvert\nearrow\rangle$	$\lvert\leftrightarrow\rangle$	$\lvert\updownarrow\rangle$	$\lvert\leftrightarrow\rangle$	$\lvert\nwarrow\rangle$	$\lvert\updownarrow\rangle$
Bob's basis	⊕	⊗	⊗	⊗	⊗	⊗	⊕	⊕	⊕	⊗
Bob's states	$\lvert\leftrightarrow\rangle$	$\lvert\nearrow\rangle$	$\lvert\nwarrow\rangle$	$\lvert\nearrow\rangle$	$\lvert\nearrow\rangle$	$\lvert\nwarrow\rangle$	$\lvert\updownarrow\rangle$	$\lvert\leftrightarrow\rangle$	$\lvert\leftrightarrow\rangle$	$\lvert\nearrow\rangle$
Check	0	–	1	0	0	–	1	–	0	–

the photons, thus changing the compared bits in a random way. An example of the transmission sequence in the presence of an eavesdropper is schematically shown in Table 11.3. When this is the case, Alice and Bob discard the shared key and have to start their procedure again, until they can be sure that no eavesdropping has occurred. Suppose that Eve also chooses one of the two bases at random. Then Eve will have a probability of $\frac{1}{2}$ to choose the same basis as that used by Alice and Bob, thus leaving the compared bits unchanged. On the other hand, when Eve chose a different basis from that used by Alice and Bob, which happens with a probability of $\frac{1}{2}$,

Bob has again a probability of $\frac{1}{2}$ to obtain a bit that matches the one sent by Alice. Therefore, in the presence of an eavesdropper the probability that two compared bits match each other is $\frac{1}{2} + \frac{1}{2} \times \frac{1}{2} = \frac{3}{4}$. A shared key will be identified as secure only if there is no mismatch at all between the compared bits, and for a n-bit comparison this happens with a probability of $\left(\frac{3}{4}\right)^n$. By increasing the number compared bits n, Alice and Bob are able to increase the probability of detecting Eve's presence, i.e., the probability that there is *at least* one mismatch in the comparison, according to the formula

$$\wp_{\text{eve}} = 1 - \left(\frac{3}{4}\right)^n. \tag{11.49}$$

For instance, if Alice and Bob would like to exclude the presence of Eve at a confidence level of 99.99%, they need to compare 33 bits. As said, if they find a mismatch, they should repeat the protocol from the very beginning, over and over again until they find a perfect match between the compared bits. Obviously, the number of bits to be sent must be sufficiently large in order to allow a sufficiently high level of reliability. This makes Eve's attempts at eavesdropping even more difficult. Needless to say, Eve would like ideally to intercept Alice's bit, make a copy of it, and resend it unperturbed to Bob. However, as we know, this is prohibited by the no-cloning theorem.

In the BB84 protocol, Alice and Bob exploit superposition to distribute a shared secret key. Another very interesting protocol for quantum key distribution is due to Ekert[a] and is based on entanglement. In the Ekert protocol (sometimes called the EPRBE protocol, short for Einstein, Podolsky, Rosen, Bell, and Ekert), Alice prepares a sequence of entangled qubit pairs, each prepared in the same maximally entangled Bell state [see Eq. (11.43b)]

$$|\Psi^-\rangle_{12} = \frac{1}{\sqrt{2}}(|\leftrightarrow\rangle_1|\updownarrow\rangle_2 - |\updownarrow\rangle_1|\leftrightarrow\rangle_2), \tag{11.50}$$

where we remark that the chosen basis is the $\oplus$ basis. It is however noted that since entanglement is basis independent [see Section 7.9], the state $|\Psi^-\rangle$ takes the same form (up to an irrelevant global phase factor) in the $\otimes$ basis and, in fact, in any basis. She keeps the first qubit for herself and sends the second to Bob. Again,

---

[a](Ekert, 1991).

when Alice and Bob use the same $\oplus$ or $\otimes$ basis to measure their qubit, they obtain the opposite results (i.e., their results are perfectly anticorrelated [see Section 10.4]). This in turn provides them with a shared key. In this case, moreover, it is possible to make use of the Bell inequality to check the security of the protocol. A measurement by Eve on the qubit sent to Bob will leave Alice's qubit and the qubit subsequently sent to Bob by Eve in the product state $|\leftrightarrow\rangle_1|\updownarrow\rangle_2$ or $|\updownarrow\rangle_1|\leftrightarrow\rangle_2$. Alice and Bob can use a third choice of basis, say the Breidbart polarization basis $B = \{|B_0\rangle, |B_1\rangle\}$, where $|B_0\rangle$ and $|B_1\rangle$ are respectively given by[a]

$$|B_0\rangle = \cos\frac{\pi}{8}|\leftrightarrow\rangle + \sin\frac{\pi}{8}|\updownarrow\rangle, \tag{11.51a}$$

$$|B_1\rangle = \cos\frac{\pi}{8}|\leftrightarrow\rangle - \sin\frac{\pi}{8}|\updownarrow\rangle, \tag{11.51b}$$

together with the classical bits correspondence

$$0 \Longleftrightarrow |B_0\rangle \quad \text{and} \quad 1 \Longleftrightarrow |B_1\rangle, \tag{11.52}$$

so that, as they establish a shared key, they also collect enough data to test the Bell inequality (10.35). If the latter is violated, then Alice and Bob can be sure that their qubit pairs are indeed entangled, thus ruling out the presence of an eavesdropper.

Finally, it should be noted that quantum cryptography may be only exploited to establish and distribute a private key, not to transmit any message. Such a key may, of course, be used with any chosen cryptographic algorithm to encrypt and to decrypt a message, which can be transmitted over standard communication channels.

Problem 11.9 Show that for the state $|\Psi^-\rangle_{12}$ and the three directions given by the $\oplus$, $\otimes$, and B bases, the Bell inequality (10.35) is indeed violated. (*Hint*: See Footnotes c, p. 371 and a, this page.)

11.7 Mutual Information and Entanglement

As we have said, quantum systems in pure states (i.e., the states showing quantum features in the maximum degree) have a zero von

---

[a]We note that the Breidbart polarization states $|B_0\rangle$ and $|B_1\rangle$ correspond respectively to the spin up and spin down states of a spin $\frac{1}{2}$ particle in the direction $\mathbf{n} = \frac{1}{\sqrt{2}}(\mathbf{e}_z + \mathbf{e}_x)$.

Neumann entropy. This is the reason why we have suggested that the universe as a whole may have zero entropy [see Sections 9.6 and 9.7]. This is due to the fact that features are the strongest form of interdependences between systems in our universe. As a consequence, quantum systems are the most ordered systems in nature. This confirms the fact that they have an infinite amount of potential information, though it can only be partially accessed by information acquisition [see Sections 6.9, 11.2, and 11.3]. Conversely, this accounts for the fact that it is necessary for the system to lose features to the environment and to increase the entropy when being measured. It is indeed impossible to extract information from a system that is perfectly ordered (as well as when it is totally disordered). Information acquisition is possible only when there is a sort of tradeoff between order and disorder.

We now use the classical treatment of information for developing considerations that will be extended to quantum information later on. Suppose that we have two systems characterized respectively by the sets J and K of elements $j \in J$ and $k \in K$, in which the probability of j is given by $\wp(j)$ and the probability of k by $\wp(k)$. It is noted that in classical information theory J and K may represents respectively the sets of characters in a signal transmitted by Alice and received by Bob in a certain communication between them. Because of possible errors induced by noise, J and K as well as $\wp(j)$ and $\wp(k)$ might not be identical. From our discussion in Section 9.6, the Shannon entropies of J and K are respectively given by

$$H(J) = -\sum_{j \in J} \wp(j) \lg \wp(j) \quad \text{and} \quad H(K) = -\sum_{k \in K} \wp(k) \lg \wp(k),$$

(11.53)

which quantify the respective potential information associated with J and K. Then the question arises naturally as to how to quantify the joint information associated with J *and* K. Indeed, such a information is measured by the *joint entropy* of J and K, which is defined by

$$H(J, K) = -\sum_{j \in J} \sum_{k \in K} \wp(j, k) \lg \wp(j, k),$$

(11.54)

where $\wp(j, k)$ is the joint probability of j and k. Let us consider the case in which J and K are uncorrelated (or statistically independent) [see Box 10.1], then we have $\wp(j, k) = \wp(j)\wp(k)$ for all $j \in J$ and $k \in K$. In this case, the joint entropy $H(J, K)$ is given by

$$
\begin{aligned}
H(J, K) &= -\sum_{j\in J}\sum_{k\in K} \wp(j)\wp(k)\lg[\wp(j)\wp(k)]\\
&= -\sum_{j\in J}\sum_{k\in K} \wp(j)\wp(k)[\lg\wp(j) + \lg\wp(k)]\\
&= -\sum_{j\in J}\sum_{k\in K} \wp(j)\wp(k)\lg\wp(j) - \sum_{j\in J}\sum_{k\in K} \wp(j)\wp(k)\lg\wp(k)\\
&= -\sum_{j\in J} \wp(j)\lg\wp(j) - \sum_{k\in K} \wp(k)\lg\wp(k)\\
&= H(J) + H(K),
\end{aligned}
\tag{11.55}
$$

where in obtaining the forth equality use has been made of the condition

$$
\sum_{j\in J} \wp(j) = \sum_{k\in K} \wp(k) = 1.
\tag{11.56}
$$

In other words, the joint potential information associated with uncorrelated J and K is the sum of the potential information associated with J and K. As a matter of fact, it can be proved that the values of $H(J)$, $H(K)$, and $H(J, K)$ are constrained by the inequality

$$
H(J) + H(K) \geq H(J, K).
\tag{11.57}
$$

In other words, the sum of the entropies of the two systems taken separately is greater than or equal to the entropy of the combined system. Indeed, when the two systems are uncorrelated they are more disordered. This property allows us to define an important non-negative quantity, called the *mutual information* of J and K, by [see Box 11.3]

$$
I(J : K) = H(J) + H(K) - H(J, K),
\tag{11.58}
$$

which is a measure of the correlation between J and K and therefore the information that is shared between the sender and the receiver in a protocol of information transmission. For this reason, in classical information theory if J and K represents respectively the sending and receiving of a signal, then $I(J : K)$ is interpreted as the information transmitted by the communication. However, the concept of mutual information is much wider and covers also situations in which no signal at all is sent, as it happens for many quantum mechanical situations and therefore, in its generality, only means to share some information (and for this reason, as we shall see, it is particularly suited to be a measure of entanglement).

Another important quantity relevant to our discussion is the *conditional entropy*

$$H(J \,|K) = -\sum_{j\in J}\sum_{k\in K} \wp(j, k) \lg \wp(j|k), \tag{11.59}$$

Box 11.3 Non-negativity of mutual information

We note that the proof of the inequality (11.57) is tantamount to proving the non-negativity of mutual information. From a pure mathematical point of view, a function is said to be *convex* on an interval if the graph of the function within the interval lies below the line segment joining the end points of the graph in the same interval. In general, a function $f(x)$ is convex on an interval $[a, b]$ if for any two points x_1 and x_2 in $[a, b]$ the following inequality holds

$$f(\lambda x_1 + (1 - \lambda)x_2) \le \lambda f(x_1) + (1 - \lambda)f(x_2), \tag{11.60}$$

where $0 < \lambda < 1$ is an arbitrary number. Examples of convex functions include the quadratic function $f(x) = x^2$ and the exponential function $f(x) = e^x$ [see Fig. 6.2]. Moreover, a function $f(x)$ is said to be *concave* on an interval $[a, b]$ if the function $-f(x)$ is convex on that interval. Examples of concave functions are the square root function $f(x) = \sqrt{x}$ and the logarithmic function $f(x) = \log_b x$ (with an arbitrary base $b > 0$).

The non-negativity of mutual information is a direct consequence of the fact that for a concave function $f(x)$ we always have the inequality (known as the Jensen inequality)

$$\sum_j a_j f(x_j) \le f\left(\sum_j a_j x_j\right), \tag{11.61}$$

where $\sum_j a_j = 1$. Using the Jensen inequality (11.61) for the logarithmic function $f(x) = \lg x$ and the fact that

$$\sum_{j \in J} \sum_{k \in K} \wp(j, k) = 1, \qquad (11.62)$$

we obtain from Eq. (11.68) that

$$-I(J : K) = \sum_{j \in J} \sum_{k \in K} \wp(j, k) \lg \frac{\wp(j)\wp(k)}{\wp(j, k)}$$

$$\leq \lg \left(\sum_{j \in J} \sum_{k \in K} \wp(j, k) \frac{\wp(j)\wp(k)}{\wp(j, k)} \right)$$

$$= \lg \left(\sum_{j \in J} \sum_{k \in K} \wp(j)\wp(k) \right)$$

$$= \lg 1$$

$$= 0, \qquad (11.63)$$

where uses has also been made of Eq. (11.56). Therefore, it follows that $I(J : K) \geq 0$ and the identity holds if and only if $\wp(j, k) = \wp(j)\wp(k)$, i.e., J and K are uncorrelated.

where $\wp(j|k)$ is the condition probability of j given k. From the above definition it follows that

$$H(J|K) = -\sum_{j \in J} \sum_{k \in K} \wp(j, k) \lg \wp(j|k)$$

$$= -\sum_{j \in J} \sum_{k \in K} \wp(j, k) \lg \frac{\wp(j, k)}{\wp(k)}$$

$$= -\sum_{j \in J} \sum_{k \in K} \wp(j, k)[\lg \wp(j, k) - \lg \wp(k)]$$

$$= -\sum_{j \in J} \sum_{k \in K} \wp(j, k) \lg \wp(j, k) + \sum_{k \in K} \wp(k) \lg \wp(k)$$

$$= H(J, K) - H(K). \qquad (11.64)$$

We note that in obtaining the above result use has been made of Eq. (10.50), the expression

$$\wp(j, k) = \wp(j|k)\wp(k), \qquad (11.65)$$

the property (9.32c) of logarithms, and the fact that [see Eq. (10.56)]

$$\sum_{j \in J} \wp(j, k) = \wp(k). \tag{11.66}$$

If in the above formalism, we interpret K as the input code and J as the output code, the conditional entropy $H(J \,|\, K)$ as expressed in the last line of Eq. (11.64), represents the measure of how much the output is different from the input and therefore is a measure of the equivocation of the signal. Indeed, by combining Eqs. (11.58) and (11.64), we can rewrite the mutual information in terms of the conditional entropy as

$$I(J : K) = H(J) + H(K) - [H(J \,|\, K) + H(K)]$$
$$= H(J) - H(J \,|\, K). \tag{11.67}$$

Upon substituting the expressions for $H(J)$ given by Eq. (11.53) and for $H(J \,|\, K)$ by Eq. (11.59) into the above equation, we can further rewrite $I(J : K)$ in terms of probability distributions as

$$I(J : K) = -\sum_{j \in J} \wp(j) \lg \wp(j) + \sum_{j \in J} \sum_{k \in K} \wp(j, k) \lg \wp(j|k)$$
$$= \sum_{j \in J} \sum_{k \in K} \wp(j, k) \lg \frac{\wp(j|k)}{\wp(j)}$$
$$= \sum_{j \in J} \sum_{k \in K} \wp(j, k) \lg \frac{\wp(j, k)}{\wp(j)\wp(k)}, \tag{11.68}$$

where use has been made use of Eq. (11.65) and the fact that [see Eq. (11.66)]

$$\wp(j) = \sum_{k \in K} \wp(j, k). \tag{11.69}$$

Note that, as can be seen from Eq. (11.68), the mutual information is a symmetric quantity, i.e., we have [see Problem 11.10]

$$I(J : K) = I(K : J), \tag{11.70}$$

which clearly is an evidence for the fact that mutual information as such represents not only information that is transmitted by communication but also information that is shared even in the absence of communication. A graphical representation of the

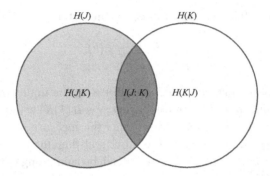

Figure 11.14 Graphic representation of the relations between the joint entropy, conditional entropy, and mutual information for two correlated systems J and K. The mutual information $I(J : K) = H(J) - H(J|K)$ is represented by the central dark gray region, where $H(J)$ is the circular region on the left (i.e., the light gray and dark gray regions) and $H(J|K)$ is the light gray region. Equivalently, $I(J : K)$ can also be expressed as $I(J : K) = H(K) - H(K|J)$, where $H(J)$ is the circular region on the right (i.e., the white and dark gray regions) and $H(K|J)$ is the white region. The joint entropy $H(J, K) = H(J) + H(K) - I(J : K)$ is represented by the whole region composed of the light gray, dark gray, and white regions.

relations between the joint entropy, conditional entropy, and mutual information is depicted in Fig. 11.14. Moreover, using Eq. (11.64) we can express $H(J)$ and $H(K)$ respectively as

$$H(J) = H(J, K) - H(K|J), \qquad (11.71a)$$

$$H(K) = H(J, K) - H(J|K). \qquad (11.71b)$$

Then, according to Eq. (11.58) we have

$$\begin{aligned}
H(J, K) &= H(J) + H(K) - I(J : K) \\
&= [H(J, K) - H(K|J)] + [H(J, K) - H(J|K)] - I(J : K) \\
&= 2H(J, K) - H(J|K) - H(K|J) - I(J : K), \qquad (11.72)
\end{aligned}$$

which in turn implies

$$H(J, K) = H(J|K) + H(K|J) + I(J : K). \qquad (11.73)$$

It is interesting to note that in this way we formulate the joint entropy of systems (codes) J an K as a sum of (i) the two conditional entropies, which express the relative independence of J on K, and vice versa, which therefore can be taken as a measure of disorder, and (ii) the correlation that they have in terms of the mutual information, representing the measure of the order of the system. Therefore, we have succeed in qualifying our statement that entropy expresses only in a first approximation the degree of disorder [see Section 9.6], since we are now able to understand that it is a kind of mix between disorder and order.

Therefore, the concept of mutual information is very important because it quantifies the interdependence between systems, or parts of a system, that contain certain information. In fact, by rewriting Eq. (11.67) as

$$H(J) = H(J|K) + I(J:K),\qquad(11.74)$$

we see that the information contained in a given system J is the information contained in the system given that another system K is known plus the mutual information between the two systems. We also stress that this equation is formally similar to Eq. (11.73). We can generalize the above equation by taking the system K as the environment E (i.e., the rest of the world), then

$$H(J) = H(J|E) + I(J:E).\qquad(11.75)$$

In other words, classically (and also quantum mechanically as we will see below) we can treat a physical system as an open system whose entropy depends on its entropic (or dynamic) relations with the environment [see Sections 9.5–9.7].

Just as we have seen in Section 9.6 that the von Neumann entropy is the extension of the classical entropy to quantum mechanics, the joint entropy, conditional entropy, and mutual information discussed above in the classical context can also be extended to quantum mechanics by means of the von Neumann entropy. In particular, the quantum mechanical counterpart of the mutual information can be used to quantify the entanglement between two quantum systems. In this way, we can easily understand that entanglement is not an absolute property (a system either has it or not) but shows degrees. For a composite system that is composed of systems 1 and 2 and described by the density matrix $\hat{\rho}_{12}$, traditionally the *entanglement* E(1, 2) of systems 1 and 2 has been calculated as

$$E(1, 2) = H_{VN}(1) + H_{VN}(2) - H_{VN}(1, 2),\qquad(11.76)$$

where $H_{VN}(1)$ and $H_{VN}(2)$ are the respective von Neumann entropies of systems 1 and 2 (calculated from the reduced density matrices $\hat{\rho}_1$ and $\hat{\rho}_2$, respectively), and $H_{VN}(1, 2)$ is the joint von Neumann entropy of systems 1 and 2 (calculated from the density matrix $\hat{\rho}_{12}$) [see Section 7.9]. In other words, entanglement is a form of quantum mutual information, in that two entangled systems are correlated because they share an amount of information that is not foreseen classically. Moreover, we can also define the entanglement of quantum systems with respect to certain measurements (as represented by observables). This is very useful for choosing the observables that optimally express the entanglement of quantum systems.[a] In this case the joint von Neumann entropy of systems 1 and 2 with respect to the measurements $\hat{O}_1$ and $\hat{O}_2$ made respectively on systems 1 and 2 is given by

$$H_{VN}(\hat{O}_1, \hat{O}_2) = -\sum_j \sum_k \langle o_{1j}, o_{2k}|\hat{\rho}_{12}|o_{1j}, o_{2k}\rangle$$

$$\times \ln\langle o_{1j}, o_{2k}|\hat{\rho}_{12}|o_{1j}, o_{2k}\rangle, \tag{11.77}$$

where $\hat{\rho}_{12}$ is the density matrix of the composite system and $|o_{1j}, o_{2k}\rangle = |o_{1j}\rangle \otimes |o_{2k}\rangle$ with $|o_{1j}\rangle$ and $|o_{2k}\rangle$ being the respective eigenstates of $\hat{O}_1$ and $\hat{O}_2$ (here for the sake of notational simplicity we have suppressed the subscripts 1 and 2 referring to the systems). The joint entropy $H_{VN}(\hat{O}_1, \hat{O}_2)$ quantifies the information contained in systems 1 and 2 that is *accessible* to the joint measurement $\hat{O}_1$ of system 1 and $\hat{O}_2$ of system 2. We also can write the von Neumann entropies of system 1 and 2 with respect to the respective observables $\hat{O}_1$ and $\hat{O}_2$ as

$$H_{VN}(\hat{O}_1) = -\sum_j \langle o_{1j}|\hat{\rho}_1|o_{1j}\rangle \ln\langle o_{1j}|\hat{\rho}_1|o_{1j}\rangle, \tag{11.78a}$$

$$H_{VN}(\hat{O}_2) = -\sum_k \langle o_{2k}|\hat{\rho}_2|o_{2k}\rangle \ln\langle o_{2k}|\hat{\rho}_2|o_{2k}\rangle, \tag{11.78b}$$

where $\hat{\rho}_1$ and $\hat{\rho}_2$ are the reduced density matrices of system 1 and 2, respectively. Similarly, the entropies $H_{VN}(\hat{O}_1)$ and $H_{VN}(\hat{O}_2)$ quantify respectively the information contained in system 1 that is accessible to the measurement $\hat{O}_1$ of system 1 and the information contained in system 2 that is accessible to the measurement $\hat{O}_2$ of system 2. Then, we may define the entanglement of systems 1 and 2 with respect to the observables $\hat{O}_1$ and $\hat{O}_2$ by

$$E(\hat{O}_1, \hat{O}_2) = H_{VN}(\hat{O}_1) + H_{VN}(\hat{O}_2) - H_{VN}(\hat{O}_1, \hat{O}_2), \tag{11.79}$$

---

[a](Zurek, 1983), (Barnett/Phoenix, 1989).

which is formally similar to the entanglement $E(1, 2)$ given by Eq. (11.76). It is interesting to note that we always have

$$E(\hat{O}_1, \hat{O}_2) \leq E(1, 2), \tag{11.80}$$

namely, the mutual information of the two systems that is accessible to two arbitrary measurements of the respective systems cannot exceed the mutual information of the two systems. This can be understood as a reformulation of the information accessibility principle in the context of quantum mutual information [see Principle 11.1].

Actually, mutual information allows us to distinguish between classical correlations and quantum features.[a] As we have seen, classically we have two equivalent expressions for mutual information, Eq. (11.58) and Eq. (11.67). Quantum mechanically it turns out that there is a difference between the two expressions (where S and A should be understood as the object system and the apparatus, respectively)

$$I(S:A) = H_{VN}(S) + H_{VN}(A) - H_{VN}(S, A), \tag{11.81a}$$
$$C(S:A) = H_{VN}(S) - H_{VN}(S|A), \tag{11.81b}$$

where $C(S:A)$ represents the part of correlations that can be attributed to classical physics and, as usual, $H_{VN}(S)$ and $H_{VN}(A)$ are computed on the reduced density matrices of the two subsystems. The conditional entropy $H_{VN}(S|A)$ requires us to specify the state of S *given* the state of A. Such a statement in quantum theory is ambiguous until the to-be-measured set of states A is selected. To this purpose, let us consider the set of one-dimensional projectors $\{\hat{P}_j^A\}$, where the label j distinguishes different outcomes of this measurement. The state of S after the outcome corresponding to $\hat{P}_j^A$ has been detected is given by [see also Eq. (9.10)]

$$\hat{\rho}_{S|\hat{P}_j^A} = \frac{\mathrm{Tr}_A(\hat{P}_j^A \hat{\rho} \hat{P}_j^A)}{\wp_j}, \tag{11.82}$$

where $\hat{\rho}$ is the density matrix of the composite system comprehending S and A, and

$$\wp_j = \mathrm{Tr}_{S,A}(\hat{P}_j^A \hat{\rho}). \tag{11.83}$$

---

[a](Olivier/Zurek, 2001).

It is clear that the entropy $H_{VN}(\hat{\rho}_{S|\hat{P}_j^A})$ represents the missing information about S. Now, the conditional entropy of S given the complete measurement $\{\hat{P}_j^A\}$ on A is represented by the probability weighted average of $H_{VN}(\hat{\rho}_{S|\hat{P}_j^A})$:

$$H_{VN}(S|\{\hat{P}_j^A\}) = \sum_j \wp_j H_{VN}(\hat{\rho}_{S|\hat{P}_j^A}), \tag{11.84}$$

which allows us to write an unambiguous expression for $C(S:A)$ as

$$C(S:A)_{\{\hat{P}_j^A\}} = H_{VN}(S) - H_{VN}(S|\{\hat{P}_j^A\}). \tag{11.85}$$

In this way, we obtain the formulation for quantum discord, i.e., the difference between the mutual information and its classical part,

$$\begin{aligned}
\mathcal{Q}(S:A)_{\{\hat{P}_j^A\}} &= I(S:A) - C(S:A)_{\{\hat{P}_j^A\}} \\
&= H_{VN}(A) - H_{VN}(S, A) + H_{VN}(S|\{\hat{P}_j^A\}). \tag{11.86}
\end{aligned}$$

This also shows that the correlation between systems that is represented by the mutual information is the sum of a classical and a quantum contribution [see Fig. 7.4 and comments]:

$$I(S:A) = C(S:A)_{\{\hat{P}_j^A\}} + \mathcal{Q}(S:A)_{\{\hat{P}_j^A\}}. \tag{11.87}$$

The quantum discord is asymmetric under exchange of S and A, since the quantity in Eq. (11.84) involves a measurement on A that allows the observer to infer the state of S. This typically implies an increase of entropy. In other words, we have

$$H_{VN}(S|\{\hat{P}_j^A\}) \geq H_{VN}(S, A) - H_{VN}(A), \tag{11.88}$$

which implies that

$$\mathcal{Q}(S:A)_{\{\hat{P}_j^A\}} \geq 0. \tag{11.89}$$

The previous formalism allows us to generalize and formulate Theorem 11.1 in mathematical terms. In fact, the classical quantity $C(S:A)$ in Eq. (11.85) is essentially the Holevo bound, i.e., the maximum amount of classical information that can be extracted from a system.[a] This also allows us to consider the complementarity between local irreversibility and global reversibility in formal terms [see Section 9.7]. In fact, each component of the object system, when coupled with the apparatus, becomes also entangled to a

---

[a](Zwolak/Zurek, 2013).

component of the environment. In other words, we establish in this way two quantum channels: one between the object system and the apparatus and the other between the object system and the environment. This shows that premeasurement, although still reversible, is a real operation, as far as it establishes external conditions for the final step of measurement [see Sections 9.3–9.4]. Now, when we get a specific measurement result, we make accessible to the latter kind of channel only a small part of the environment (the local one), while all other channels are switched off. This means that we indeed lose this part of information into the environment [see Section 9.6]. In other words, there is here a complementarity between quantum discord (global reversibility) and the Holevo bound (local irreversibility):[a] Increasing the redundant (classical) information stored in a small fragment of the universe decreases the quantum information in the much larger fragment represented by the rest of the universe. For this reason, there is also a minimum fragment size needed by an observer to learn about the object system.

Since $C(S:A)$ depends on the set $\{\hat{P}_j^A\}$, we shall usually be concerned with the set $\{\hat{P}_j^A\}$ that minimizes the discord given a certain $\hat{\rho}$. Minimizing the discord over the possible measurements on A corresponds to finding the measurement that disturbs least the overall quantum state and that, at the same time, allows one to extract the most information about S. When the quantum discord is not zero, it indicates the presence of correlations that are ultimately due to non-commutativity of quantum observables [see Section 6.9]. This may be also considered the quantumness of correlations. We remark that, in accordance with what is said in Section 7.9, entanglement is defined by the expression in Eq. (11.76) that is in general different from the discord precisely because entanglement also contains classical correlations. Note also that although classically the conditional entropy is always positive, quantum mechanically it can be negative, and its negativity is a sufficient condition of entanglement. A final remark concerns the use of terms and concepts. Both discord and quantumness hint at the same issue as the notion of features. However, the discord

---

[a] (Zwolak/Zurek, 2013).

is a measure while the quantumness expresses a character. It is therefore suitable to have a specific term that can stand on the same foot as the notions of event and property but displaying simultaneously the specificity (the quantumness) of quantum mechanics, and this is the specific advantage of the notion of features.

Problem 11.10 Prove Eq. (11.70), i.e., the mutual information is a symmetric quantity.

Problem 11.11 Let J and K have the following joint probability:

	$J = 1$	$J = 2$
$K = 1$	0	$\frac{3}{4}$
$K = 2$	$\frac{1}{8}$	$\frac{1}{8}$

Find $H(J, K)$ $H(J)$, $H(K)$, $H(J \mid K)$, $H(K \mid J)$, and $I(J : K)$.

11.8 Information and Non-Separability

We are now able to reformulate the Bell inequality in terms of the mutual information and entropy [see Section 10.4]. Indeed, Braunstein and Caves[a] derived an information-theoretic Bell inequality of the form

$$H(\hat{O}_a | \hat{O}_b) \leq H(\hat{O}_a | \hat{O}_{b'}) + H(\hat{O}_{b'} | \hat{O}_{a'}) + H(\hat{O}_{a'} | \hat{O}_b), \qquad (11.90)$$

where $\hat{O}_a$ and $\hat{O}_{a'}$ are observables of system 1 while $\hat{O}_b$ and $\hat{O}_{b'}$ are observables of system 2. This inequality can also be considered an information-theoretic reformulation of the CHSH inequality (10.38). We also note that the observables $\hat{O}_a$, $\hat{O}_b$, and alike do not necessarily represent spin components but are quite general. In the above expression, all the conditional entropies have the same form of Eq. (11.64). In particular, $H(\hat{O}_a | \hat{O}_b)$ is the conditional entropy of system 1 accessible to the observable $\hat{O}_a$ given that system 2 is

---

[a](Braunstein/Caves, 1988), (Braunstein/Caves, 1990).

measured with the observable $\hat{O}_b$. This quantity is defined by

$$H(\hat{O}_a|\hat{O}_b) = H(\hat{O}_a, \hat{O}_b) - H(\hat{O}_b), \qquad (11.91)$$

where it is noted that the entropies appear in the above equation (and similar ones) are understood to be calculated in a local hidden variable theory. The information-theoretic Bell inequality (11.90) can be derived by using the following inequalities:

(i) The inequality

$$H(\hat{O}_a|\hat{O}_b) \le H(\hat{O}_a), \qquad (11.92)$$

which means that the information contained in one of the systems that is accessible to a measurement decreases given that the other system has been already measured. This is quite intuitive, especially considering that any measurement is a selection act.

(ii) The inequalities

$$H(\hat{O}_a) \le H(\hat{O}_a, \hat{O}_b) \quad \text{and} \quad H(\hat{O}_b) \le H(\hat{O}_a, \hat{O}_b), \qquad (11.93)$$

meaning that the information contained in two systems that is accessible to two measurements of the respective systems is no less than the information contained in either system that is accessible to individual measurement.

(iii) The following generalization of the inequality (11.93)

$$\begin{aligned} H(\hat{O}_a, \hat{O}_b) &\le H(\hat{O}_a, \hat{O}_{a'}, \hat{O}_b, \hat{O}_{b'}) \\ &= H(\hat{O}_a|\hat{O}_{a'}, \hat{O}_b, \hat{O}_{b'}) + H(\hat{O}_{b'}|\hat{O}_{a'}, \hat{O}_b) \\ &\quad + H(\hat{O}_{a'}|\hat{O}_b) + H(\hat{O}_b), \end{aligned} \qquad (11.94)$$

where the equality is obtained by a recursion of the definition of the conditional entropy analogous to Eq. (11.64).

Now the information-theoretic Bell inequality (11.90) can be obtained from the inequality (11.94) by noting that

$$H(\hat{O}_a|\hat{O}_{a'}, \hat{O}_b, \hat{O}_{b'}) \le H(\hat{O}_a|\hat{O}_{b'}), \qquad (11.95a)$$

$$H(\hat{O}_{b'}|\hat{O}_{a'}, \hat{O}_b) \le H(\hat{O}_{b'}|\hat{O}_{a'}), \qquad (11.95b)$$

which follow directly from the inequality (11.92). Substituting the above two equalities into the equality (11.94) and using Eq. (11.91) to rewrite the joint entropy $H(\hat{O}_a, \hat{O}_b)$ on the left-hand side of latter equality in terms of the conditional entropy $H(\hat{O}_a|\hat{O}_b)$, we find that the unwanted $H(\hat{O}_b)$ terms on both sides cancel and the desired inequality follows. Braunstein and Caves succeeded in showing that in the context of Bell's spin-like experiments the quadrilateral inequality (11.90) is violated by quantum mechanics.

It is possible to arrive at a result formally similar to the previous one by using a different conceptual instrument.[a] Recalling the definition of the mutual information, we now define the *information distance* $D(\hat{O}_a, \hat{O}_b)$ of two observables $\hat{O}_a$ and $\hat{O}_b$ by

$$
\begin{aligned}
D(\hat{O}_a, \hat{O}_b) &= H(\hat{O}_a|\hat{O}_b) + H(\hat{O}_b|\hat{O}_a) \\
&= H(\hat{O}_a, \hat{O}_b) - I(\hat{O}_a : \hat{O}_b) \\
&= 2H(\hat{O}_a, \hat{O}_b) - H(\hat{O}_a) - H(\hat{O}_b),
\end{aligned}
\tag{11.96}
$$

which measures the lack of correlation between the observables $\hat{O}_a$ and $\hat{O}_b$. Obviously, the information distance is positive semidefinite and symmetric. It can also be proved that for three arbitrary observables $\hat{O}_a$, $\hat{O}_b$, and $\hat{O}_c$ the classical information distance satisfies the following triangular inequality

$$
D(\hat{O}_a, \hat{O}_b) + D(\hat{O}_b, \hat{O}_c) \geq D(\hat{O}_a, \hat{O}_c).
\tag{11.97}
$$

The so-called classical quadrilateral information-distance Bell inequality can be formulated similarly and is given by

$$
D(\hat{O}_a, \hat{O}_b) + D(\hat{O}_b, \hat{O}_{a'}) + D(\hat{O}_{a'}, \hat{O}_{b'}) \geq D(\hat{O}_a, \hat{O}_{b'}).
\tag{11.98}
$$

To construct a counterexample of the above quadrilateral inequality in quantum mechanics, we consider the situation in which the observables $\hat{O}_a$ and $\hat{O}_b$ represent components of spin in directions separated by an angle θ. The quantum information distance of $\hat{O}_a$ and $\hat{O}_b$ for the Bell state $|\Psi^-\rangle$ [see Eq. (11.43b)] is given by

$$
D_{\Psi^-}(\hat{O}_a, \hat{O}_b) = 2 f\left(\frac{\theta}{2}\right),
\tag{11.99}
$$

where

$$
f(\phi) = -\cos^2\phi \ln\cos^2\phi - \sin^2\phi \ln\sin^2\phi.
\tag{11.100}
$$

Now choose the four observable $\hat{O}_a$, $\hat{O}_b$, $\hat{O}_{a'}$, and $\hat{O}_{b'}$ as depicted in Fig. 11.15, such that the angles between the coplanar directions

---

[a](Schumacher, 1990), (Schumacher, 1991).

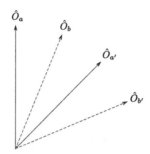

Figure 11.15 Spin measurement yielding a violation of the information-distance quadrilateral inequality for the Bell state $|\Psi^-\rangle$. The angles between the coplanar directions represented by $\hat{O}_a$ and $\hat{O}_b$, by $\hat{O}_b$ and $\hat{O}_{a'}$, as well as by $\hat{O}_{a'}$ and $\hat{O}_{b'}$ are all equal to $\frac{\pi}{8}$.

represented by $\hat{O}_a$ and $\hat{O}_b$, by $\hat{O}_b$ and $\hat{O}_{a'}$, as well as by $\hat{O}_{a'}$ and $\hat{O}_{b'}$ are all equal to $\frac{\pi}{8}$. Then, we have

$$D_{\Psi^-}(\hat{O}_a, \hat{O}_b) = D_{\Psi^-}(\hat{O}_b, \hat{O}_{a'}) = D_{\Psi^-}(\hat{O}_{a'}, \hat{O}_{b'}) = 2\,f\left(\frac{\pi}{16}\right) \simeq 0.323,$$

$$\tag{11.101a}$$

$$D_{\Psi^-}(\hat{O}_a, \hat{O}_{b'}) = 2\,f\left(\frac{3\pi}{16}\right) \simeq 1.236. \tag{11.101b}$$

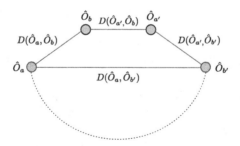

Figure 11.16 Schematic information-distance representation of quantum non-separability. The gray circles represent observables, and the solid line represents the information distance of two observables in a local hidden variable theory. The hidden-variable information distance $D(\hat{O}_a, \hat{O}_{b'})$ satisfies the information quadrilateral inequality (11.98), which however is violated by the quantum mechanical result (dotted line) for the Bell state $|\Psi^-\rangle$.

Since $0.323 + 0.323 + 0.323 < 1.236$, it follows that

$$D_\Psi\text{-}(\hat{O}_a, \hat{O}_b) + D_\Psi\text{-}(\hat{O}_b, \hat{O}_{a'}) + D_\Psi\text{-}(\hat{O}_{a'}, \hat{O}_{b'}) \leq D_\Psi\text{-}(\hat{O}_a, \hat{O}_{b'}).$$

$$(11.102)$$

Therefore, it is evident that the quadrilateral inequality (11.98) is violated. A schematic representation of the counterexample is depicted in Fig. 11.16. In quantum mechanics, $\hat{O}_a$ and $\hat{O}_{b'}$ may be "further apart" in information distance than the information quadrilateral inequality permits, given the "closeness" of $\hat{O}_a$ to $\hat{O}_b$, $\hat{O}_b$ to $\hat{O}_{a'}$, and $\hat{O}_{a'}$ to $\hat{O}_{b'}$. Since the information distance measures the lack of correlation between two observables, we can say that $\hat{O}_a$ and $\hat{O}_{b'}$ may be less correlated in quantum mechanics than it would be possible in a local hidden variable theory.

11.9 Summary

In this chapter, we have

- Explained which kind of problems arise when considering quantum systems as information processors and have solved them.
- Proved the no-cloning theorem, which states that it is not possible to create identical copies of an arbitrary unknown quantum state.
- Introduced the fundamental concept of potential information and shown that it has to do with encoding and sharing of information.
- Discussed the information accessibility principle according to which we can cannot have access to the whole potential information.
- Learned the very basic of quantum gates and quantum computation.
- Analyzed the quantum algorithm for solving the Deutsch problem, which however cannot be solved by employing any classical algorithm.
- Assimilated the amazing phenomenon called quantum teleportation.

- Analysed the Bennett and Brassard (BB84) protocol and the Ekert protocol for quantum key distribution.
- Made general considerations about the classical and quantum information, shown how to deal with entanglement by means of quantum mutual information and introduced the concept of quantum discord that allows us to distinguish between classical and quantum correlations.
- Expressed entanglement in terms of the mutual information and considered two information-theoretic formulations of the Bell inequalities.

Chapter 12

Interpretation

In this chapter, we shall collect and generalize the results of the previous three chapters. We shall come back to the issue of quantum correlations in terms of information. We shall consider now the second aspect of quantum information [see Section 11.3], that is, information selection leading to a local growth of entropy (as it is the case during a measurement). This will be the basis for examining again the whole process of measurement and finally assign an ontological status to fundamental entities of quantum theory like states, observables, and properties.

12.1 Information Acquisition

We recall that quantum mechanically, in order to recover the information about an object system, we need the coupling with an apparatus [see Sections 9.3–9.4]. Classically, we have a similar situation. Here, we have an unknown parameter k whose value we wish to know and some data d pertaining to a set D at our disposal. This is a very important point, since we *never* have direct access to things (whose properties are described by k) but always

Quantum Mechanics for Thinkers
Gennaro Auletta and Shang-Yung Wang
Copyright © 2014 Pan Stanford Publishing Pte. Ltd.
ISBN 978-981-4411-71-4 (Hardcover), 978-981-4411-72-1 (eBook)
www.panstanford.com

to things *through* data.[a] These data can be represented by the position of the pointer of our measuring apparatus or simply by the impulse our sensory system has received, or even by the way we receive information about the position of the pointer through our sensory system. It does not matter how long this chain may be. The important point is a matter of principle.

Principle 12.1 (Information Acquisition) *We can receive information about objects and events only conditionally on the data at our disposal.*

Let us consider a classical example. Suppose that we wished to know exactly what the distribution of matter was in the early universe. We can know this by collecting data about the cosmic microwave background radiation we receive now. This again shows a very important common point between quantum and classical physics that is not well understood, and which has been pointed out by Wheeler's delayed choice experiment [see Section 5.6]. We cannot receive any information about past events unless they are received through *present effects* (data). This is an equivalent formulation of what we have said before, since any event, represented by a parameter k, can be known only through a *later* effect due to the finite speed of light. As a matter of fact, all of our perceptual experience is mediated and slightly delayed in time. Moreover, since we always have experience only of a part of the possible effects produced by events, this is again an application of the principle of information accessibility [see Principle 11.1].

Obviously, once we have observed or acquired data, we must perform an information extrapolation that allows us to have an "informed guess" about the value of the parameter k. This is the process of information selection. As we know, the joint probability $\wp(j, k)$ that we select the event j while having an event represented by an unknown parameter k (i.e., the probability that both event k and event j occur) is given by [see Eq. (10.50)]

$$\wp(j, k) = \wp(j|k)\wp(k), \tag{12.1}$$

where $\wp(j|k)$ is the conditional probability of the selection event j given the source event represented by k. Now, by taking into account

---

[a](Zurek, 2004).

the data d that are somehow the interface between the source event k and our selection event j we may express the probability $\wp(j|k)$ as[a]

$$\wp(j|k) = \sum_{d \in D} \wp(j|d)\wp(d|k), \tag{12.2}$$

where the summation is over all the data d pertaining to the set D. By substituting the above expression into Eq. (12.1) we obtain

$$\wp(j, k) = \sum_{d \in D} \wp(j|d)\wp(d|k)\wp(k)$$

$$= \sum_{d \in D} \wp(j|d)\wp(d, k). \tag{12.3}$$

A sample probability tree illustrating the use of the above equation is dictated in Fig. 12.1. We note that Eq. (12.3) can be considered a generalization of the well-known formula [see Eq. (10.57)]

$$\wp(j) = \sum_{d \in D} \wp(j|d)\wp(d), \tag{12.4}$$

and it reduces to the latter when $\wp(k) = 1$, i.e., when the value of the parameter k is known with certainty. It is important to stress that the two conditional probabilities $\wp(j|d)$ and $\wp(d|k)$ are quite different. The probability $\wp(d|k)$ represents how *faithful* our data are given the event k, that is, how reliable our apparatus (or sensory system) is. Instead, the probability $\wp(j|d)$ represents our ability to select a *single* event j which can be used to interpret the data d in the best way. Moreover, using the Bayes theorem (10.53) we express $\wp(k|j)$ in terms of $\wp(j|d)$ and $\wp(d|k)$ as

$$\wp(k|j) = \frac{\wp(k)\wp(j|k)}{\wp(j)}$$

$$= \frac{\wp(k)}{\wp(j)} \sum_{d \in D} \wp(j|d)\wp(d|k). \tag{12.5}$$

In other words, we can invert the kind of question we pose and try to infer the unknown parameter k conditioned on having selected the event j.

Having made these considerations, we immediately see that Eq. (12.2) or (12.3) represents the classical analogue of the quantum

---

[a](Helstrom, 1976).

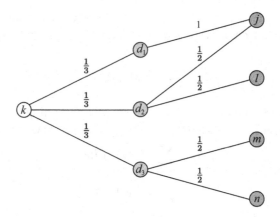

Figure 12.1 A tree diagram for calculating conditional probabilities (from left to right). Suppose that we have an event k, occurring with a probability $\wp(k) = \frac{1}{2}$, and a particular set of effects (data at our disposal) d_1, d_2, and d_3, each one occurring with conditional probability $\frac{1}{3}$ given k. It is easy to see then that the probability to select events j and k is $\wp(j, k) = [\wp(j|d_1)\wp(d_1|k) + \wp(j|d_2)\wp(d_2|k)]\wp(k) = \left(1 \times \frac{1}{3} + \frac{1}{2} \times \frac{1}{3}\right) \times \frac{1}{2} = \frac{1}{4}$, while the probability to select events n and k is $\wp(n, k) = \wp(n|d_3)\wp(d_3|k)\wp(k) = \frac{1}{2} \times \frac{1}{3} \times \frac{1}{2} = \frac{1}{12}$. Note that when $\wp(k) = 1$, we have $\wp(d_i|k) = \wp(d_i)$ for any of the data and $\wp(j, k) = \wp(j|k) = \wp(j)$. Note also that the sum of the probabilities leaving any node is equal to 1.

measurement process analyzed in Chapter 9. The conditional probability $\wp(d|k)$ corresponds to the coupling between the object system and apparatus in quantum mechanics. Obviously, the difference between the classical and the quantum case is that, when we have an entanglement, we can have a perfect correlation between the apparatus and object system, which is difficult to obtain in classical situations. According to the analysis developed in Section 11.7, the quantity

$$H(J|K) = -\sum_{j \in J}\sum_{k \in K}\wp(j, k)\lg\wp(j|k), \qquad (12.6)$$

given by Eq. (11.59) but whose conditional probability $\wp(j|k)$ is given now by Eq. (12.2), expresses the incertitude of the selection (or detection) $j \in J$ when the parameter $k \in K$ varies. On the other hand, the surprisal $I(j|k) = -\lg[\wp(j|k)]$ represents the surprisal of the result j given k and therefore the amount of the information that we have effectively gained through the selection of j.

We would like now to show that the counterpart of the conditional probabilitiy $\wp(j|k)$ in quantum mechanics is the probability of a final detection event, which, given a certain experimental context (a premeasurement), allows us to finally ascribe a property to the object system (the system that has been measured). Let us for the moment consider only the object system and the apparatus (with the exclusion of the environment), and in particular the premesurement that describes the type of transformation given by Eq. (9.13). Suppose that the initial state of the apparatus is some ready state $|A_0\rangle$ while the state of the object system is some superposition state $|\psi_S\rangle$. For the sake of simplicity, we shall consider a two-state system as the object system. Then, the transformation (9.13) takes the form

$$|\psi_S\rangle|A_0\rangle \longrightarrow c_0|0\rangle|a_0\rangle + c_1|1\rangle|a_1\rangle, \tag{12.7}$$

where $|0\rangle$ and $|1\rangle$ are the system states representing the eigenstates of some system observable to be measured, while $|a_0\rangle$ and $|a_1\rangle$ are respectively the apparatus states corresponding to $|0\rangle$ and $|1\rangle$. In the density matrix formalism the initial state of the subject system and the apparatus appearing on the left-hand side of the transformation (12.7) may be described by the (factorized) density matrix $\hat{\rho}_S\hat{\rho}_A$, where

$$\hat{\rho}_S = |\psi_S\rangle\langle\psi_S| \quad \text{and} \quad \hat{\rho}_A = |A_0\rangle\langle A_0|. \tag{12.8}$$

This is in agreement with the assumption that the entanglement between the object system and the apparatus created during the premeasurement step is the result of a unitary transformation (for instance through a Hadamard gate, as shown in the previous chapter). Indeed, we remind the reader that only the final step of selection or detection is not unitary. Therefore, we have the following unitary transformation [see Eq. (7.121)]

$$\hat{\rho}_S\hat{\rho}_A \longrightarrow \hat{U}_t\hat{\rho}_S\hat{\rho}_A\hat{U}_t^\dagger, \tag{12.9}$$

where $\hat{U}_t$ is the time evolution operator whose form depends on the coupling of the system and the apparatus, and for the sake of notational simplicity the time dependence in the above expression has been suppressed. It is easy to find that in the case under consideration we have

$$\hat{U}_t\hat{\rho}_S\hat{\rho}_A\hat{U}_t^\dagger = |c_0|^2|0\rangle\langle 0| \otimes |a_0\rangle\langle a_0| + |c_1|^2|1\rangle\langle 1| \otimes |a_1\rangle\langle a_1|$$
$$+ c_0c_1^*|0\rangle\langle 1| \otimes |a_0\rangle\langle a_1| + c_0^*c_1|1\rangle\langle 0| \otimes |a_1\rangle\langle a_0|. \tag{12.10}$$

Just before the detection, the probability that the apparatus will read the value a_j ($j = 0, 1$) is given by

$$\wp(a_j) = \text{Tr}_A\left[\hat{P}_{a_j} \text{Tr}_S(\hat{U}_t \hat{\rho}_S \hat{\rho}_A \hat{U}_t^\dagger)\right], \tag{12.11}$$

where $\hat{P}_{a_j} = |a_j\rangle\langle a_j|$ is the projector onto the apparatus state $|a_j\rangle$. Note that the previous probability only takes into account correlated terms due to the entanglement between the object system and apparatus. The above result can be verified quite easily. Indeed, the partial trace over the system will sum only the diagonal terms in the system Hilbert space $\mathcal{H}_S$, while the subsequent application of the projector $\hat{P}_{a_j}$ projects out a single term in the apparatus Hilbert space $\mathcal{H}_A$, yielding

$$\hat{P}_{a_j} \text{Tr}_S(\hat{U}_t \hat{\rho}_S \hat{\rho}_A \hat{U}_t^\dagger) = |c_j|^2 |a_j\rangle\langle a_j|. \tag{12.12}$$

Finally tracing out the apparatus, we shall get the probability, which, in our case, is simply given by $|c_j|^2$. Using the cyclic property of the trace (7.115c) and the fact that Tr_S does not act on the apparatus, we may rewrite Eq. (12.11) as

$$\begin{aligned}
\wp(a_j) &= \text{Tr}_A\left[\hat{P}_{a_j} \text{Tr}_S(\hat{U}_t \hat{\rho}_S \hat{\rho}_A \hat{U}_t^\dagger)\right] \\
&= \text{Tr}_A\left[\text{Tr}_S(\hat{U}_t^\dagger \hat{P}_{a_j} \hat{U}_t \hat{\rho}_S \hat{\rho}_A)\right] \\
&= \text{Tr}_A\left[\text{Tr}_S(\hat{U}_t^\dagger |a_j\rangle\langle a_j| \hat{U}_t \hat{\rho}_S) \hat{\rho}_A\right].
\end{aligned} \tag{12.13}$$

It is convenient to define a new Hermitian operator $\hat{E}_j$ in the apparatus Hilbert space $\mathcal{H}_A$ by

$$\hat{E}_j = \text{Tr}_S(\hat{U}_t^\dagger \hat{P}_{a_j} \hat{U}_t \hat{\rho}_S), \tag{12.14}$$

in terms of which the above equation can be written in a form formally similar to Eq. (9.9) as

$$\wp(a_j) = \text{Tr}_A(\hat{E}_j \hat{\rho}_A). \tag{12.15}$$

The projection-like operator $\hat{E}_j$ is called the *effect operator* (or effect for short), which plays an important role in the theory of generalized measurement. From the definition (12.14), it follows that

$$\sum_j \hat{E}_j = \hat{I}, \tag{12.16}$$

where $\hat{I}$ is the identity operator (in the apparatus Hilbert space $\mathcal{H}_A$) and use has been made of the completeness relation for the apparatus states

$$\sum_j \hat{P}_{a_j} = \hat{I}. \tag{12.17}$$

It can be shown that $\hat{E}_j$ is positive semidefinite, i.e.,

$$\langle \phi_A | \hat{E}_j | \phi_A \rangle \geq 0 \quad \text{for all} \quad |\phi_A\rangle \in \mathcal{H}_A. \tag{12.18}$$

However, unlike the projectors, the effect operators in general does not satisfy the requirement of orthogonality [see Eq. (3.62)], namely,

$$\hat{E}_j \hat{E}_k \neq \delta_{jk} \hat{E}_k. \tag{12.19}$$

This can be used to generalize the concept of property, which however is an issue that goes beyond the scope of this book. Moreover, substituting $\hat{\rho}_S = |\psi_S\rangle\langle\psi_S|$ into Eq. (12.14), we can explicitly calculate the trace over the system to obtain

$$\begin{aligned}
\hat{E}_j &= \mathrm{Tr}_S\left(\hat{U}_t^\dagger |a_j\rangle\langle a_j| \hat{U}_t |\psi_S\rangle\langle\psi_S|\right) \\
&= \sum_{k=0,1} \langle k| \hat{U}_t^\dagger |a_j\rangle\langle a_j| \hat{U}_t |\psi_S\rangle\langle\psi_S|k\rangle \\
&= \langle\psi_S| \hat{U}_t^\dagger |a_j\rangle\langle a_j| \hat{U}_t |\psi_S\rangle \\
&= \hat{\vartheta}_j^\dagger \hat{\vartheta}_j, \tag{12.20}
\end{aligned}$$

where

$$\hat{\vartheta}_j = \langle a_j| \hat{U}_t |\psi_S\rangle, \quad \hat{\vartheta}_j^\dagger = \langle\psi_S| \hat{U}_t^\dagger |a_j\rangle. \tag{12.21}$$

Note that $\hat{\vartheta}_j$ and $\hat{\vartheta}_j^\dagger$ do *not* represent probability amplitudes because the time evolution operator $\hat{U}_t$ describes the coupling of the apparatus and the system, whereas the kets $|a_j\rangle$ and $|\psi_S\rangle$ represent respectively only the apparatus state and the system state. As a result, $\hat{\vartheta}_j$ is called the *amplitude operator*. From its definition, it is clear that the amplitude operator $\hat{\vartheta}_j$ describes the three steps of the measurement of a given observable [see Section 9.5]:

(i) Preparation of the initial state of the system (i.e., the input $|\psi_S\rangle$),

(ii) Unitary time evolution (i.e., the coupling or premeasurement) that entangles the system with the apparatus and allows us to select an observable (i.e., the process represented by $\hat{U}_t$).

(iii) Detection by the apparatus (i.e., the output $|a_j\rangle$) that allows us to assign a property to the system.

We can also summarize what is said here by writing [see Eq. (9.10)]

$$\hat{\rho}_A^{(f)} = \frac{1}{\wp(a_j)} \hat{\vartheta}_j \hat{\rho}_A \hat{\vartheta}_j^\dagger, \tag{12.22}$$

where $\hat{\rho}_A^{(f)}$ is the state of the apparatus after the detection corresponding to the value a_j. Let the parameter j be associated with one of the detection outcomes a_j that corresponds to the apparatus state $|a_j\rangle$ and the parameter k with one of the state vectors $|\psi_k\rangle$ in a given orthonormal basis for the object system. Then, following a similar analysis as above and using Eqs. (12.15) and (12.20), we have

$$\wp(j|k) = \text{Tr}_A(\hat{\vartheta}_{jk}\hat{\rho}_A\hat{\vartheta}_{jk}^\dagger), \tag{12.23}$$

where

$$\hat{\vartheta}_{jk} = \langle a_j|\hat{U}_t|\psi_k\rangle. \tag{12.24}$$

Therefore, we have shown that the amplitude operator $\hat{\vartheta}_{jk}$ is the quantum mechanical counterpart of the classical conditional probability $\wp(j|k)$ given by Eq. (12.2).

Problem 12.1 Show that the effect operator $\hat{E}_j$ is Hermitian and positive semidefinite.

12.2 Bounds on Information Acquisition

The general question that we like to address now is whether there are specific quantum mechanical bounds on information acquisition. Is the bound set by the Bell inequality (10.35) a necessity or are there more stringent bounds? And if there are, what is their meaning? The present section is devoted to the exploration of these issues. Let us take advantage of the CHSH inequality (10.38). Since, as can be seen from Eq. (10.22), each of the terms in Eq. (10.38) lies between -1 and $+1$, the natural upper bound for the entire expression is $+4$. This is precisely the case if we demand that the probabilities satisfy only the causal communication requirement, i.e., that they do not violate relativistic (or Einstein's) locality (what is called the no-signaling requirement).[a] In this case, we have

$$|\langle \mathbf{a}, \mathbf{b}\rangle + \langle \mathbf{a}, \mathbf{b}'\rangle + \langle \mathbf{a}', \mathbf{b}\rangle - \langle \mathbf{a}', \mathbf{b}'\rangle| \leq 4. \tag{12.25}$$

This can be proved as follows. The only requirement of relativistic locality is that the operations one perform locally here are not

---

[a](Popescu/Rohrlich, 1994), (Hillery/Yurke, 1995).

influenced by the operations someone performs far away elsewhere [see Section 10.6]. This implies in particular that the probability that one obtains a certain outcome (say 1) when choosing the direction **a** is independent of the outcomes (either $+1$ or -1) when someone elsewhere choses a direction **b** or **b**$'$, that is,

$$\wp(1, 1|\mathbf{a}, \mathbf{b}) + \wp(1, -1|\mathbf{a}, \mathbf{b}) = \wp(1, 1|\mathbf{a}, \mathbf{b}') + \wp(1, -1|\mathbf{a}, \mathbf{b}').$$
(12.26)

Similar considerations hold for arbitrary directions. If we consider only this requirement, we are allowed to build the set of probabilities

$$\wp(1, 1|\mathbf{a}, \mathbf{b}) = \wp(-1, -1|\mathbf{a}, \mathbf{b}) = \frac{1}{2}, \tag{12.27a}$$

$$\wp(1, 1|\mathbf{a}, \mathbf{b}') = \wp(-1, -1|\mathbf{a}, \mathbf{b}') = \frac{1}{2}, \tag{12.27b}$$

$$\wp(1, 1|\mathbf{a}', \mathbf{b}) = \wp(-1, -1|\mathbf{a}', \mathbf{b}) = \frac{1}{2}, \tag{12.27c}$$

$$\wp(1, -1|\mathbf{a}', \mathbf{b}') = \wp(-1, 1|\mathbf{a}', \mathbf{b}') = \frac{1}{2}, \tag{12.27d}$$

while all other probabilities are zero. In fact, when we measure particle 1 along the direction **a** and particle 2 along the direction **b**, we can obtain both 1 or both -1. We also note that here only the probabilities $\wp(1, -1|\mathbf{a}', \mathbf{b}')$ and $\wp(1, -1|\mathbf{a}', \mathbf{b}')$ show anticorrelation. Abstractly speaking, we could write the expectation values occurring in inequality (12.25) as

$$\langle \mathbf{a}, \mathbf{b} \rangle = \wp(1, 1|\mathbf{a}, \mathbf{b}) + \wp(-1, -1|\mathbf{a}, \mathbf{b})$$
$$- \wp(1, -1|\mathbf{a}, \mathbf{b}) - \wp(-1, 1|\mathbf{a}, \mathbf{b}), \tag{12.28}$$

where the negative sign of the latter two probabilities is due to the fact that both represent anticorrelations. However, this expectation value in the paramount case in which all the four probabilities in the above equation are equal mirrors the separability condition (10.16), that is, the absence of correlations (or $\langle \mathbf{a}, \mathbf{b} \rangle = 0$) between the two systems. This can be easily acknowledged when considering that all possible combinations would occur with equal probability [see, for instance, the state given by Eq. (7.128)]. However, in the model of the no-signalling case that we have chosen, the latter two probabilities in the above equation are zero. Therefore, the expectation value $\langle \mathbf{a}, \mathbf{b} \rangle$ reduces to

$$\langle \mathbf{a}, \mathbf{b} \rangle = \wp(1, 1|\mathbf{a}, \mathbf{b}) + \wp(-1, -1|\mathbf{a}, \mathbf{b}). \tag{12.29a}$$

Similarly, for the other three expectation values we have

$$\langle \mathbf{a}, \mathbf{b}' \rangle = \wp(1, 1 | \mathbf{a}, \mathbf{b}') + \wp(-1, -1 | \mathbf{a}, \mathbf{b}'), \qquad (12.29b)$$

$$\langle \mathbf{a}', \mathbf{b} \rangle = \wp(1, 1 | \mathbf{a}', \mathbf{b}) + \wp(-1, -1 | \mathbf{a}', \mathbf{b}), \qquad (12.29c)$$

$$\langle \mathbf{a}', \mathbf{b}' \rangle = -\wp(1, -1 | \mathbf{a}', \mathbf{b}') - \wp(-1, 1 | \mathbf{a}', \mathbf{b}'). \qquad (12.29d)$$

In this way, taking into account the probabilities given by Eqs. (12.27), we obtain the upper bound 4 set by Eq. (12.25).

We have already seen that quantum mechanical correlations violate the much stronger bound (i.e., 2) imposed by Eq. (10.38). As a consequence of this situation, the question arises naturally as to whether quantum mechanical correlations fill the gap between 2 and 4 or, in other words, whether there is an upper bound for quantum mechanical correlations smaller than 4? Tsirelson proved the following theorem:[a]

Theorem 12.1 (Tsirelson) Let $\hat{O}_a$, $\hat{O}_{a'}$, $\hat{O}_b$, $\hat{O}_{b'}$ be arbitrary Hermitian operators on a two-dimensional Hilbert space, each having eigenvalues 1 and -1 and satisfying the conditions $[\hat{O}_a, \hat{O}_b] = 0$, and so on, for the other pairs (a, b'), (a', b), and (a', b'), then the following equality holds in quantum mechanics:

$$\left| \langle \hat{O}_a \hat{O}_b \rangle + \langle \hat{O}_{a'} \hat{O}_b \rangle + \langle \hat{O}_a \hat{O}_{b'} \rangle - \langle \hat{O}_{a'} \hat{O}_{b'} \rangle \right| \leq 2\sqrt{2}. \qquad (12.30)$$

To prove the theorem, let us define the operator

$$\hat{B} = \hat{O}_a \hat{O}_b + \hat{O}_{a'} \hat{O}_b + \hat{O}_a \hat{O}_{b'} - \hat{O}_{a'} \hat{O}_{b'}, \qquad (12.31)$$

which is analogous to the Bell operator $\hat{B}$ defined for the spin observables in Section 10.3 [see Eq. (10.36)]. Since each of the Hermitian operators $\hat{O}_a$, $\hat{O}_{a'}$, $\hat{O}_b$, and $\hat{O}_{b'}$ has eigenvalues 1 and -1, it follows that their squares $\hat{O}_a^2$, $\hat{O}_{a'}^2$, $\hat{O}_b^2$, and $\hat{O}_{b'}^2$ are equal to the identity operator $\hat{I}$. This allows us to define a new Hermitian operator $\hat{A}$ by

$$\hat{A} = 2\sqrt{2}\hat{I} - \hat{B}$$

$$= \frac{1}{\sqrt{2}} (\hat{O}_a^2 + \hat{O}_{a'}^2 + \hat{O}_b^2 + \hat{O}_{b'}^2) - \hat{B}, \qquad (12.32)$$

which in turn can be rewritten as [see Problem 12.2]

$$\hat{A} = \frac{1}{\sqrt{2}} \left[\left(\hat{O}_a - \frac{\hat{O}_b + \hat{O}_{b'}}{\sqrt{2}} \right)^2 + \left(\hat{O}_{a'} - \frac{\hat{O}_b - \hat{O}_{b'}}{\sqrt{2}} \right)^2 \right]. \qquad (12.33)$$

---

[a](Tsirelson, 1980).

Since $\hat{A}$ is in the form of the sum of squares of Hermitian operators, it is evidently positive semidefinite and has a non-negative expectation value [see Eq. (6.158) and the discussion thereafter], that is,

$$\langle \hat{A} \rangle = 2\sqrt{2} - \langle \hat{B} \rangle \geq 0, \qquad (12.34)$$

which leads to

$$\langle \hat{B} \rangle \leq 2\sqrt{2}. \qquad (12.35)$$

A similar argument with the Hermitian operator $\hat{A}' = 2\sqrt{2}\hat{I} + \hat{B}$ leads to [see Problem 12.3]

$$\langle \hat{B} \rangle \geq -2\sqrt{2}. \qquad (12.36)$$

Therefore, we conclude that

$$\left| \langle \hat{B} \rangle \right| \leq 2\sqrt{2}, \qquad (12.37)$$

which proves the theorem.

The importance of the Tsirelson theorem lies in the fact that it proves that quantum mechanics does *not* fill the entire gap between the two bounds set by Eqs. (10.38) and (12.25). The former inequality sets bound 2 for classical separable theories while quantum mechanics satisfy the bound $2\sqrt{2}$, which is still stricter than the bound 4 imposed by Eq. (12.25). In other words, quantum mechanics certainly allows correlations that are not allowed by local hidden variable theories. However, there is a wide spectrum of "hyper-correlations" that satisfy the bound imposed by Eq. (12.25) but are nevertheless not allowed by quantum mechanics, since they do not satisfy the Tsirelson inequality (12.30). Therefore, we need still to clarify the relations between these different bounds. To examine this point, let us reformulate the CHSH inequality (10.38) in terms of the numerical parameter

$$D = \frac{B}{2} - 1. \qquad (12.38)$$

where B is the expectation value of the Bell operator $\hat{B}$ evaluated in a local hidden variable theory [see Eq. (10.37)]. It is clear that when $D = 0$, we have $B = 2$ and the upper bound of the CHSH inequality is attained. In other words, the parameter D is a measure of the deviation of B from the bound set by classical separability, in that classical separability is satisfied for $D \leq 0$ but violated for

$D > 0$. Moreover, it proves convenient to formulate the problem under current consideration in information-theoretic terms. In specific, we consider the situation in which Alice and Bob respectively choose the inputs j and k (where $j, k = 0, 1$) with the possible outputs of each input given by 0 and 1. Given the respective inputs j and k of Alice and Bob, the correlation of their outputs C_{jk} for the no-signaling case can be expressed as the sum of two joint conditional probabilities as

$$C_{00} = \wp(11|00) + \wp(00|00), \qquad (12.39a)$$

$$C_{01} = \wp(11|01) + \wp(00|01), \qquad (12.39b)$$

$$C_{10} = \wp(11|10) + \wp(00|10), \qquad (12.39c)$$

$$C_{11} = -\wp(10|11) - \wp(01|11), \qquad (12.39d)$$

where the two digits following the vertical lines are the respective inputs of Alice and Bob, while those preceding the vertical line are the corresponding outputs. It is evident that C_{00}, C_{01}, C_{10}, and C_{11} are respectively reformulation of the expectation values $\langle \mathbf{a}, \mathbf{b} \rangle$, $\langle \mathbf{a}, \mathbf{b}' \rangle$, $\langle \mathbf{a}', \mathbf{b} \rangle$, and $\langle \mathbf{a}', \mathbf{b}' \rangle$ [see Eqs. (12.29)]. Again, we remark that only the latter one is an anticorrelation (expressed by the anticorrelated outputs 10 and 01). The expectation value B can be written in term of C_{jk} as

$$B = C_{00} + C_{00} + C_{10} - C_{11}, \qquad (12.40)$$

which yields[a]

$$D = \frac{B}{2} - 1$$

$$= \frac{1}{2}(C_{00} + C_{01} + C_{10} - C_{11}) - 1. \qquad (12.41)$$

Let us now consider the simplest case in which

$$C_{00} = C_{01} = C_{10} = -C_{11} = C > 0, \qquad (12.42)$$

which implies $B = 4C$. Then, we can rewrite the expression (12.41) as

$$D(C) = 2C - 1. \qquad (12.43)$$

It is easy to distinguish, in terms of the value of C, the following three cases:

---

[a](Masanes *et al.*, 2006).

(i) Classical case: for $0 < C \leq \frac{1}{2}$ we have $-1 < D \leq 0$ and $0 < B \leq 2$. Indeed, in this case we have classical separability.

(ii) Quantum case: for $\frac{1}{2} < C \leq \frac{1}{\sqrt{2}}$ we have $0 < D \leq \sqrt{2} - 1$ and $2 < B \leq 2\sqrt{2}$. This is the case in which correlations are purely quantum mechanical and we have quantum non-separability.

(iii) Hyper-correlation case: for $\frac{1}{\sqrt{2}} < C \leq 1$ we have $\sqrt{2} - 1 < D \leq 1$ and $2\sqrt{2} < B \leq 4$. This is precisely the case in which the causal no-signaling requirement is respected, but there exist correlations stronger than the strongest quantum correlations.

The quantum and the hyper-correlation cases can be further distinguished by using the recently proposed principle of information causality,[a] which can be formulated as follows.

Principle 12.2 (Information Causality) *The information gain that Bob can reach about a previously unknown to him data set of Alice, by using all his local resources (which may be correlated with Alice's resources) and m classical bits communicated by Alice, is at most m bits.*

To illustrate this principle, let us consider the quantum teleportation protocol discussed in Section 11.5. As said it suffices that two bits of classical information is sent by Alice (according to a previously established code, like the binary numbers 00, 01, 10, and 11) so that Bob may perform the required operation as per the prearranged rules and to recover the encoded one qubit of quantum information. However, Bob can gain at most one bit of information [see Theorem 11.1] but not the whole amount of potential information contained in the qubit. Otherwise, Bob would be able to know exactly the state of the qubit teleported by Alice [see also Section 7.8]. In this simple case, the information gain I of Bob is bound by the two bits of information communicated by Alice, i.e.,

$$I \leq 2. \tag{12.44}$$

The amazing results found by Pawłowski *et al.* are that the principle of information causality is respected by both classical and quantum mechanics, whereas it is violated by all hypothetical theories that

---

[a](Pawlowski *et al.*, 2009).

fulfill the no-signaling requirements (known as the no-signaling theories) but are endowed with correlations that are stronger than the strongest quantum correlations. In other words, all non–local theories that exceed the Tsirelson bound also violate the principle of information causality. Therefore, the Tsirelson bound sets a limit on the possibility of acquiring information that is stronger than the no-signaling requirement.

Now, an important question arises naturally as to what would happen in a world in which the Tsirelson bound is violated but the no-signaling requirement is satisfied? Let us come back to the Eberhard theorem and in particular let us reformulate Eqs. (10.48) and (10.49) in analogy with Eq. (12.26) as

$$\wp(1|\mathbf{a}) = \wp(1, 1|\mathbf{a}, \mathbf{b}) + \wp(1, -1|\mathbf{a}, \mathbf{b}), \qquad (12.45a)$$

$$\wp(1|\mathbf{b}) = \wp(1, 1|\mathbf{a}, \mathbf{b}) + \wp(-1, 1|\mathbf{a}, \mathbf{b}), \qquad (12.45b)$$

respectively, and similarly for the other outcomes. This clearly confirms that quantum mechanics requires a full independence between apparatus settings (here represented by $\mathbf{a}$ and $\mathbf{b}$). These settings correspond to local operations performed in complete separation from other operations that could be performed far away elsewhere.[a] As we have stressed in Sections 10.6 and 11.3, quantum correlations represent indeed interdependences between possible measurement outcomes, but not interdependences between possible apparatus settings. In other words, violation of the quantum mechanical bound (and of the information causality principle) would imply that there are correlations between possible apparatus settings (although in the absence of any signal exchange). Let us consider the correlations C_{jk} given by Eqs. (12.39) that enter the CHSH inequality. We have expressed these correlations in terms of the probabilities $\wp(11|00)$, $\wp(00|00)$, etc. Following the standard approach in quantum mechanics (and our physical experience) we have naturally interpreted those probabilities as the conditional probabilities that both Alice and Bob get the outcome 1 given that they have both chosen the setting 0, both Alice and Bob get the outcome 0 given that they have both chosen the setting 0,

---

[a](Auletta, 2011b).

etc. This is displayed by the fact that we find quite natural in a quantum framework to say that, if Bob knows which was the setting of Alice (whether 0 or 1), he is able to infer which was her outcome (whether 0 or 1). This is another way to say that quantum entanglement is a non-local interdependence between possible measurement outcomes. This is indeed the quantum-information resource that is used in quantum cryptography [see Section 11.6]. Moreover, this fully justifies our analysis of Schrödinger's point of view [Section 10.2] that correlations forbid ascription of reality only to properties but not to observables (which are singled out through measurement settings). However, nothing forbids us to interpret such a probability $\wp(11|00)$ as meaning that, in the context of Bayesian inferences (10.53) and (12.5), Bob is able to predict the probability that Alice has chosen the setting 0, once he knows that Alice and him have obtained the outcome 1. This is still allowed by the no-signaling condition (12.26). In such a case we would have hyper-correlations that are quantum mechanically (and physically) forbidden by the principle of information causality.

Now, in such a world in which settings (and not only outcomes) are shared, this would imply that also the encoding of information is shared. Indeed, we have shown in Section 11.2 that the latter deals with the choice of a basis, which in a measurement context is the choice of apparatus settings. In other words, in a world endowed with hyper-correlations associated with the sharing of settings, the encoding of information would be no longer a local procedure. Because quantum mechanics satisfies and saturates the bound imposed by the principle of information causality, and because in so doing it also sets specific constraints on both the possible correlations and the possible interactions (also causal interconnections) in our universe, the fact that non-local encoding of information is forbidden in quantum mechanics justifies quantum information as both a general theory of information and a general theory of causality.

Problem 12.2 Verify the operator $\hat{A}$ defined by Eq. (12.32) indeed can be written in the form given by Eq. (12.33).

Problem 12.3 Complete the proof of the Tsirelson theorem by explicitly working out the lower bound of the inequality (12.30).

Problem 12.4 Consider an observable $\hat{A}$ of a two-state system defined by

$$\hat{A} = \frac{1}{\sqrt{2}}(\hat{O}^2 + \hat{O}'^2),$$

where $\hat{O} = |1\rangle\langle 1| - |2\rangle\langle 2|$ and $\hat{O}' = |3\rangle\langle 3| - |4\rangle\langle 4|$, with $\{|1\rangle, |2\rangle\}$ begin an orthonormal basis and

$$|3\rangle = \frac{1}{\sqrt{2}}(|1\rangle + |2\rangle), \quad |4\rangle = \frac{1}{\sqrt{2}}(|1\rangle - |2\rangle).$$

Show that the expectation value $\langle \hat{A} \rangle_\psi$ of $\hat{A}$ on an arbitrary state $|\psi\rangle$ is non-negative.

12.3 Operations

As a consequence of the previous two sections, we are authorized to consider the quantum mechanical process of information acquisition as the most general way in which information is exchanged in our universe. Following the analysis of measurement, which is a good model of information acquisition as well as of dynamic interactions between open systems, we can affirm that any information acquisition can be thought of as a three-system and three-step process.[a] Indeed, as discussed in Sections 9.5 and 12.1, the whole measurement process can be divided into three steps: A first step in which we prepare the system, then a second step in which the premeasurement (i.e., coupling or entanglement of the system and the apparatus) is established, and finally a third step in which the selection is made by the detector. Hence, the preparation, premeasurement, and measurement (or detection) constitute the three fundamental local operations a system can undergo in quantum mechanics. Such operations should be thought of as concrete interventions on the physical world that can somehow affect the system at hand.[b] Again, this does not imply at all any subjectivism, since analogues of the three operations can happen spontaneously in nature. The only difference between the

---

[a] (Auletta, 2011a, Chapter 2).
[b] (Braginsky/Khalili, 1992).

operations and their spontaneous analogues is that the former, being instantiated in a controlled way, allow us to make an inference about the object system (and therefore making ascription of reality possible).

A state of the form $|\psi\rangle = c_0|0\rangle + c_1|1\rangle$ can always be *prepared*. For instance, the state $|0\rangle$ may represent horizontal photon polarization state, while the state $|1\rangle$ may represent vertical photon polarization state. Then, the state $|\psi\rangle$ may represent certain polarization orientation, determined by the coefficients c_0 and c_1. A preparation can be understood as the *determination* of the state of a system (indeed, it is also called determinative measurement). It is the procedure through which only systems in a certain (previously theoretically defined) state are selected and delivered for further procedures, that is, allowed to undergo subsequent operations (premeasurement and measurement). A premeasurement consists in an *interrogation* of a quantum system with respect to some degree of freedom (such as position, momentum, energy, angular momentum, and so on). Quantum mechanics seems to imply that the specific basis used for the expansion of the system–apparatus composite state is irrelevant, and therefore that premeasurement is not about a specific observable. The fact is that, at a rather abstract level, different bases for the same system–apparatus composite state are possible [see Section 9.4]. This also reflects the equivalence of the Schrödinger and Heisenberg pictures in describing the time evolution of the composite system [see Section 7.5].

However, we should not mix measurement procedures with algorithms. When we consider a specific physical situation (that is, once a particular setting of the apparatus is chosen), we introduce a further degree of determination and are no longer authorized to treat different experimental contexts as equivalent [see Sections 10.2 and 11.3]. It would be highly unphysical to consider all observables as equivalent in a concrete experimental context since changing the apparatus basis (i.e., the settings) means a concrete change in the apparatus as such, so that we may no longer assume to have the same or an *equivalent* measurement process. For this reason, choosing a certain experimental context univocally individuates a certain observable. This is exactly the reason why we said that *actual* external conditions are needed to obtain an event. That an experimental context individuates a certain observable (better, a certain degree of freedom) is also true, to a certain extent, from a classical point of view, since each apparatus is better suited for measuring a certain observable and not others. Finally, we stress that, as features are an intrinsic characteristic of entanglement and since premeasurement essentially consists in entangling the object system with the apparatus, features play an important role in premeasurement operations.

A detection (or a measurement in the strict sense of the word) is an answer to our interrogation. When we establish an entanglement, we are actually also entangling the object system with some detector. Although a detector is in general considered part of an apparatus, they have here two conceptually very different functions. An apparatus is a *coupling* device, while a detector is a *selection* device (we may then use the term apparatus to cover both functions but we should avoid any confusion on this point). This justifies the fact that, properly speaking, the apparatus is an interface between the detector and the object system. When a suitable selection is made, the detection apparatus is in one of its basis states and, through the coupling with the object system, it tells information about the latter, and therefore allow us to ascribe a property to the object system. In other words, this connection allows for a certain random outcome that tells us something about the input state. On this basis, as already announced in Section 10.2, we may consider that each step here (from preparation through premeasurement up to detection) can be understood as a further degree of determination, or that the whole process can be seen as a dynamical process through which, starting from some potential reality and a suitable context, the actual reality (the event) is activated.

12.4 Theoretical Entities

In conclusion of this examination, we ask about the ontological meaning of terms like state, observable, and property. Since a state can be prepared using different procedures, we can understand the state as an equivalence class of preparations. Indeed, different preparations can lead to the same state (they can be considered equivalent) and also different systems may be prepared in the same state.[a] It is true that when a system is in a given eigenstate of an observable (say, $|x\rangle$), we often assume that it has the property associated with this state. However, in so doing, we are mixing two different issues:

(i) One issue is that the state is the complete catalogue of all probabilities that may be calculated (the algorithm level).

(ii) Another issue is that whether or not this state is subjected to measurement in order to establish the associated property (the operational aspect).

This distinction is very important and we have already stressed it in the previous section when speaking of premeasurement. Indeed, any output state, from a pure formal point of view, can be also considered a superposition [see Section 11.2]. In other words, if the system would undergo another experimental procedure, the output state would no longer instantiate the property we assumed to be real. It would be indeed a weird situation if properties did appear and disappear depending on the expansion of the state we are considering. For this reason, we suggest that properties are *never acquired* but *only inferred* given certain detection events. They can also be ascribed in the preparation or premeasurement step, but only in a conditional and probabilistic sense when taking into account the whole measurement procedure the system is submitted to. However, in this case we probabilistically associate possible properties to the components of the prepared system but not to the prepared state itself. This obviously does not mean that the state is a pure formal entity either. However, it is an interpreted piece of ontology referring, in a non-mirroring way, to the deep and hidden dynamical interplay between features and events that is involved in any dynamic process between open systems. On the other hand, an observable is a physical magnitude, namely, a collection of possible properties and each of them can be represented by or associated with a projector. Following EPR,[b] we can say that an observable is an interpreted element of reality [see Section 10.2]. Indeed, it

---

[a] (Auletta/Torcal, 2010).
[b] (Einstein *et al.*, 1935).

represents a "dimension" of the system through which the latter is defined. Moreover, since when measuring a certain observable a number of slightly different concrete physical contexts (setups) could be equally good, an observable is an equivalence class of premeasurements. This sheds light on our previous examination. The fact that an observable is an equivalence class of operations clearly confirms that it is an element of reality, although interpreted. We finally stress that properties may be operationally understood as an equivalence class of detection events. This also shows that properties cannot be identified with events, nor with any actual form of reality, and, as anticipated, are rather inferred given a certain event and a certain premeasurement.

Both classically and quantum mechanically, states are equivalence classes of preparations, observables are equivalence classes of premeasurements, and properties are equivalence classes of events. The main difference is that quantum systems present non-local features that deeply affect the way in which we can define and treat theoretical entities. Since in classical mechanics all observables commute, there is no fundamental conceptual distinction between event, property, observable, and state. Moreover, this gives the illusion that properties (and also physical magnitudes and states) can be directly identified with events and therefore are actual forms of reality. For this reason, quantum mechanics teaches us a quite general lesson, since the existence of features deeply affects the notion of the state (which is not simply a sum but a *combination* of properties), the definition of observables (they may not commute), and the conception of properties (they cannot be all compatible, and the possibility to infer some of them excludes the simultaneous consideration of the other ones). Obviously, the specific aspects mentioned are typical of quantum systems. However, the general lesson is that quantum theory prevents the illusion of identifying these physical concepts with directly experienceable, actual realities. In fact, quantum realities are very elusive. Non-local features cannot be experienced at all but they frame possible local events, while these local events can be experienced but they happen randomly or in an uncontrolled way and precisely for this reason they do not allow an access to the hidden correlations. We discover in this way that reality is something not only much more complex but also much more beautiful than previously thought of.

12.5 Fundamental Information Triad

The final choice that happens at the detectors can be random or not (in the quantum mechanical case it is random). However, there is always some sort of incertitude affecting the final selection act. Viewing the whole from the point of view of the involved systems, we have that the object (measured) system which represents, as we have stressed, encoded information evolving reversibly in time can be considered an information *processor* [see Section 11.2]. Its time evolution, which could be considered a change in some informational content (a change that can be either random or according to a program) provides the starting point of the whole process. The measuring device that is coupled to the object system is a *regulator*, while the final operation of detection is done through a *decider*. The regulator owes its name to the fact that choosing a certain apparatus with a certain experimental setup and a certain pointer indeed contributes to determine the conditions in which information is acquired. However, since these conditions cannot provide the necessary variety of bits (which is guaranteed at the source by the processor), this determination is rather a tuning of the measurement process. The only encoded activity that is necessary is in the processor and in the final detection, since the regulator connects previously independent systems. Obviously, the decider can provide a selection that, thanks to the indirect connection with the processor through the regulator, will finally consist in an option within a set of alternative possibilities.

The whole process can be represented in Fig. 12.2 as the fundamental information triad. The relation established between the regulator and the processor is coupling, which allows information to be subsequently acquired. The relation between the decider and the regulator is information selection. Finally, the decider can acquire information about the processor (or the event resulting from processing) by performing in this way the analogue of inferring (the

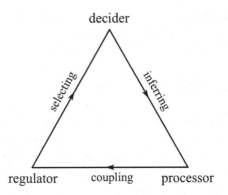

Figure 12.2 The fundamental information triad is consists of three elements (processor, regulator, and decider) and three relations (coupling, selection, and inferring). Adapted from (Auletta, 2011a, p. 48).

experimenter performs in this way precisely such an inference). In other words, when a decider (even randomly) selects and eventually stores some information which through an appropriate coupling reveals something about another system, we have something that at a pure physical level bears some structural relation to what we are authorized to call a true inference. Therefore, this inferring, or the whole measuring process, must not be understood merely in subjective terms. It is also important to realize that this inferring could be considered part of a further preparation procedure in which a processor is determined. For instance, we may decide to measure our system again (when it is not annihilated in the detection event) starting from its output state. The reason could be that we are not sure of our inference because of some doubts about the apparatus reliability. In this way, the whole process presents a certain circularity, as shown in Fig. 12.2.

Also Shannon understood very well that information is concerned with a reduction of incertitude (a choice between alternative possibilities) and that, in order to have an exchange of information, we need both a variety and an interdependence.[a] However, he mainly dealt with engineering problems of communications, in which the task is to increase the fidelity between an input and

---

[a](Shannon, 1948).

an output in controlled situations. In this case, the reduction of incertitude already happens at the source (by the sender who chooses a certain message among many possible ones). This is rather a limiting case in ordinary life, and the problem was that Shannon's followers took this specific model as a general model of information. The worry consists in the fact that, in the most general case, the reduction of incertitude is *only at the output* and not at the input. This is evident for quantum systems since qubits are not selected messages, otherwise they could be cloned, which is impossible due to non-local features [see Theorem 11.2]. A good example is provided by the delayed choice experiment discussed in Section 5.6. The reason is that, in most (even classical) situations, nobody has the control on the source, and therefore, even in the case in which a particular message has been selected by a sender, it remains unknown to the receiver (it is as if it were undetermined).[a] In such a situation, one is obliged to make an inference about the input starting from a certain selection at the output (again in the context of Bayesian inference). This is also sometimes true for the senders; if they desired to be certain about the message that has been sent, they need to process it again, and in this way reduce the incertitude that may affect (their knowledge about) the input message. On the contrary, the selection operated by the receiver is the received message (it is the event that has happened). Obviously, the receivers may also try to verify again whether their understanding is correct. However, this understanding concerns the inference about the input message, not the act of selection, i.e., the event itself by which a reduction of incertitude at the output has been produced forever. This is an irreversible event and therefore an ultimate fact. We have already generalized this by saying that it is the final act of information selection that introduces irreversibility in any information exchange or processing.

Summarizing, there are only two ways to deal with information apart from processing it: either by sharing it (which represents the non-local aspect) or by selecting it (which represents the local aspect).[b] Quantum mechanics allows us to deeply understand this

---

[a] (Auletta *et al.*, 2008).
[b] (Auletta, 2005).

point. Here, the input information is intrinsically uncertain although shared and the only event we have is at the output. Again, we see how helpful quantum mechanics is in order to really understand how the world is.

12.6 Summary

In this chapter, we have

- Studied the classical and quantum processes of information acquisition and introduced the effect and amplitude operators.
- Introduced the Tsirelson bound as a limit of information acquisition as such.
- Explained that violation of the Tsirelson bound by a non–local theory would imply a world in which the settings and the encoding of information are shared.
- Shown that the three steps of measurement (preparation, premeasurement, and detection) correspond to the three stages in the determination of a system.
- Interpreted states, observables, and properties as equivalence classes of preparations, premeasurements, and events, respectively.
- Provided a very general model for information acquisition in which the three elements (processor, regulator, and decider) and three relations (coupling, selection, and inferring) constitute the fundamental information triad.

Bibliography

[Arnold, 1978]   Arnold, Vladimir I., *Mathematical Methods of Classical Mechanics* (trans. from Russian), New York, Springer, 1978; 2nd ed., 1989.

[Aspect *et al.*, 1982]   Aspect, A., Dalibard, J., and Roger, G., "Experimental Test of Bell's Inequalities Using Time-Varying Analyzers," *Physical Review Letters* **49**: 1804–1807.

[Atkins/De Paula, 2006]   Atkins, P. W. and De Paula, J., *Physical Chemistry*, Oxford, University Press.

[Atkins/Friedman, 2005]   Atkins, P. W. and Friedman, R., *Molecular Quantum Mechanics*, Oxford, University Press, 4th ed.

[Auletta, 2000]   Auletta, G., *Foundations and Interpretation of Quantum Mechanics: In the Light of a Critical-Historical Analysis of the Problems and of a Synthesis of the Results*, Singapore, World Scientific, 2000; rev. ed., 2001.

[Auletta, 2004a]   ——, "Critical Examination of the Conceptual Foundations of Classical Mechanics in the Light of Quantum Physics," *Epistemologia* **27**: 55–82.

[Auletta, 2004b]   ——, "Decoherence and Triorthogonal Decomposition," *International Journal of Theoretical Physics* **43**: 2263–2274.

[Auletta, 2004c]   ——, "Interpretazioni soggettivistiche della misurazione in meccanica quantistica," *Teorie e modelli* **9**: 43–61.

[Auletta, 2005]   ——, "Quantum Information as a General Paradigm," *Foundations of Physics* **35**: 787–815.

[Auletta, 2006a] ——, "The Ontology Suggested by Quantum Mechanics," in Valore, Paolo (Ed.), *Topics on General and Formal Ontology*, Milan, Polimetrica: 161–179.

[Auletta, 2006b] ——, "The Problem of Information," in Auletta, Gennaro (Ed.), *The Controversial Relations Between Science and Philosophy: A Critical Assessment*, Vatican City, Libreria Editrice Vaticana: 109–127.

[Auletta, 2009] ——, "What About the Three Forms of Inference?" *Acta Philosophica* **1.18**: 59–74.

[Auletta, 2011a] ——, *Cognitive Biology: Dealing with Information from Bacteria to Minds*, Oxford University Press, 2011.

[Auletta, 2011b] Auletta, G., "Correlations and Hyper-Correlations," *Journal of Modern Physics* **2**: 958–961.

[Auletta/Torcal, 2010] Auletta, G. and Torcal, L., "From Wave–Particle to Features–Event Complementarity," *International Journal of Theoretical Physics* **50**: 3654–3668.

[Auletta *et al.*, 2008] Auletta, G., Ellis, G., and Jaeger, L., "Top-Down Causation by Information Control: From a Philosophical Problem to a Scientific Research Program," *Journal of the Royal Society: Interface* **5**: 1159–1172.

[Auletta *et al.*, 2009] Auletta, G., Fortunato, M., and Parisi, G., *Quantum Mechanics: A Modern Perspective*, Cambridge, University Press, 2009.

[Ball, 1999] Ball, P., *The Self-Made Tapestry: Pattern Formation in Nature*, Oxford, University Press, 1999.

[Barnett/Phoenix, 1989] Barnett, S. M. and Phoenix, S. J. D., "Entropy as a Measure of Quantum Optical Correlation," *Physical Review* **A40**: 2404–2409.

[Barrow *et al.*, 2004] Barrow, J. D., Davies, P. C. W., and Harper, C. L. Jr. (Eds.), *Science and Ultimate Reality: Quantum Theory, Cosmology and Complexity*, Cambridge, Cambridge University Press, 2004.

[Baumgarten, 1739]   Baumgarten, A. G., *Metaphysica*, 1739; 7th ed., Halle 1779; rep. Hildesheim-New York, Georg Olms, 1982.

[Bell, 1964]   Bell, J. S., "On Einstein Podolsky Rosen Paradox," *Physics* **1**: 195–200; rep. in (Bell, 1994, 14–21).

[Bell, 1973]   ——, "Subject and Object," in J. Mehra (Ed.), *The Physicist's Conception of Nature*, Dordrecht, Reidel, 1973; rep. in (Bell, 1994, 40–44).

[Bell, 1976]   ——, "The Theory of Local Beables," *Epistemological Letters* (1976); rep. in (Bell, 1994, 52–62).

[Bell, 1994]   ——, *Speakable and Unspeakable in Quantum Mechanics*, Cambridge, University Press, 1994.

[Bell, 1990]   ——, "Against 'Measurement'," *Physics World* **3**: 33–40.

[Bennett, 1973]   Bennett, C. H., "Logical Reversibility of Computation," *IBM Journal of Research and Development* **17**: 525–532.

[Bennett, 1982]   ——, "The Thermodynamics of Computation. A Review," *International Journal of Theoretical Physics* **21**: 905–940.

[Bennett/Brassard, 1984]   Bennett, C. H. and Brassard, G., "Quantum Cryptography: Public Key Distribution and Coin Tossing," *Proceedings IEEE International Conference on Computer Systems*, New York, IEEE, 1984: 175–179.

[Bennett/Landauer, 1985]   Bennett, C. H. and Landauer, R., "The Fundamental Physical Limits of Computation," *Scientific American* **253**: 48–56.

[Bennett/Wiesner, 1992]   Bennett, C. H. and Wiesner, S. J., "Communication via One- and Two-Particle Operators on EPR States," *Physical Review Letters* **69**: 2881–2884.

[Bennett *et al.*, 1993]   Bennett, C. H., Brassard, G., Crepeau, C., Jozsa, R., Peres, A., and Wootters, W. K., "Teleporting an Unknown Quantum State via Dual Classical and EPR Channels," *Physical Review Letters* **70**: 1895–1899.

[Berthiaume *et al.*, 2000]   Berthiaume, A., van Dam, W., and Laplante, S., "Quantum Kolmogorov Complexity," *Journal of Computer and System Sciences* **63**: 201–221.

[Bohm, 1951]   Bohm, D., *Quantum Theory*, New York, Prentice-Hall, 1951.

[Bohr, 1913]   Bohr, N., "On the Constitution of Atoms and Molecules," *Philosophical Magazine* **26**: 1–25, 476–502, 857–875.

[Bohr, 1920]   ——, "Über die Serienspektra der Element," *Zeitschrift für Physik* **2**: 423–478; rep. in (Bohr CW, III, 241–282).

[Bohr, 1928]   ——, "The Quantum Postulate and the Recent Development of Atomic Theory," *Nature* **121**: 580–590.

[Bohr, 1935a]   ——, "Quantum Mechanics and Physical Reality," *Nature* **136**: 65.

[Bohr, 1935b]   ——, "Can Quantum-Mechanical Description of Physical Reality Be Considered Complete?" *Physical Review* **48**: 696–702.

[Bohr CW]   ——, *Niels Bohr Collected Works*, Vols. I–XIII (Gen. Ed. Finn Aaserud), Amsterdam, North Holland, 1972–2008.

[Born/Jordan, 1925]   Born, M. and Jordan, P., "Zur Quantenmechanik," *Zeitschrift für Physik* **34**: 858–888.

[Born *et al.*, 1926]   Born, M., Heisenberg, W., and Jordan, P., "Zur Quantenmechanik II," *Zeitschrift für Physik* **35**: 557–615.

[Boscovich, 1754]   Boscovich, R. J., *De continuitatis lege and ejus consectariis, pertinentibus ad prima materiae elementa eorumque vires*, Roma, Collegio Romano.

[Bose, 1924]   Bose, S., "Plancks Gesetz und Lichtquantenhypothese," *Zeitschrift für Physik* **26**: 178–181.

[Bouwmeester *et al.*, 1997]   Bouwmeester, D., Pan, J.–W., Mattle, K., Eibl, M., Weinfurter, H., and Zeilinger, A., "Experimental Quantum Teleportation," *Nature* **390**: 575–579.

[Braginsky/Khalili, 1992] Braginsky, V. B. and Khalili, F. Y., *Quantum Measurement*, Cambridge, University Press, 1992.

[Braunstein/Caves, 1988] Braunstein, S. L. and Caves, C. M., "Information-Theoretic Bell Inequalities," *Physical Review Letters* **61**: 662–665.

[Braunstein/Caves, 1990] ——, "Wringing Out Better Bell Inequalities," *Annals of Physics* **202**: 22–56.

[Braunstein *et al.*, 1992] Braunstein, S. L., Mann, A., and Revzen, M., "Maximal Violation of Bell Inequalities for Mixed States," *Physical Review Letters* **68**: 3259–3261.

[Brune *et al.*, 1996] Brune, M., Hagley, E., Dreyer, J., Maître, X., Maali, A., Wunderlich, C., Raimond, J. M., and Haroche, S., "Observing the Progressive Decoherence of the 'Meter' in a Quantum Measurement," *Physical Review Letters* **77**: 4887–4890.

[Cerf *et al.*, 1998] Cerf, N. J., Adami, C., and Kwiat, P. G., "Optical Simulation of Quantum Logic," *Physical Review* **A57**: R1477–R1480.

[Chiribella *et al.*, 2011] Chiribella, G., D'Ariano, G. M., and Perinotti, P., "Informational Derivation of Quantum Theory," *Physical Review* **A84**: 012311.

[Cini, 1983] Cini, M., "Quantum Theory of Measurement Without Wave Packet Collapse," *Nuovo Cimento* **73B**: 27–54.

[Clauser/Shimony, 1978] Clauser, J. F. and Shimony, A., "Bell's Theorem: Experimental Tests and Implications," *Reporting Progress Physics* **41**: 1881–1927.

[Clauser *et al.*, 1969] Clauser, J. F., Horne, M. A., Shimony, A., and Holt, R. A., "Proposed Experiment to Test Local Hidden-Variable Theories," *Physical Review Letters* **23**: 880–884.

[Compton, 1923] Compton, A. H., "A Quantum Theory of the Scattering of X-rays by Light Elements," *Physical Review* **21**: 483–502.

[Cullerne/Machacek, 2008]   Cullerne, J. P. and Machacek, A., *The Language of Physics: A Foundation for University Study*, Oxford, Oxford University Press.

[D'Ariano, 2012]   D'Ariano, G. M., "A Quantum Digital Universe: Quantum Information Helps Foundations of Physics," *Il Nuovo Saggiatore* **28. 3–4**: 13–22.

[D'Ariano/Yuen, 1996]   D'Ariano, G. M. and Yuen, H. P., "Impossibility of Measuring the Wave Function of a Single Quantum System," *Physical Review Letters* **76**: 2832–2835.

[Davisson/Germer, 1927]   Davisson, C. J. and Germer, L. H., "Diffraction of Electrons By a Crystal of Nickel," *Physical Review* **30**: 705–740.

[De Broglie, 1924]   de Broglie, L., "Sur la définition générale de la correspondance entre onde et mouvement," *Comptes Rendus à l'Academie des Sciences* **179**: 39–40.

[Deutsch, 1985]   Deutsch, D., "Quantum Theory, the Church-Turing Principle and the Universal Quantum Computer," *Proceedings of the Royal Society of London* **A400**: 97–117.

[Dewitt, 1970]   DeWitt, B. S., "Quantum Mechanics and Reality," *Physics Today* **23**: 30–40; rep. in (Dewitt/Graham, 1973, 155–65).

[Dewitt/Graham, 1973]   DeWitt, B. S. and Graham, N., (Eds.) *The Many World Interpretation of Quantum Mechanics*, Princeton, University Press, 1973.

[Dieks, 1982]   Dieks D., "Communication by EPR Devices," *Physics Letters A* **92**: 271–272.

[Dirac, 1958]   Dirac, P. A. M., *The Principles of Quantum Mechanics*, 4th ed., Oxford, Clarendon, 1958.

[Eberhard, 1978]   Eberhard, P. H., "Bell's Theorem and the Different Concepts of Locality," *Nuovo Cimento* **46B**: 392–419.

[Einstein, 1905]   Einstein, A., "Über einen die Erzeugung und Verwandlung des Lichtes betreffenden heuristischen Gesichtspunkt," *Annalen der Physik* **17**: 132–148.

[Einstein *et al.*, 1935] Einstein, A., Podolsky, B., and Rosen N., "Can Quantum-Mechanical Description of Physical Reality be Considered Complete?" *Physical Review* **47**: 777–780.

[Ekert, 1991]   Ekert, A. K., "Quantum Cryptography Based on Bell's Theorem," *Physical Review Letters* **67**: 661–663.

[Elitzur/Vaidman, 1993]   Elitzur, A. C. and Vaidman, L., "Quantum Mechanical Interaction–Free Measurements," *Foundations of Physics* **23**: 987–997.

[Everett, 1957]   Everett, H. III, "'Relative State' Formulation of Quantum Mechanics," *Review of Modern Physics* **29**: 454–462; rep. in (Dewitt/Graham, 1973, 141–49).

[Fermi, 1926]   Fermi, E., "Zur Quantelung des Idealen Einatomigen Gases," *Zeitschrift für Physik* **36**: 902–912.

[Feynman, 1982]   Feynman, R. P., "Simulating Physics with Computers," *International Journal of Theoretical Physics* **21**: 467–488.

[Feynman, 1986]   Feynman, Richard P., "Quantum Mechanical Computers," *Foundations of Physics* **16**: 507–531.

[Finkelstein, 1996]   Finkelstein, D. R., *Quantum Relativity*, Berlin, Springer.

[Fredkin/Toffoli, 1982]   Fredkin, E. and Toffoli, T., "Conservative Logic," *International Journal of Theoretical Physics* **21**: 219–253.

[Freedman/Clauser, 1972] Freedman, S. J. and Clauser, J. F., "Experimental Test of Local Hidden-Variable Theories," *Physical Review Letters* **28**: 938–41.

[Freimund *et al.*, 2001] Freimund, D. L., Aflatooni, K., and Batelaan, H., "Observation of the Kapitza–Dirac Effect," *Nature* **413**: 142–143.

[Friedrich/Herschbach, 2003]   Friedrich, B. and Herschbach, D., "Stern and Gerlach: How a Bad Cigar Helped Reorient Atomic Physics," *Physics Today* **56**: 53–59.

[Furusawa *et al.*, 1998]   Furusawa, A., Sørensen, J. L., Braunstein, S. L., Fuchs, C. A., Kimble, H. J., and Polzik, E. S., "Unconditional Quantum Teleportation," *Science* **282**: 706–709.

[Gerlach/Stern, 1922a]   Gerlach, W. and Stern, O., "Der experimentelle Nachweis des magnetischen Moments des Silberatoms," *Zeitschrift für Physik* **8**: 110–111.

[Gerlach/Stern, 1922b]   ——, "Der experimentelle Nachweis der Richtungsquantelung im Magnetfeld," *Zeitschrift für Physik* **9**: 349–352.

[Gerlach/Stern, 1922c]   ——, "Das magnetische Moment des Silberatoms," *Zeitschrift für Physik* **9**: 353–355.

[Greenberger/Yasin, 1988]   Greenberger, D. M. and Yasin, A., "Simultaneous Wave and Particle Knowledge in a Neutron Interferometer," *Physics Letters* **128A**: 391–394.

[Gröblacher *et al.*, 2007]   Gröblacher, S., Paterek, T., Kaltenbaek, R., Brukner, Č, Żukowski, M., Aspelmeyer, M., and Zeilinger, A., "An Experimental Test of Non-Local Realism," *Nature* **446**: 871–875.

[Hasegawa *et al.*, 2006]   Hasegawa, Y., Loidl, R., Badurek, G., Baron, M., and Rauch, H., "Quantum Contextuality in a Single–Neutron Optical Experiment," *Physical Review Letters* **97**: 230401.

[Heisenberg, 1925]   Heisenberg, W., "Über quantentheoretische Umdeutung kinematischer und mechanischer Beziehungen," *Zeitschrift für Physik* **33**: 879–893.

[Heisenberg, 1927]   ——, "Über den anschaulichen Inhalt der quantentheoretischen Kinematik und Mechanik," *Zeitschrift für Physik* **43**: 172–198.

[Heller, 2006]   Heller, M., *Some Mathematical Physics for Philosophers*, Rome, Pontifical Council for

Culture and Pontifical Gregorian University, 2006.

[Helstrom, 1976] Helstrom, C. W., *Quantum Detection and Estimation Theory*, New York, Academic, 1976.

[Hempel, 1953] Hempel, C. G., *Methods of Concept Formation Science*, Chicago, University of Chicago Press, 1953.

[Hertz, 1887] Hertz, H. R. "Ueber sehr schnelle elektrische Schwingungen," *Annalen der Physik* **267**: 421–448.

[Hillery/Yurke, 1995] Hillery, M. and Yurke, B., "Bell's Theorem and Beyond," *Quantum Optics* **7**: 215–227.

[Holevo, 1998] Holevo, A. S., "The Capacity of the Quantum Channel with General Signal States," *IEEE Transactions on Information Theory* **44**: 269–273.

[Horodecki *et al.*, 2005] Horodecki, M., Oppenheim, J., and Winter, A., "Partial Quantum Information," *Nature* **436**: 673–76.

[Jammer, 1974] Jammer, M., *The Philosophy of Quantum Mechanics. The Interpretation of Quantum Mechanics in Historical Perspective*, New York, Wiley, 1974.

[Joos/Zeh, 1985] Joos, E. and Zeh, H. D., "The Emergence of Classical Properties Through Interaction with the Environment," *Zeitschrift für Physik* **B59**: 223–243.

[Kim, 1984] Kim, J., "Epiphenomenal and Supervenient Causation," *Midwest Studies in Philosophy* **9**: 257–270; rep. in (Kim, 1993, 92–108).

[Kim, 1993] ——, *Supervenience and Mind: Selected Philosophical Essays*, Cambridge, University Press, 1993, 1995.

[Khinchin, 1957] Khinchin, A. I., *Mathematical Foundations of Information Theory* (Engl. Trans.), New York, Dover, 1957.

[Kochen/Specker, 1967] Kochen, S. and Specker, E., "The Problem of Hidden Variables in Quantum Mechanics,"

Journal of Mathematics and Mechanics **17**: 59–87.

[Kolmogorov, 1963]   Kolmogorov, A. N., "On Tables of Random Numbers," *Sankhya Ser* **A. 25**: 369–375; rep. in *Theoretical Computer Science* **207**: 387–395.

[Küppers, 1990]   Küppers, B.-O., *Information and the Origin of Life*, Cambridge, MA, MIT Press.

[Kwiat *et al.*, 1996]   Kwiat, P., Weinfurter, H., and Zeilinger, A., "Quantum Seeing in the Dark," *Scientific American* (Nov. *1996*), 72–78.

[Landauer, 1961]   Landauer, R., "Irreversibility and Heat Generation in the Computing Process," *IBM Journal Res. Dev.* **5**: 183–191.

[Landauer, 1991]   ——, "Information is physical," *Physics Today* **44**: 23–29.

[Landauer, 1996]   ——, "Minimal Energy Requirements in Communication," *Science* **272**: 1914–1919.

[Laplace, 1796]   Laplace, P.-S., *Exposition du système du monde*, Paris 1796; 1835; Fayard, 1984.

[Le Bellac, 2006]   Le Bellac, M., *Quantum Physics*, Cambridge, University Press, 2006.

[Leibniz MS]   Leibniz, G. W., *Mathematische Scriften* (Ed. Gerhardt), Halle, 1860; rep. Hildesheim, Olms, 1971.

[Leibniz PS]   ——, *Philosophische Scriften* (Ed. Gerhardt), Halle, 1875; rep. Hildesheim, Olms, 1978.

[Leibfried *et al.*, 2005]   Leibfried, D., Knill, E., Seidelin, S., Britton, J., Blakestad, R. B., Chiaverini, J., Hume, D. B., Itano, W. M., Jost, J. B., Langer, C., Ozeri, R., Reichle, R., and Wineland, D. J., "Creation of a Six–Atom 'Schrödinger Cat' State," *Nature* **438**: 639–642.

[Lindblad, 1983]   Lindblad, G., *Non-Equilibrium Entropy and Irreversibility*, Dordrecht, Reidel, 1983.

[Lockwood, 1996]   Lockwood, M., "'Many Minds' Interpretations of Quantum Mechanics," *British Journal for Philosophy of Science* **47**: 159–188.

[London/Bauer, 1939]   London, F. and Bauer, E., *La théorie de l'observation en mécanique quantique*, Paris, Hermann, 1939; Eng. trans. in (Wheeler/Zurek, 1983, 217–259).

[Masanes *et al.*, 2006]   Masanes, L., Acin, A., and Gisin, N., "General Properties of Nonsignaling Theories," *Physical Review* **A73**: 012112.

[Maxwell, 1873]   Maxwell, J. C., *A Treatise on Electricity and Magnetism*, Oxford, Clarendon Press, 1873.

[Mo *et al.*, 2010]   Mo, H., van den Bosch, F., and White, S., *Galaxy Formation and Evolution*, Cambridge, Cambridge University Press.

[Monroe *et al.*, 1996]   Monroe, C., Meekhof, D. M., King, B. E., and Wineland, D. J., "A 'Schrödinger Cat' Superposition State of an Atom," *Science* **272**: 1131–1136.

[Neder *et al.*, 2007]   Neder, I., Ofek, N., Chung, Y., Heiblum, M., Mahalu, D., Umansky, V., "Interference between Two Indistinguishable Electrons from Independent Sources," *Nature* **448**: 333–337.

[Nielsen/Chuang, 2011]   Nielsen, M. A. and Chuang, I. L., *Quantum Computation and Quantum Information*, Cambridge, University Press, 2011.

[Olivier/Zurek, 2001]   Ollivier, H. and Zurek, W. H., "Quantum Discord: A Measure of the Quantumness of Correlations," *Physics Review Letters* **88**: 017901.

[Pauli, 1925]   Pauli, W., "Über den Zusammenhang des Abschlusses von Elektronengruppen im Atom mit der Komplexstruktur der Felder," *Zeitschrift für Physik* **31**: 765–783.

[Pauli, 1927]   ——, "Zur Quantenmechanik des magnetischen Elektrons," *Zeitschrift für Physik* **43**: 601–623.

[Pawlowski *et al.*, 2009]   Pawłowski, M., Paterek, T., Kaszlikowski, D., Scarani, V., Winter, A., Zukowski, M. Z., "Information Causality as a Physical Principle," *Nature* **461**: 1101–1104.

[Peirce, 1866]   Peirce, C. S., "The Logic of Science or Induction and Hypothesis: Lowell Lectures," in (Peirce W, I, 357–504).

[Peirce, 1877]   ——, "The Fixation of Belief," *Popular Science Monthly* **12**: 1–15; rep. in (Peirce W, III, 242–257).

[Peirce, 1878]   ——, "Deduction, Induction, and Hypothesis," *Popular Science Monthly* **13**: 470–482; rep. in (Peirce W, III, 323–38).

[Peirce CP]   ——, *The Collected Papers*, Vols. I–VI (Eds. Charles Hartshorne/Paul Weiss), Cambridge, MA, Harvard University Press, 1931–1935; Vols. VII–VIII (Ed. Arthur W. Burks), Cambridge, MA, Harvard University Press, 1958.

[Peirce W]   ——, *Writings of Charles S. Peirce*, A Chronological Edition, Vol. I–VI (Eds. Peirce Edition Project), Bloomington, Indiana University Press, 1982–1999.

[Pitowsky, 1989]   Pitowsky, I., *Quantum Probability Quantum Logic*, Berlin, Springer, 1989.

[Planck, 1900a]   Planck, M., "Über die Verbesserung der Wien'schen Spektralgleichung," *Verhandlungen der Deutschen Physikalischen Gesellschaft* **2**: 202–204.

[Planck, 1900b]   ——, "Zur Theorie des Gesetzes der Energieverteilung im Normalspektrum," *Verhandlungen der Deutschen Physikalischen Gesellschaft* **2**: 237–245.

[Plenio/Vitelli, 2001]   Plenio, M. B., and Vitelli, V., "The Physics of Forgetting: Landauer's Erasure Principle and Information Theory," *Contemporary Physics* **42**: 25–60.

[Popescu/Rohrlich, 1994]   Popescu, S. and Rohrlich, D., "Quantum Nonlocality as an Axiom," *Foundations of Physics* **24**: 379–385.

[Popper, 1934]	Popper, K. R., *Logik der Forschung*, Wien, Springer, 1934, 8th ed. Tübingen, Mohr, 1984.
[Renninger, 1960]	Renninger, M. K., "Messungen ohne Störung des Meßobjekts," *Zeitschrift für Physik* **158**: 417–421.
[Santos, 1991]	Santos, Emilio, "Does Quantum Mechanics Violate the Bell Inequalities?" *Physical Review Letters* **66**: 1388–90; Errata: 3227.
[Savage/Walls, 1985]	Savage, C. M. and Walls, D. F., "Damping of Quantum Coherence: The Master-Equation Approach," *Physical Review* **A32**: 2316–2323.
[Schrödinger, 1926a]	Schrödinger, E., "Quantisierung als Eigenwertproblem I," *Annalen der Physik* **79**: 361–376; rep. in (Schrödinger GA, III, 82–97).
[Schrödinger, 1926b]	——, "Quantisierung als Eigenwertproblem II," *Annalen der Physik* **79**: 489–527; rep. in (Schrödinger GA, III, 98–136).
[Schrödinger, 1926c]	——, "Quantisierung als Eigenwertproblem III," *Annalen der Physik* **80**: 437–90; rep. in (Schrödinger GA, III, 166–219).
[Schrödinger, 1926d]	——, "Quantisierung als Eigenwertproblem IV," *Annalen der Physik* **81**: 109–39; rep. in (Schrödinger GA, III, 220–250).
[Schrödinger, 1935a]	——, "Die gegenwärtige Situation in der Quantenmechanick. I–III," *Naturwissenschaften* **23**: 807–12, 823–28, 844–49; rep. in (Schrödinger GA, III, 484–501).
[Schrödinger, 1935b]	Schrödinger, Erwin, "Discussion of Probability Relations Between Separated Systems," *Proceedings of the Cambridge Philosophical Society* **32**: 446–52; rep. in (Schrödinger GA, I, 433–39).
[Schrödinger GA]	——, *Gesammelte Abhandlungen*, Vols. I–IV, Wien, Verlag der Österreichischen Akademie der Wissenschaften Wien, Vieweg, Braunschweig & Wiesbaden, 1984.

[Schumacher, 1990]   Schumacher, Benjamin W., "Information from Quantum Measurements," in (Zurek, 1990, 29–37).

[Schumacher, 1991]   ——, "Information and Quantum Nonseparability," *Physical Review* **A44**: 7047–7052.

[Selleri, 1988]   Selleri, Franco (Ed.), *Quantum Mechanics versus Local Realism*, New York, Plenum.

[Shannon, 1948]   Shannon, C. E., "A Mathematical Theory of Communication," *Bell System Technical Journal* **27**: 379–423; 623–656.

[Summhammer *et al.*, 1983]   Summhammer, J., Badurek, G., Rauch, H., Kischko, U., and Zeilinger A., "Direct Observation of Fermion Spin Superposition by Neutron Interferometry," *Physical Review* **A27**: 2523–2532.

[Tarozzi, 1996]   Tarozzi, G., "Quantum Measurements and Macroscopical Reality: Epistemological Implications of a Proposed Paradox," *Foundations of Physics* **26**: 907–917.

[Tsirelson, 1980]   Tsirelson, B. S., "Quantum Generalizations of Bell's Inequality," *Letters in Mathematical Physics* **4**: 93–100.

[Turing, 1936]   Turing, A.M., "On Computable Numbers, with an Application to the Entscheidungsproblem," *Proceedings of the London Mathematical Society*, **42**: 230–265.

[Turing, 1937]   ——, "On Computable Numbers, with an Application to the Entscheidungsproblem: A correction," *Proceedings of the London Mathematical Society*, **43**: 544–546.

[Uhlenbeck/Goudsmit, 1925]   Uhlenbeck, G. E. and Goudsmit, S. A., "Ersetzung der Hypothese vom unmechanischen Zwang durch eine Forderung bezüglich des inneren Verhaltens jedes einzelnen Elektrons," *Naturwissenschaften* **47**: 953–954.

[Uhlenbeck/Goudsmit, 1926]   ——, "Spinning Electrons and the Structure of Spectra," *Nature* **117**: 264–265.

[Vedral *et al.*, 1997a]   Vedral, V., Plenio, M. B., Rippin, M. A., and Knight, P. L., "Quantifying Entanglement," *Physical Review Letters* **78**: 2275–2279.

[Von Neumann, 1932]   von Neumann, J., *Mathematische Grundlagen der Quantenmechanik*, Berlin, Springer, 1932, 1968, 1996.

[Von Weizsäcker, 1972]   von Weizsäcker, C. F., "Evolution und Entropiewachstum," *Nova Acta Leopoldina* **37**: 515.

[Wang *et al.*, 1991]   Wang, L. J., Zou, X. Y., and Mandel, L., "Induced Coherence Without Induced Emission," *Physical Review* **A44**: 4614–4622.

[Watson/Crick, 1953]   Watson J. D. and Crick F. H. C., "A Structure for Deoxyribose Nucleic Acid," *Nature* **171**: 737–738.

[Wheeler, 1983]   Wheeler, J. A., "Law Without Law," in (Wheeler/Zurek, 1983, 182–213).

[Wheeler, 1990]   ——, "Information, Physics, Quantum: The search for Links," in: (Zurek, 1990, 3–28).

[Wheeler/Zurek, 1983]   Wheeler, J. A. and Zurek, W. (Eds.), *Quantum Theory and Measurement*, Princeton, University Press, 1983.

[Wiesner, 1983]   Wiesner, Stephen J., "Conjugate Coding," *Sigact News* **15**: 78–88.

[Wigner, 1961]   Wigner, E. P., "Remarks on the Mind–Body Question," in I. J. Good (Ed.), *The Scientist Speculates*, London, Heinemann, 1961: 284–302; rep. in (Wheeler/Zurek, 1983, 168–181).

[Wootters/Zurek, 1982]   Wootters, W. K. and Zurek, W. H., "A Single Quantum Cannot be Cloned," *Nature* **299**: 802–803.

[Zeh, 1970]   Zeh, H. D, "On the Interpretation of Measurement in Quantum Theory," *Foundations of Physics* **1**: 69–76; rep. in (Wheeler/Zurek, 1983, 342–349).

[Zukowski *et al.*, 1993]   Zukowski, M., Zeilinger, A., Horne, M. A., and Ekert, A. K., "Event-Ready-Detectors' Bell Experiment via Entanglement Swapping," *Physical Review Letters* **71**: 4287–4290.

[Zurek, 1981]   Zurek, W. H., "Pointer Basis of Quantum Apparatus: Into What Mixture Does the Wave Packet Collapse?" *Physical Review* **D24**: 1516–1525.

[Zurek, 1982]   ——, "Environment-Induced Superselection Rules," *Physical Review* **D26**: 1862–1880.

[Zurek, 1983]   ——, "Information Transfer in Quantum Measurements: Irreversibility and Amplification," in P. Meystre, P. and M. O. Scully (Eds.), *Quantum Optics, Experimental Gravity, and Measurement Theory*, New York, Plenum, 1983: 87–116.

[Zurek, 1990]   Zurek, W. H. (Ed.), *Complexity, Entropy and the Physics of Information*, Redwood City, Addison-Wesley, 1990.

[Zurek, 2004]   Zurek, W. H., "Quantum Darwinism and Envariance," in (Barrow *et al.*, 2004, 121–137).

[Zurek, 2007]   ——, "Quantum Origin of Quantum Jumps: Breaking of Unitary Symmetry Induced by Information Transfer in the Transition from Quantum to Classical," *Physical Review* **A76**: 052110.

[Zurek, 2013]   ——, "Wave-Packet Collapse and the Core Quantum Postulates: Discreteness of Quantum Jumps from Unitarity, Repeatability, and Actionable Information," *Physical Review* **A87**: 052111.

[Zwolak/Zurek, 2013]   Zwolak, M. and Zurek, W. H., "Complementarity of Quantum Discord and Classically Accessible Information," *Scientific Reports* **3**: 1729.

Author Index

Subject Index

Solutions to Selected Problems

Chapter 2

2.1 This is straightforward.

$$(2+i)^* = 2-i, \quad (1+3i)^* = 1-3i,$$
$$|2+i|^2 = (2+i)(2-i) = 5,$$
$$|1+3i|^2 = (1+3i)(1-3i) = 10.$$

2.2 Using the Euler formula (2.12) and Table 2.1, we have

$$e^{i0} = \cos 0 + i \sin 0 = 1,$$
$$e^{i\frac{\pi}{4}} = \cos \frac{\pi}{4} + i \sin \frac{\pi}{4} = \frac{1}{\sqrt{2}}(1+i),$$
$$e^{i\frac{\pi}{2}} = \cos \frac{\pi}{2} + i \sin \frac{\pi}{2} = i,$$
$$e^{i\frac{3\pi}{4}} = \cos \frac{3\pi}{4} + i \sin \frac{3\pi}{4} = -\frac{1}{\sqrt{2}}(1-i),$$
$$e^{i\pi} = \cos \pi + i \sin \pi = -1.$$

2.5 The square of a complex number is in general another complex number but probabilities should be real non-negative numbers and less than or equal to 1 [see Box 2.1]. Therefore, we need to compute the *square modulus* of the corresponding coefficient.

2.6 This is straightforward.

$$\wp_2 = \left| \frac{1}{2}\left(1 - e^{i\phi}\right) \right|^2 = \frac{1}{4}\left(1 - e^{i\phi}\right)\left(1 - e^{-i\phi}\right)$$
$$= \frac{1}{4}\left(2 - e^{i\phi} - e^{-i\phi}\right) = \frac{1}{2}\left(1 - \cos\phi\right).$$

2.7 Since

$$c_1^2 = \left[\frac{1}{2}\left(1 + e^{i\phi}\right)\right]^2 = \frac{1}{4}\left(1 + e^{i\phi}\right)\left(1 + e^{i\phi}\right)$$

$$= \frac{1}{4}\left(1 + e^{2i\phi} + 2e^{i\phi}\right),$$

$$c_2^2 = \left[\frac{1}{2}\left(1 - e^{i\phi}\right)\right]^2 = \frac{1}{4}\left(1 - e^{i\phi}\right)\left(1 - e^{i\phi}\right)$$

$$= \frac{1}{4}\left(1 + e^{2i\phi} - 2e^{i\phi}\right),$$

their sum is

$$c_1^2 + c_2^2 = \frac{1}{4}\left(1 + e^{2i\phi} + 2e^{i\phi}\right) + \frac{1}{4}\left(1 + e^{2i\phi} - 2e^{i\phi}\right)$$

$$= \frac{1}{2}\left(1 + e^{2i\phi}\right),$$

which for almost all values of ϕ is not equal to 1.

2.8 The detection probabilities at D1 and D2 are

$$\wp_1 = \frac{1}{2}\left(1 + \cos\phi\right) \quad \text{and} \quad \wp_2 = \frac{1}{2}\left(1 - \cos\phi\right),$$

respectively. Since for $\phi = 0$ (in radians) we have $\cos\phi = 1$, it is clear that for this value D2 never clicks. Moreover, since for $\phi = \pi$ we have $\cos\phi = -1$, it follows that for this value D1 to never clicks.

Chapter 3

3.1 This is straightforward.

$$|\psi\rangle = \frac{1}{\sqrt{2}}(|a\rangle + |a'\rangle) = \frac{1}{\sqrt{2}}\begin{pmatrix} 1 \\ 1 \end{pmatrix},$$

$$|\psi'\rangle = \frac{1}{\sqrt{2}}(|a\rangle - |a'\rangle) = \frac{1}{\sqrt{2}}\begin{pmatrix} 1 \\ -1 \end{pmatrix}.$$

3.2 We have

$$|a\rangle = \begin{pmatrix} 1 \\ 0 \end{pmatrix}, \quad |a'\rangle = \begin{pmatrix} 0 \\ 1 \end{pmatrix},$$

therefore

$$\langle a|a'\rangle = (1\ 0)\begin{pmatrix} 0 \\ 1 \end{pmatrix} = \begin{pmatrix} 0 \\ 0 \end{pmatrix} = 0.$$

3.3 This is straightforward.

$$\langle \psi | a \rangle = \begin{pmatrix} c_a^* & c_{a'}^* \end{pmatrix} \begin{pmatrix} 1 \\ 0 \end{pmatrix} = c_a^*, \quad \langle \psi | a' \rangle = \begin{pmatrix} c_a^* & c_{a'}^* \end{pmatrix} \begin{pmatrix} 0 \\ 1 \end{pmatrix} = c_{a'}^*.$$

3.4 Let us first expand the state vector $|\phi\rangle$ as

$$|\phi\rangle = d|b\rangle + d'|b'\rangle$$
$$= \frac{1}{\sqrt{2}} \left[d(|a\rangle + |a'\rangle) + d'(|a\rangle - |a'\rangle) \right]$$
$$= \frac{1}{\sqrt{2}} \left[(d + d')|a\rangle + (d - d')|a'\rangle \right].$$

Then, we have

$$\langle \psi | \varphi \rangle = \frac{1}{\sqrt{2}} \begin{pmatrix} c^* & c'^* \end{pmatrix} \begin{pmatrix} d + d' \\ d - d' \end{pmatrix}$$
$$= \frac{1}{\sqrt{2}} \left[c^*(d + d') + c'^*(d - d') \right].$$

3.5 This is straightforward.

$$\left(\langle a | + \langle a' | \right) \left(|a\rangle + |a'\rangle \right) = \langle a | a \rangle + \langle a | a' \rangle + \langle a' | a \rangle + \langle a' | a' \rangle$$
$$= \langle a | a \rangle + 0 + 0 + \langle a' | a' \rangle$$
$$= \langle a | a \rangle + \langle a' | a' \rangle.$$

3.6 This is straightforward.

$$\hat{A} = \begin{bmatrix} 3 & 9 \\ 9 & 9 \end{bmatrix}, \qquad \hat{B} = \begin{bmatrix} 21 & 28 \\ 49 & 64 \end{bmatrix}.$$

3.8 This is straightforward.

$$\hat{P}_a |a\rangle = |a\rangle \langle a | a \rangle = |a\rangle, \qquad \hat{P}_{a'} |a'\rangle = |a'\rangle \langle a' | a' \rangle = |a'\rangle,$$

$$\hat{P}_a |a'\rangle = |a\rangle \langle a | a' \rangle = 0, \qquad \hat{P}_{a'} |a\rangle = |a'\rangle \langle a' | a \rangle = 0.$$

3.9 We have

$$\hat{P}_a |\psi\rangle = \begin{bmatrix} 1 & 0 \\ 0 & 0 \end{bmatrix} \begin{pmatrix} c_a \\ c_{a'} \end{pmatrix} = \begin{bmatrix} 1 & 0 \\ 0 & 0 \end{bmatrix} \begin{pmatrix} \langle a | \psi \rangle \\ \langle a' | \psi \rangle \end{pmatrix}$$
$$= \begin{pmatrix} \langle a | \psi \rangle \\ 0 \end{pmatrix} = \langle a | \psi \rangle |a\rangle,$$

$$\hat{P}_{a'} |\psi\rangle = \begin{bmatrix} 0 & 0 \\ 0 & 1 \end{bmatrix} \begin{pmatrix} c_a \\ c_a' \end{pmatrix} = \begin{bmatrix} 0 & 0 \\ 0 & 1 \end{bmatrix} \begin{pmatrix} \langle a | \psi \rangle \\ \langle a' | \psi \rangle \end{pmatrix}$$
$$= \begin{pmatrix} 0 \\ \langle a' | \psi \rangle \end{pmatrix} = \langle a' | \psi \rangle |a'\rangle.$$

3.12 We have

$$
\hat{I}\hat{P}_a = \begin{bmatrix} 1 & 0 \\ 0 & 1 \end{bmatrix} \begin{bmatrix} 1 & 0 \\ 0 & 0 \end{bmatrix} = \begin{bmatrix} 1 & 0 \\ 0 & 0 \end{bmatrix} = \hat{P}_a,
$$

$$
\hat{I}\hat{P}_{a'} = \begin{bmatrix} 1 & 0 \\ 0 & 1 \end{bmatrix} \begin{bmatrix} 0 & 0 \\ 0 & 1 \end{bmatrix} = \begin{bmatrix} 0 & 0 \\ 0 & 1 \end{bmatrix} = \hat{P}_{a'},
$$

$$
\hat{I}|\psi\rangle = \begin{bmatrix} 1 & 0 \\ 0 & 1 \end{bmatrix} \begin{pmatrix} c_a \\ c_{a'} \end{pmatrix} = \begin{pmatrix} c_a \\ c_{a'} \end{pmatrix} = |\psi\rangle.
$$

3.11 This is straightforward.

$$
\hat{P}_a^2 = (|a\rangle\langle a|)(|a\rangle\langle a|) = (|a\rangle\langle a|)\langle a|a\rangle = |a\rangle\langle a| = \hat{P}_a,
$$

$$
\hat{P}_{a'}^2 = (|a'\rangle\langle a'|)(|a'\rangle\langle a'|) = (|a'\rangle\langle a'|)\langle a'|a'\rangle = |a'\rangle\langle a'| = \hat{P}_{a'},
$$

$$
\hat{P}_a\hat{P}_{a'} = (|a\rangle\langle a|)(|a'\rangle\langle a'|) = (|a\rangle\langle a'|)\langle a|a'\rangle = 0,
$$

$$
\hat{P}_{a'}\hat{P}_a = (|a'\rangle\langle a'|)(|a\rangle\langle a|) = (|a'\rangle\langle a|)\langle a'|a\rangle = 0.
$$

Chapter 4

4.1 This is straightforward.

$$
\hat{P}_{a'}|a'\rangle = |a'\rangle = +1\,(|a'\rangle),
$$

$$
\hat{P}_{a'}|a\rangle = 0 = 0\,(|a\rangle).
$$

4.2 Since

$$
\hat{O}'|a\rangle = \begin{bmatrix} -1 & 0 \\ 0 & +1 \end{bmatrix} \begin{pmatrix} 1 \\ 0 \end{pmatrix} = -\begin{pmatrix} 1 \\ 0 \end{pmatrix} = -|a\rangle,
$$

$$
\hat{O}'|a'\rangle = \begin{bmatrix} -1 & 0 \\ 0 & +1 \end{bmatrix} \begin{pmatrix} 0 \\ 1 \end{pmatrix} = \begin{pmatrix} 0 \\ 1 \end{pmatrix} = +|a'\rangle,
$$

then the eigenvalues of the eigenstates $|a\rangle$ and $|a'\rangle$ are -1 and $+1$, respectively.

4.3 This is straightforward.

$$
\hat{P}_a^\dagger = (|a\rangle\langle a|)^\dagger = |a\rangle\langle a| = \hat{P}_a,
$$

$$
\hat{P}_{a'}^\dagger = (|a'\rangle\langle a'|)^\dagger = |a'\rangle\langle a'| = \hat{P}_{a'}.
$$

4.4 This is straightforward.
$$\hat{P}_h|\psi\rangle = |h\rangle\langle h|\langle h|\psi\rangle|h\rangle + |h\rangle\langle h|\langle v|\psi\rangle|v\rangle = \langle h|\psi\rangle|h\rangle,$$

$$\hat{P}_v|\psi\rangle = |v\rangle\langle v|\langle h|\psi\rangle|h\rangle + |v\rangle\langle v|\langle v|\psi\rangle|v\rangle = \langle v|\psi\rangle|v\rangle.$$

4.5 The reason is that probabilities are calculated as square moduli of the corresponding probability amplitudes.

4.6 This is straightforward.
$$c_a = \langle a|\hat{I}|\psi\rangle$$
$$= \langle a| (|v\rangle\langle v| + |h\rangle\langle h|) |\psi\rangle$$
$$= \langle a|v\rangle c_v + \langle a|h\rangle c_h,$$
$$c_{a'} = \langle a'|\hat{I}|\psi\rangle$$
$$= \langle a'| (|v\rangle\langle v| + |h\rangle\langle h|) |\psi\rangle$$
$$= \langle a'|v\rangle c_v + \langle a'|h\rangle c_h.$$

4.7 This is straightforward.
$$\hat{U}^\dagger\hat{U} = \begin{bmatrix} \langle h|a\rangle^* & \langle v|a\rangle^* \\ \langle h|a'\rangle^* & \langle v|a'\rangle^* \end{bmatrix} \begin{bmatrix} \langle h|a\rangle & \langle h|a'\rangle \\ \langle v|a\rangle & \langle v|a'\rangle \end{bmatrix}$$
$$= \begin{bmatrix} \langle a|h\rangle\langle h|a\rangle + \langle a|v\rangle\langle v|a\rangle & \langle a|h\rangle\langle h|a'\rangle + \langle a|v\rangle\langle v|a'\rangle \\ \langle a'|h\rangle\langle h|a\rangle + \langle a'|v\rangle\langle v|a\rangle & \langle a'|h\rangle\langle h|a'\rangle + \langle a'|v\rangle\langle v|a'\rangle \end{bmatrix}$$
$$= \begin{bmatrix} \langle a| (|h\rangle\langle h| + |v\rangle\langle v|) |a\rangle & \langle a| (|h\rangle\langle h| + |v\rangle\langle v|) |a'\rangle \\ \langle a'| (|h\rangle\langle h| + |v\rangle\langle v|) |a\rangle & \langle a'| (|h\rangle\langle h| + |v\rangle\langle v|) |a'\rangle \end{bmatrix}$$
$$= \begin{bmatrix} \langle a|a\rangle & \langle a|a'\rangle \\ \langle a'|a\rangle & \langle a'|a'\rangle \end{bmatrix}$$
$$= \begin{bmatrix} 1 & 0 \\ 0 & 1 \end{bmatrix} = \hat{I}.$$

4.9 If we write
$$|1\rangle = \begin{pmatrix} 1 \\ 0 \end{pmatrix} \quad \text{and} \quad |2\rangle = \begin{pmatrix} 0 \\ 1 \end{pmatrix},$$
then
$$\hat{U}_{BS}\left[\frac{1}{\sqrt{2}}(|d\rangle + e^{i\phi}|u\rangle)\right] = \frac{1}{\sqrt{2}}\begin{bmatrix} 1 & 1 \\ 1 & -1 \end{bmatrix}\frac{1}{\sqrt{2}}\begin{pmatrix} 1 \\ e^{i\phi} \end{pmatrix}$$
$$= \frac{1}{2}\begin{pmatrix} 1 + e^{i\phi} \\ 1 - e^{i\phi} \end{pmatrix},$$
which is the same as the result in Eq. (2.29).

4.10 Since the phase shifter does not affect the state $|d\rangle$ and produces a phase shift ϕ to the state $|u\rangle$, we can express $\hat{U}_\phi$ in the basis $\{|d\rangle, |u\rangle\}$ as

$$\hat{U}_\phi = \begin{bmatrix} 1 & 0 \\ 0 & e^{i\phi} \end{bmatrix}.$$

Moreover, we have

$$\hat{U}_\phi^\dagger \hat{U}_\phi = \begin{bmatrix} 1 & 0 \\ 0 & e^{-i\phi} \end{bmatrix} \begin{bmatrix} 1 & 0 \\ 0 & e^{i\phi} \end{bmatrix} = \begin{bmatrix} 1 & 0 \\ 0 & 1 \end{bmatrix} = \hat{I},$$

and also $\hat{U}_\phi \hat{U}_\phi^\dagger = \hat{I}$, hence $\hat{U}_\phi$ is unitary.

4.11 Assuming that $|a'\rangle = c_v|v\rangle - c_h|h\rangle$, we have

$$\hat{P}_{a'} = (c_v|v\rangle - c_h|h\rangle)\left(c_v^*\langle v| - c_h^*\langle h|\right)$$
$$= |c_v|^2\,|v\rangle\langle v| + |c_h|^2\,|h\rangle\langle h| - c_v c_h^*|v\rangle\langle h| - c_v^* c_h|h\rangle\langle v|.$$

Applying it to the state $c_v|v\rangle$, we obtain

$$\hat{P}_{a'}\left(c_v|v\rangle\right) = c_v\Big(\,|c_v|^2\,|v\rangle\langle v| + |c_h|^2\,|h\rangle\langle h| - c_v c_h^*|v\rangle\langle h|$$
$$- c_v^* c_h|h\rangle\langle v|\Big)|v\rangle$$
$$= c_v\left(|c_v|^2\,|v\rangle - c_v^* c_h|h\rangle\right)$$
$$= |c_v|^2\left(c_v|v\rangle - c_h|h\rangle\right)$$
$$= |c_v|^2\,|a'\rangle,$$

which is formally similar to that obtained in the text. This means that it does not matter in which direction the intermediate polarization filter is (the only requirement is that it is different from both the vertical and horizontal directions) because the output state will always have a nonzero probability along that direction.

4.12 We have

$$\hat{P}_h\left(\hat{P}_v\left(\hat{P}_{a'}|a\rangle\right)\right) = \begin{bmatrix} 1 & 0 \\ 0 & 0 \end{bmatrix} \left\{ \begin{bmatrix} 0 & 0 \\ 0 & 1 \end{bmatrix} \left\{ \begin{bmatrix} |c_h'|^2 & c_h' c_v'^* \\ c_v' c_h'^* & |c_v'|^2 \end{bmatrix} \begin{pmatrix} c_h \\ c_v \end{pmatrix} \right\} \right\}$$

$$= \begin{bmatrix} 1 & 0 \\ 0 & 0 \end{bmatrix} \left\{ \begin{bmatrix} 0 & 0 \\ 0 & 1 \end{bmatrix} \begin{pmatrix} c_h\,|c_h'|^2 + c_h' c_v'^* c_v \\ c_v' c_h'^* c_h + c_v\,|c_v'|^2 \end{pmatrix} \right\}$$

$$= \begin{bmatrix} 1 & 0 \\ 0 & 0 \end{bmatrix} \begin{pmatrix} 0 \\ c_v' c_h'^* c_h + c_v\,|c_v'|^2 \end{pmatrix}$$

$$= 0$$

and

$$\hat{P}_h\left(\hat{P}_{a'}\left(\hat{P}_v|a\rangle\right)\right) = \begin{bmatrix} 1 & 0 \\ 0 & 0 \end{bmatrix} \left\{ \begin{bmatrix} |c_h'|^2 & c_h'c_v'^* \\ c_v'c_h'^* & |c_v'|^2 \end{bmatrix} \left\{ \begin{bmatrix} 0 & 0 \\ 0 & 1 \end{bmatrix} \begin{pmatrix} c_h \\ c_v \end{pmatrix} \right\} \right\}$$

$$= \begin{bmatrix} 1 & 0 \\ 0 & 0 \end{bmatrix} \left\{ \begin{bmatrix} |c_h'|^2 & c_h'c_v'^* \\ c_v'c_h'^* & |c_v'|^2 \end{bmatrix} \begin{pmatrix} 0 \\ c_v \end{pmatrix} \right\}$$

$$= \begin{bmatrix} 1 & 0 \\ 0 & 0 \end{bmatrix} \begin{pmatrix} c_h'c_v'^*c_v \\ |c_v'|^2 c_v \end{pmatrix}$$

$$= \begin{pmatrix} c_h'c_v'^*c_v \\ 0 \end{pmatrix}$$

$$= c_h'c_v'^*c_v|h\rangle.$$

Chapter 5

5.1 This is straightforward. For the state $|f\rangle$ in Eq. (5.4)

$$|f\rangle = \frac{e^{i\phi}}{2}(|1\rangle - |2\rangle),$$

we have

$$\wp_1 = |\langle 1|f\rangle|^2 = \left|\frac{e^{i\phi}}{2}\right|^2 = \frac{1}{4}, \quad \wp_2 = |\langle 2|f\rangle|^2 = \left|\frac{e^{i\phi}}{2}\right|^2 = \frac{1}{4}.$$

For the state $|f\rangle$ in Eq. (5.5)

$$|f\rangle = \frac{1}{2}(|1\rangle - |2\rangle),$$

we have

$$\wp_1 = |\langle 1|f\rangle|^2 = \left|\frac{1}{2}\right|^2 = \frac{1}{4}, \quad \wp_2 = |\langle 2|f\rangle|^2 = \left|\frac{1}{2}\right|^2 = \frac{1}{4}.$$

Therefore state vectors that differ by a global phase factor represent the same physical state.

5.2 In the case in which the upper path is blocked, we have

$$\frac{1}{\sqrt{2}}(|d\rangle + |u\rangle) \xrightarrow{S} \frac{1}{\sqrt{2}}|d\rangle.$$

Then, the final state after BS2 is

$$|f\rangle = \frac{1}{2}(|1\rangle + |2\rangle),$$

where we have used the transformation (4.36). The detection probabilities at D1 and D2 are, respectively,

$$\wp_1 = \wp_2 = \frac{1}{4},$$

in which case we have no dark output (the two detectors receive on average an equal number of photons independent of the phase). Since these probabilities are precisely the same when it is the lower path that is blocked, in both cases we can only infer that one of the paths is blocked but cannot tell which one.

5.3 In this case, we use the probability $\wp_2$ in Eq. (2.34):

$$\wp_2 = \frac{1}{2}(1 - \cos\phi).$$

Also in this case we have violation of the inequality (5.6) whenever $\wp_2 > 1/2$, although the violation is maximum when $\phi = \pi$.

5.5 Since T^2 and R^2 are respectively the probabilities of a photon being transmitted and reflected by a beam splitter and since transmission and reflection are the only two possible mutually exclusive cases, the sum of whose probabilities has to be one.

5.6 Making use of the unitary operator $\hat{U}_\phi$ representing the phase shifter that is obtained in Prob. 4.10, we have

$$
\begin{aligned}
\hat{U}_{BS2}\hat{U}_\phi\hat{U}_{BS1}|i\rangle &= \frac{1}{\sqrt{2}}\begin{bmatrix} 1 & 1 \\ 1 & -1 \end{bmatrix}\begin{bmatrix} 1 & 0 \\ 0 & e^{i\phi} \end{bmatrix}\begin{bmatrix} T & R \\ R & -T \end{bmatrix}\begin{pmatrix} 1 \\ 0 \end{pmatrix} \\
&= \frac{1}{\sqrt{2}}\begin{bmatrix} 1 & 1 \\ 1 & -1 \end{bmatrix}\begin{bmatrix} 1 & 0 \\ 0 & e^{i\phi} \end{bmatrix}\begin{pmatrix} T \\ R \end{pmatrix} \\
&= \frac{1}{\sqrt{2}}\begin{bmatrix} 1 & 1 \\ 1 & -1 \end{bmatrix}\begin{pmatrix} T \\ e^{i\phi}R \end{pmatrix} \\
&= \frac{1}{\sqrt{2}}\begin{pmatrix} T + e^{i\phi}R \\ T - e^{i\phi}R \end{pmatrix},
\end{aligned}
$$

which is the same as the result in Eq. (5.11).

5.7 We have

$$\wp_3 = \frac{1}{2} \left(T + e^{i\phi}R\right) \left(T + e^{-i\phi}R\right)$$

$$= \frac{1}{2} \left(T^2 + e^{i\phi}TR + e^{-i\phi}TR + R^2\right)$$

$$= \frac{1}{2} \left(T^2 + R^2\right) + TR \cos\phi$$

$$= \frac{1}{2} + TR \cos\phi,$$

$$\wp_4 = \frac{1}{2} \left(T - e^{i\phi}R\right) \left(T - e^{-i\phi}R\right)$$

$$= \frac{1}{2} \left(T^2 - e^{i\phi}TR - e^{-i\phi}TR + R^2\right)$$

$$= \frac{1}{2} \left(T^2 + R^2\right) - TR \cos\phi$$

$$= \frac{1}{2} - TR \cos\phi,$$

where use has been made of the formula (2.20). A comparison with the probabilities in Eqs. (2.33) and (2.34) shows that the latter represent a special case of the above result for $T = R = \frac{1}{\sqrt{2}}$.

5.8 Recall that we have assumed T and R to be real and non-negative, hence

$$\hat{U}_{BS1}\hat{U}^\dagger_{BS1} = \begin{bmatrix} T & R \\ R & -T \end{bmatrix} \begin{bmatrix} T & R \\ R & -T \end{bmatrix}$$

$$= \begin{bmatrix} T^2 + R^2 & TR - TR \\ TR - TR & T^2 + R^2 \end{bmatrix}$$

$$= \begin{bmatrix} 1 & 0 \\ 0 & 1 \end{bmatrix} = \hat{I},$$

which shows that $\hat{U}_{BS1}$ is unitary.

5.9 In the case in which the detectors are located after BS2, the experiment is the same as that described in Secs. 2.3–2.6. In the case in which we decide to place the detectors before BS2, the final state of the photon is [see, e.g., Eq. (2.25)]

$$|f\rangle = \frac{1}{\sqrt{2}} \left(|d\rangle + e^{i\phi} |u\rangle \right).$$

It is evident that the probabilities of detecting the photon at D_A and D_B are the same and equal to $\frac{1}{2}$. Most importantly, since the component $|u\rangle$ is detected at detector D_A while the component $|d\rangle$ at D_B, we can tell unequivocally from which path the photon came once one of the two detectors clicks.

Chapter 6

6.1 The eigenbasis of $\hat{x}$ can be written as

$$\left\{ |x_0\rangle = \begin{pmatrix} 1 \\ 0 \\ 0 \\ 0 \end{pmatrix}, |x_1\rangle = \begin{pmatrix} 0 \\ 1 \\ 0 \\ 0 \end{pmatrix}, |x_2\rangle = \begin{pmatrix} 0 \\ 0 \\ 1 \\ 0 \end{pmatrix}, |x_3\rangle = \begin{pmatrix} 0 \\ 0 \\ 0 \\ 1 \end{pmatrix} \right\}.$$

The state $|\psi\rangle$ can then be expanded as

$$|\psi\rangle = \begin{pmatrix} \langle x_0|\psi\rangle \\ \langle x_1|\psi\rangle \\ \langle x_2|\psi\rangle \\ \langle x_3|\psi\rangle \end{pmatrix} = \begin{pmatrix} c_0 \\ c_1 \\ c_2 \\ c_3 \end{pmatrix}.$$

6.2 The first two integrals are straightforward.

$$\frac{1}{4} \int_{-2}^{3} x^3 dx = \frac{1}{4} \frac{x^4}{4} \Big|_{-2}^{3} = \frac{1}{16} \left[3^4 - (-2)^4 \right]$$

$$= \frac{1}{16} (81 - 16) = \frac{65}{16},$$

$$\int_{0}^{\pi} \sin x \, dx = -\cos x \Big|_{0}^{\pi} = 1 - (-1) = 2.$$

The third one can be computed by observing that if we rescale $kx = y$, then the area under the graph of e^{kx} between $x = 0$ and $x = t$ is scaled by a factor of k (in the horizontal direction) to become the area under the graph of e^y between $y = 0$ and $y = kt$. Hence

$$\int_{0}^{t} e^{kx} dx = \frac{1}{k} \int_{0}^{kt} e^y dy = \frac{e^y}{k} \Big|_{0}^{kt} = \frac{e^{kt} - 1}{k}.$$

Indeed, this observation is very general and the corresponding method of evaluating integrals is called the *change of variables*.

6.3 Partial fraction expansion and the property that integrals are invariant under translations (6.22) are used here. Write

$$\frac{3x}{x^2 + x - 2} = \frac{1}{x - 1} + \frac{2}{x + 2},$$

then we have

$$\int_2^4 \frac{3x}{x^2 + x - 2} dx = \int_2^4 \left(\frac{1}{x - 1} + \frac{2}{x + 2} \right) dx$$

$$= \int_1^3 \frac{1}{x} dx + \int_4^6 \frac{2}{x} dx$$

$$= (\ln 3 - \ln 1) + 2(\ln 6 - \ln 4)$$

$$= \ln 3 + 2\ln 3 - 2\ln 2$$

$$= 3\ln 3 - 2\ln 2,$$

where we recall that $\ln(ab) = \ln a + \ln b$.

6.4 It is easy to see that we have

$$1 = \langle \psi | \psi \rangle = \langle \psi | \hat{I} | \psi \rangle = \int_{-\infty}^{+\infty} \langle \psi | x \rangle \langle x | \psi \rangle dx = \int_{-\infty}^{+\infty} |\psi(x)|^2 dx$$

and

$$1 = \langle \psi | \psi \rangle = \langle \psi | \hat{I} | \psi \rangle = \int_{-\infty}^{+\infty} \langle \psi | \mathbf{r} \rangle \langle \mathbf{r} | \psi \rangle d^3 r = \int_{-\infty}^{+\infty} |\psi(\mathbf{r})|^2 d^3 r.$$

It is also clear that the above two expressions correspond to the normalization condition for the discrete case that can be written as

$$\sum_j \wp_j = 1,$$

which in turn shows that the true probabilities are $|\psi(x)|^2 dx$ and $|\psi(\mathbf{r})|^2 d^3 r$.

6.5 From the normalization condition

$$1 = \int_{-\infty}^{+\infty} dx \, |\psi(x)|^2$$

$$= \int_{-\infty}^{+\infty} dx (\mathcal{N} e^{-\lambda|x|/\hbar})^2$$

$$= \mathcal{N}^2 \int_{-\infty}^{0} dx e^{-2\lambda|x|/\hbar} + \mathcal{N}^2 \int_{0}^{+\infty} dx e^{-2\lambda|x|/\hbar}$$

$$= 2\mathcal{N}^2 \int_{0}^{\infty} dx \, e^{-2\lambda x/\hbar}$$

$$= \frac{\mathcal{N}^2 \hbar}{\lambda},$$

we find

$$\mathcal{N} = \sqrt{\frac{\lambda}{\hbar}}.$$

6.6 The product rule, quotient rule, and the chain rule are used here.

$$\frac{d}{dx}\sqrt{x} = \frac{1}{2}x^{-\frac{1}{2}} = \frac{1}{2\sqrt{x}},$$

$$\frac{d}{dx}(x \cos x) = \cos x - x \sin x,$$

$$\frac{d}{dx}\left(\frac{\sin x}{x}\right) = \frac{x \cos x - \sin x}{x^2} = \frac{\cos x}{x} - \frac{\sin x}{x^2},$$

$$\frac{d}{dx}\left(\frac{1}{1+\sqrt{x}}\right) = -\left(1 + x^{\frac{1}{2}}\right)^{-2}\left(\frac{1}{2}x^{-\frac{1}{2}}\right) = -\frac{1}{2\sqrt{x}(1+\sqrt{x})^2}.$$

6.7 Assuming term-by-term differentiation is permissible (which can be justified), we have

$$\frac{de^x}{dx} = \sum_{n=1}^{\infty} \frac{x^{n-1}}{(n-1)!} = \sum_{n=0}^{\infty} \frac{x^n}{n!}.$$

Hence we conclude that the derivative of the exponential is the exponential itself.

6.8 The quotient rule and the chain rule are used here.

$$\frac{\partial}{\partial x}\left(\frac{x-y}{x+y}\right) = \frac{1}{x+y} - \frac{x-y}{(x+y)^2},$$

$$\frac{\partial}{\partial y}\left(\frac{x-y}{x+y}\right) = -\frac{1}{x+y} - \frac{x-y}{(x+y)^2},$$

$$\frac{\partial^2}{\partial y \partial x}\left(\frac{x-y}{x+y}\right) = \frac{\partial}{\partial y}\left(\frac{1}{x+y} - \frac{x-y}{(x+y)^2}\right) = \frac{2(x^2-y^2)}{(x+y)^4},$$

$$\frac{\partial^2}{\partial x \partial y}\left(\frac{x-y}{x+y}\right) = -\frac{\partial}{\partial x}\left(\frac{1}{x+y} + \frac{x-y}{(x+y)^2}\right) = \frac{2(x^2-y^2)}{(x+y)^4}.$$

6.9 This is straightforward.

$$\frac{\partial f}{\partial t} = 2t - 8x = 2t - 8\sin t,$$

$$\frac{df}{dt} = \frac{\partial f}{\partial t} + \frac{\partial f}{\partial x}\frac{dx}{dt} = 2t - 8x + (-8t - 2x)\cos t$$

$$= 2t - 8\sin t - (8t + 2\sin t)\cos t.$$

6.11 From Eq. (6.114), we have

$$\hat{U}_x^\dagger(a) = \left(e^{-\frac{i}{\hbar}a\hat{p}_x}\right)^\dagger = e^{\frac{i}{\hbar}a\hat{p}_x}.$$

To prove the unitarity of $\hat{U}_x(a)$, it suffices to show that

$$\hat{U}_x(a)\hat{U}_x^\dagger(a) = \hat{U}_x^\dagger(a)\hat{U}_x(a) = \hat{I}.$$

We prove the first part:

$$\hat{U}_x(a)\hat{U}_x^\dagger(a) = e^{-\frac{i}{\hbar}a\hat{p}_x}e^{\frac{i}{\hbar}a\hat{p}_x} = e^{-\frac{i}{\hbar}a\hat{p}_x + \frac{i}{\hbar}a\hat{p}_x} = e^0 = \hat{I}.$$

6.12 From Eq. (6.131) and the unitarity of $\hat{U}_x(a)$, we have

$$\hat{U}_x^\dagger(a)|x\rangle = \hat{U}_x^{-1}(a)|x\rangle = |x - a\rangle,$$

which implies

$$\langle x|\hat{U}_x^\dagger(a)\hat{x}\hat{U}_x(a)|x'\rangle = \langle x|\hat{U}_x^\dagger(a)\hat{x}|x' + a\rangle$$

$$= (x' + a)\langle x|\hat{U}_x^\dagger(a)|x' + a\rangle$$

$$= (x' + a)\langle x|x'\rangle$$

$$= \langle x|(\hat{x} + a)|x'\rangle.$$

Since the above equation is valid for arbitrary $\langle x|$ and $|x'\rangle$, it follows that $\hat{U}_x^\dagger(a)\hat{x}\hat{U}_x(a) = \hat{x} + a$.

6.13 We have

$$1 = \langle \psi | \psi \rangle = \int_{-\infty}^{+\infty} \langle \psi | p_x \rangle \langle p_x | \psi \rangle dp_x = \int_{-\infty}^{+\infty} |\widetilde{\psi}(p_x)|^2 dp_x.$$

We can proceed in a similar way if we like to obtain the three-dimensional counterpart of this proof.

6.14 We have

$$\begin{aligned}
\widetilde{\psi}(p_x) &= \frac{1}{\sqrt{2\pi\hbar}} \int_0^1 \psi(x) e^{-\frac{i}{\hbar} p_x x} dx \\
&= \frac{1}{\sqrt{2\pi\hbar}} \int_0^1 e^{-\frac{i}{\hbar} p_x x} dx \\
&= \frac{i\hbar}{\sqrt{2\pi\hbar} p_x} e^{-\frac{i}{\hbar} p_x x} \Big|_{x=0}^{x=1} \\
&= \frac{i\hbar}{\sqrt{2\pi\hbar} p_x} \left(e^{-\frac{i}{\hbar} p_x} - 1 \right).
\end{aligned}$$

6.15 The properties of the commutators (6.151) and the commutation relation $[\hat{x}, \hat{p}_x] = i\hbar\hat{x}$ are used here.

$$[\hat{x}^2, \hat{p}_x] = \hat{x}[\hat{x}, \hat{p}_x] + [\hat{x}, \hat{p}_x]\hat{x} = 2i\hbar\hat{x},$$

$$[\hat{x}, \hat{p}_x^2] = \hat{p}_x[\hat{x}, \hat{p}_x] + [\hat{x}, \hat{p}_x]\hat{p}_x = 2i\hbar\hat{p}_x,$$

$$[\hat{x}\hat{p}_x, \hat{p}_x\hat{x}] = [\hat{p}_x\hat{x} + i\hbar, \hat{p}_x\hat{x}] = 0.$$

6.17 The expression for $\Delta_\psi^2 \hat{O}$ in Eq. (6.158) can be simplified to

$$\Delta_\psi^2 \hat{O} = \langle \hat{O}^2 \rangle_\psi - \langle \hat{O} \rangle_\psi^2,$$

where use has been made of the properties of the expectation value given by Eqs. (6.157). Since

$$\hat{O}^2 = (|h\rangle\langle h| - |v\rangle\langle v|)(|h\rangle\langle h| - |v\rangle\langle v|) = |h\rangle\langle h| + |v\rangle\langle v| = \hat{I},$$

we have $\langle \hat{O}^2 \rangle_\psi = 1$ and $\Delta_\psi^2 \hat{O} = \langle \hat{O}^2 \rangle_\psi - \langle \hat{O} \rangle_\psi^2 = 1.$

Chapter 7

7.1 Let the displacement of the object be x, then the force acting on the subject is $F(x) = -kx$, where the minus sign means the direction of the force is opposite to the of the displacement. It is straightforward to show that $F(x)$ is a conservative force and the corresponding potential energy is $V(x) = \frac{1}{2}kx^2$ [see Eq. (7.1)]. Hence, the Hamiltonian of the object is

$$H(p_x, x) = \frac{p_x^2}{2m} + \frac{kx^2}{2},$$

where p_x is the momentum of the object. The Hamilton equations can be found by direct substitution [see Eq. (7.3)]

$$\frac{dx}{dt} = \frac{p_x}{m}, \qquad \frac{dp_x}{dt} = -kx.$$

7.2 From Eqs. (7.4) and (7.13), we obtain

$$i\hbar \frac{\partial}{\partial t} \psi(x, t) = -\frac{\hbar^2}{2m} \frac{\partial^2}{\partial x^2} \psi(x, t) + mgx\, \psi(x, t),$$

where the gravitational field is chosen to point in the negative x direction.

7.4 The states $|E_1\rangle$ and $|E_2\rangle$ are energy eigenstates, so they are stationary states whose time evolution are respectively given by

$$|E_1(t)\rangle = e^{-\frac{i}{\hbar}E_1 t}|E_1\rangle, \qquad |E_2(t)\rangle = e^{-\frac{i}{\hbar}E_2 t}|E_2\rangle.$$

By expanding the initial state $|\psi(0)\rangle$ as a superposition of the energy eigenstates $|E_1\rangle$ and $|E_2\rangle$ as

$$|\psi(0)\rangle = |1\rangle = \frac{1}{\sqrt{2}}(|E_1\rangle + |E_2\rangle),$$

we can express the state of the system at time t as

$$|\psi(t)\rangle = \frac{1}{\sqrt{2}}(e^{-\frac{i}{\hbar}E_1 t}|E_1\rangle + e^{-\frac{i}{\hbar}E_2 t}|E_2\rangle).$$

Hence, probability of finding the system in the state $|1\rangle$ is

$$\wp_1(t) = |\langle 1|\psi(t)\rangle|^2 = \frac{1}{4}\left| e^{-\frac{i}{\hbar}E_1 t} + e^{-\frac{i}{\hbar}E_2 t} \right|^2$$

$$= \frac{1}{2}\left[1 + \cos\frac{(E_1 - E_2)t}{\hbar} \right].$$

7.5 From Eq. (7.40) and the similar ones for $\hat{Y}$ and $\hat{Z}$, we have

$$[\hat{X}, \hat{Y}] \xrightarrow{\hat{U}} [\hat{X}', \hat{Y}'] = [\hat{U}^\dagger \hat{X}\hat{U}, \hat{U}^\dagger \hat{Y}\hat{U}]$$

$$= \hat{U}^\dagger \hat{X}\hat{U}\, \hat{U}^\dagger \hat{Y}\hat{U} - \hat{U}^\dagger \hat{Y}\hat{U}\, \hat{U}^\dagger \hat{X}\hat{U}$$

$$= \hat{U}^\dagger \hat{X}\hat{Y}\hat{U} - \hat{U}^\dagger \hat{Y}\hat{X}\hat{U}$$

$$= \hat{U}^\dagger [\hat{X}, \hat{Y}]\hat{U}$$

$$= \hat{U}^\dagger \hat{Z}\hat{U}$$

$$= \hat{Z}',$$

where in the third line use has been made of the property $\hat{U}\hat{U}^\dagger = \hat{I}$.

7.6 From Eqs. (7.40) and (7.44), we have

$$\langle\psi'|\hat{X}|\psi'\rangle = \langle\hat{U}\psi'|\hat{X}|\hat{U}\psi'\rangle = \langle\psi'|\hat{U}^\dagger\hat{X}\hat{U}|\psi'\rangle = \langle\psi|\hat{X}'|\psi\rangle,$$

where we note that $\langle\hat{U}\psi'| = |\hat{U}\psi'\rangle^\dagger = (\hat{U}|\psi'\rangle)^\dagger = \langle\psi'|\hat{U}^\dagger$.

7.7 The Hamiltonian in the Heisenberg picture is given by [see Eq. (7.54)]

$$\hat{H}(t) = \hat{U}_t^\dagger(t)\,\hat{H}\,\hat{U}_t(t) = \hat{H}\,\hat{U}_t^\dagger(t)\,\hat{U}_t(t) = \hat{H},$$

where $\hat{H}$ (without time dependence) is the Hamiltonian in the Schrödinger picture. In the derivation above use has been made of the fact that $\hat{H}$ and $\hat{U}_t(t)$ commute.

7.8 From Eq. (7.54), we have

$$\begin{aligned}
\hat{O}(t) &= \hat{U}_t^\dagger(t)\,\hat{O}\,\hat{U}_t(t) \\
&= \hat{U}_t^\dagger(t)\,[\hat{O}_1,\,\hat{O}_2]\,\hat{U}_t(t) \\
&= \hat{U}_t^\dagger(t)\,\hat{O}_1\hat{O}_2\,\hat{U}_t(t) - \hat{U}_t^\dagger(t)\,\hat{O}_2\hat{O}_1\,\hat{U}_t(t) \\
&= [\hat{U}_t^\dagger(t)\,\hat{O}_1\,\hat{U}_t(t)]\,[\hat{U}_t^\dagger(t)\,\hat{O}_2\,\hat{U}_t(t)] \\
&\quad - [\hat{U}_t^\dagger(t)\,\hat{O}_2\,\hat{U}_t(t)]\,[\hat{U}_t^\dagger(t)\,\hat{O}_1\,\hat{U}_t(t)] \\
&= \hat{O}_1(t)\hat{O}_2(t) - \hat{O}_2(t)\hat{O}_1(t) \\
&= [\hat{O}_1(t),\,\hat{O}_2(t)],
\end{aligned}$$

where in the forth line the identity operator $\hat{I} = \hat{U}_t(t)\hat{U}_t^\dagger(t)$ has been inserted and the square parentheses have only the purpose to help the reader to single out the relevant parts of the expression.

7.10 We have

$$\begin{aligned}
[\hat{a},\,\hat{a}^\dagger] &= \frac{m}{2\hbar\omega}\left(\omega^2\,[\hat{x},\,\hat{x}] + \mathrm{i}\frac{\omega}{m}\,[\hat{p}_x,\,\hat{x}]\right. \\
&\qquad\left. - \mathrm{i}\frac{\omega}{m}\,[\hat{x},\,\hat{p}_x] + \frac{1}{m^2}\,[\hat{p}_x,\,\hat{p}_x]\right) \\
&= -\frac{\mathrm{i}}{\hbar}\,[\hat{x},\,\hat{p}_x] = \hat{I},
\end{aligned}$$

where we have made use of the commutation relation $[\hat{x},\,\hat{p}_x] = \mathrm{i}\hbar$.

7.11 We perform a similar derivation for the annihilation operator. First, let us write

$$\hat{N}\hat{a}^{\dagger}|n\rangle = \left(\hat{N}\hat{a}^{\dagger} - \hat{a}^{\dagger}\hat{N} + \hat{a}^{\dagger}\hat{N}\right)|n\rangle$$
$$= \left([\hat{N}, \hat{a}^{\dagger}] + \hat{a}^{\dagger}\hat{N}\right)|n\rangle$$
$$= -\hat{a}^{\dagger}|n\rangle + n\hat{a}^{\dagger}|n\rangle$$
$$= (n+1)\hat{a}^{\dagger}|n\rangle,$$

where in the first line we have simply added and subtracted the same quantity and in the third line we have made use of both Eq. (7.81) and the following commutation relation [see Eq. (7.78)]:

$$[\hat{N}, \hat{a}^{\dagger}] = [\hat{a}^{\dagger}\hat{a}, \hat{a}^{\dagger}] = \hat{a}^{\dagger}[\hat{a}, \hat{a}^{\dagger}] = \hat{a}^{\dagger}.$$

From the previous it follows that $\hat{a}^{\dagger}|n\rangle = |\hat{a}^{\dagger}n\rangle$ is an eigenstate of $\hat{N}$ with eigenvalue $n + 1$, that is, we have

$$\hat{N}|\hat{a}^{\dagger}n\rangle = (n+1)|\hat{a}^{\dagger}n\rangle,$$

but from Eq. (7.81) it follows that

$$\hat{N}|n+1\rangle = (n+1)|n+1\rangle.$$

Since the eigenstates $\hat{a}|n\rangle$ and $|n+1\rangle$ correspond to the same eigenvalue, they have to be proportional to each other. Hence we have

$$\hat{a}|n\rangle = c'_n|n+1\rangle,$$

where c'_n is a proportional constant that is formally given by $c'_n = \langle n+1|\hat{a}^{\dagger}n\rangle$. To find c'_n, we left multiply both sides of the previous equation by $\langle \hat{a}n|$ and take the norm squared of the state vector $\hat{a}^{\dagger}|n\rangle$

$$\langle \hat{a}n|\langle n+1|\hat{a}^{\dagger}n\rangle|n+1\rangle = \left|c'_n\right|^2 = \langle n|\hat{a}\hat{a}^{\dagger}|n\rangle$$
$$= \langle n|(\hat{N} + \hat{I})|n\rangle = n+1,$$

since $\hat{a}\hat{a}^{\dagger} = \hat{a}\hat{a}^{\dagger} - \hat{a}^{\dagger}\hat{a} + \hat{a}^{\dagger}\hat{a} = \hat{N} + \hat{I}$. Therefore, we find $c'_n = \sqrt{n+1}$ up to a global phase factor, which can be absorbed into $|n+1\rangle$. Then, the desired result follows.

7.12 We prove this relation by induction. From Eq. (7.90) we immediately obtain

$$|1\rangle = \hat{a}^\dagger|0\rangle.$$

Assuming that the relation holds for a given n, we must prove that it holds for $n+1$ as well, that is

$$|n+1\rangle = \frac{\left(\hat{a}^\dagger\right)^{n+1}}{\sqrt{(n+1)!}}|0\rangle.$$

In fact, we have

$$\frac{\left(\hat{a}^\dagger\right)^{n+1}}{\sqrt{(n+1)!}}|0\rangle = \frac{\hat{a}^\dagger}{\sqrt{n+1}}\frac{\left(\hat{a}^\dagger\right)^n}{\sqrt{n!}}|0\rangle$$

$$= \frac{\hat{a}^\dagger}{\sqrt{n+1}}|n\rangle = |n+1\rangle.$$

7.13 Taking advantage of Eqs. (6.52) and (6.117), we have

$$0 = \sqrt{\frac{m}{2\hbar\omega}}\left(\omega x + \frac{\hbar}{m}\frac{d}{dx}\right)\langle x|n\rangle$$

$$= \left(\omega x + \frac{\hbar}{m}\frac{d}{dx}\right)\psi_0(x)$$

$$= \frac{\hbar}{m}\left(\frac{d}{dx}\psi_0(x) + \frac{m\omega}{\hbar}x\psi_0(x)\right),$$

where we have used ordinary derivative in the place of partial derivative due to the considerations expressed in Footnote a, p. 139.

7.14 From Eq. (7.98) and the normalization condition, it follows that

$$|\mathcal{N}|^2\int_{-\infty}^{+\infty} dx\, e^{-\frac{m\omega}{\hbar}x^2} = 1.$$

Using the known mathematical formula

$$\int_{-\infty}^{+\infty} dy\, e^{-ay^2} = \sqrt{\frac{\pi}{a}},$$

and setting $a = m\omega/\hbar$, we obtain

$$|\mathcal{N}|^2\sqrt{\frac{\pi\hbar}{m\omega}} = 1.$$

Taking $\mathcal{N}$ real for simplicity and without loss of generality, we finally have

$$\mathcal{N} = \left(\frac{m\omega}{\pi\hbar}\right)^{\frac{1}{4}}.$$

7.15 (a) This part is straightforward.

$$[\hat{x}, \hat{H}] = \frac{i\hbar\hat{p}_x}{m}, \qquad [\hat{p}_x, \hat{H}] = -i\hbar m\omega^2\hat{x}.$$

(b) From the commutators in (a), the Heisenberg equations for $\hat{x}(t)$ and $\hat{p}_x(t)$ are given by [see Eq. (7.55)]

$$\frac{d\hat{x}(t)}{dt} = \frac{\hat{p}_x(t)}{m},$$

$$\frac{d\hat{p}_x(t)}{dt} = -m\omega^2\hat{x}(t).$$

Taking time derivative of the first equation and using the second equation to eliminate $d\hat{p}_x/dt$, we obtain

$$\frac{d^2\hat{x}(t)}{dt^2} + \omega^2 x(t) = 0.$$

The above equation has the solution [see Box 7.1 and in particular Eq. (7.72)]

$$\hat{x}(t) = c_1 \cos\omega t + c_2 \sin\omega t,$$

which yields

$$\hat{p}_x(t) = m\omega(c_2 \cos\omega t - c_1 \sin\omega t).$$

Substituting the initial condition $\hat{x}(0) = \hat{x}$ and $\hat{p}_x(0) = \hat{p}_x$, we finally obtain

$$\hat{x}(t) = \hat{x} \cos\omega t + \frac{\hat{p}_x}{m\omega} \sin\omega t,$$

$$\hat{p}_x(t) = \hat{p}_x \cos\omega t - m\omega\hat{x} \sin\omega t.$$

7.16 In the basis

$$|h\rangle = \begin{pmatrix} 1 \\ 0 \end{pmatrix} \quad \text{and} \quad |v\rangle = \begin{pmatrix} 0 \\ 1 \end{pmatrix},$$

we have

$$\hat{\rho} = |c_h|^2 \begin{pmatrix} 1 \\ 0 \end{pmatrix} (1\ 0) + |c_v|^2 \begin{pmatrix} 0 \\ 1 \end{pmatrix} (0\ 1)$$

$$+ c_h c_v^* \begin{pmatrix} 1 \\ 0 \end{pmatrix} (0\ 1) + c_h^* c_v \begin{pmatrix} 0 \\ 1 \end{pmatrix} (1\ 0).$$

Then the desired result is easily obtained.

7.17 In both cases, $\mathrm{Tr}\,\hat{\rho}$ is a restatement of the normalization condition that the sum of all probabilities of a set of mutually exclusive events be equal to one. In the basis where the density matrix is diagonal, the corresponding diagonal elements are precisely the probabilities that the system may be found in a certain set of mutually exclusive pure states. Therefore, the sum of the diagonal elements has to be one because the trace of a matrix is invariant under a change of basis [see Eq. (7.116)].

7.18 For the mixture (7.110), we have

$$\hat{\rho}'^2 = \begin{bmatrix} |c_h|^2 & 0 \\ 0 & |c_v|^2 \end{bmatrix} \begin{bmatrix} |c_h|^2 & 0 \\ 0 & |c_v|^2 \end{bmatrix} = \begin{bmatrix} |c_h|^4 & 0 \\ 0 & |c_v|^4 \end{bmatrix} \neq \hat{\rho},$$

Instead, for the pure state (7.108) we have

$$
\begin{aligned}
\hat{\rho}^2 &= \begin{bmatrix} |c_h|^2 & c_h c_v^* \\ c_h^* c_v & |c_v|^2 \end{bmatrix} \begin{bmatrix} |c_h|^2 & c_h c_v^* \\ c_h^* c_v & |c_v|^2 \end{bmatrix} \\
&= \begin{bmatrix} |c_h|^4 + |c_h|^2 |c_v|^2 & |c_h|^2 c_h c_v^* + c_h c_v^* |c_v|^2 \\ c_h^* c_v |c_h|^2 + |c_v|^2 c_h^* c_v & |c_h|^2 |c_v|^2 + |c_v|^4 \end{bmatrix} \\
&= \begin{bmatrix} |c_h|^2 \left(|c_h|^2 + |c_v|^2\right) & c_h c_v^* \left(|c_h|^2 + |c_v|^2\right) \\ c_h^* c_v \left(|c_h|^2 + |c_v|^2\right) & |c_v|^2 \left(|c_h|^2 + |c_v|^2\right) \end{bmatrix} \\
&= \begin{bmatrix} |c_h|^2 & c_h c_v^* \\ c_h^* c_v & |c_v|^2 \end{bmatrix} = \hat{\rho},
\end{aligned}
$$

where use has been made of $|c_h|^2 + |c_v|^2 = 1$.

7.19 This is straightforward but tedious.

$$
\begin{aligned}
\hat{\rho}'_{12} &= |\Psi\rangle\langle\Psi|_{12} \\
&= \frac{1}{4} \left(|h_1 \otimes h_2\rangle + |h_1 \otimes v_2\rangle + |v_1 \otimes h_2\rangle + |v_1 \otimes v_2\rangle \right) \\
&\quad \left(\langle h_1 \otimes h_2| + \langle h_1 \otimes v_2| + \langle v_1 \otimes h_2| + \langle v_1 \otimes v_2| \right) \\
&= \frac{1}{2} \left(|h_1\rangle\langle h_1| + |h_1\rangle\langle v_1| + |v_1\rangle\langle h_1| + |v_1\rangle\langle v_1| \right) \\
&\quad \otimes \frac{1}{2} \left(|h_2\rangle\langle h_2| + |h_2\rangle\langle v_2| + |v_2\rangle\langle h_2| + |v_2\rangle\langle v_2| \right).
\end{aligned}
$$

In other words, the total density matrix $\hat{\rho}_{12}$ can be factorized into a direct product of two density matrices, one describing system 1 and the other one system 2, that is, we have

$$\hat{\rho}_{12} = \hat{\rho}_1 \otimes \hat{\rho}_2.$$

This is what we mean by separable.

7.20 Let us make us of the transformations

$$|v\rangle = \frac{1}{\sqrt{2}}(|a\rangle + |a'\rangle), \quad |h\rangle = \frac{1}{\sqrt{2}}(|a\rangle - |a'\rangle),$$

which allows us to rewrite the entangled state $|\Phi\rangle_{12}$ as [see Eq. (7.131)]

$$\begin{aligned}
|\Phi\rangle_{12} &= \frac{1}{2\sqrt{2}}\Big[\left(|a\rangle - |a'\rangle\right)_1 \otimes \left(|a\rangle - |a'\rangle\right)_2 \\
&\quad + \left(|a\rangle + |a'\rangle\right)_1 \otimes \left(|a\rangle + |a'\rangle\right)_2 \Big] \\
&= \frac{1}{2\sqrt{2}}\big(|a\rangle_1|a\rangle_2 - |a\rangle_1|a'\rangle_2 - |a'\rangle_1|a\rangle_2 + |a'\rangle_1|a'\rangle_2 \\
&\quad + |a\rangle_1|a\rangle_2 + |a\rangle_1|a'\rangle_2 + |a'\rangle_1|a\rangle_2 + |a'\rangle_1|a'\rangle_2\big) \\
&= \frac{1}{2\sqrt{2}}\big(|a\rangle_1|a\rangle_2 + |a'\rangle_1|a'\rangle_2 + |a\rangle_1|a\rangle_2 + |a'\rangle_1|a'\rangle_2\big) \\
&= \frac{1}{\sqrt{2}}\big(|a\rangle_1|a\rangle_2 + |a'\rangle_1|a'\rangle_2\big),
\end{aligned}$$

where in the last lines we have omitted the symbol $\otimes$ for the sake of simplicity. The above state is clearly entangled, since it pairs the states $|a\rangle_1$ and $|a\rangle_2$ on the one hand, and the states $|a'\rangle_1$ and $|a'\rangle_2$ on the other.

7.21 This is straightforward. We have

$$\hat{\rho}_2 = \mathrm{Tr}_1\,\hat{\rho}_{12} = \langle h|\hat{\rho}_{12}|h\rangle_1 + \langle v|\hat{\rho}_{12}|v\rangle_1 = \frac{1}{2}(|h\rangle\langle h|_2 + |v\rangle\langle v|_2),$$

which also describes a mixture.

Chapter 8

8.2 This is straightforward.

$$
\begin{aligned}
\left[\hat{L}_y, \hat{L}_z\right] &= [\hat{z}\hat{p}_x - \hat{x}\hat{p}_z, \hat{x}\hat{p}_y - \hat{y}\hat{p}_x] \\
&= [\hat{z}\hat{p}_x, \hat{x}\hat{p}_y] + [\hat{x}\hat{p}_z, \hat{y}\hat{p}_x] \\
&= \hat{z}[\hat{p}_x, \hat{x}]\hat{p}_y + \hat{y}[\hat{x}, \hat{p}_x]\hat{p}_z \\
&= \hat{y}[\hat{x}, \hat{p}_x]\hat{p}_z - \hat{z}[\hat{x}, \hat{p}_x]\hat{p}_y \\
&= i\hbar(\hat{y}\hat{p}_z - \hat{z}\hat{p}_y) \\
&= i\hbar\hat{L}_x.
\end{aligned}
$$

$$
\begin{aligned}
\left[\hat{L}_z, \hat{L}_x\right] &= [\hat{x}\hat{p}_y - \hat{y}\hat{p}_x, \hat{y}\hat{p}_z - \hat{z}\hat{p}_y] \\
&= [\hat{x}\hat{p}_y, \hat{y}\hat{p}_z] + [\hat{y}\hat{p}_x, \hat{z}\hat{p}_y] \\
&= \hat{x}[\hat{p}_y, \hat{y}]\hat{p}_z + \hat{z}[\hat{y}, \hat{p}_y]\hat{p}_x \\
&= \hat{z}[\hat{y}, \hat{p}_y]\hat{p}_x - \hat{x}[\hat{y}, \hat{p}_y]\hat{p}_z \\
&= i\hbar(\hat{z}\hat{p}_x - \hat{x}\hat{p}_z) \\
&= i\hbar\hat{L}_y.
\end{aligned}
$$

8.3 Writing $\hat{\mathbf{L}}^2 = \hat{L}_x^2 + \hat{L}_y^2 + \hat{L}_z^2$, we have $\left[\hat{L}_x, \hat{\mathbf{L}}^2\right] = \left[\hat{L}_x, \hat{L}_y^2 + \hat{L}_z^2\right]$ since $\left[\hat{L}_x, \hat{L}_x^2\right] = 0$. Moreover, we have

$$
\begin{aligned}
\left[\hat{L}_x, \hat{L}_y^2\right] &= \hat{L}_y[\hat{L}_x, \hat{L}_y] + [\hat{L}_x, \hat{L}_y]\hat{L}_y = i\hbar(\hat{L}_y\hat{L}_z + \hat{L}_z\hat{L}_y), \\
\left[\hat{L}_x, \hat{L}_z^2\right] &= \hat{L}_z[\hat{L}_x, \hat{L}_z] + [\hat{L}_x, \hat{L}_z]\hat{L}_z = -i\hbar(\hat{L}_z\hat{L}_y + \hat{L}_y\hat{L}_z).
\end{aligned}
$$

Thus, we obtain $\left[\hat{L}_x, \hat{\mathbf{L}}^2\right] = 0$. It is easy to verify that this holds true also for $\hat{L}_y$ and $\hat{L}_z$, which proves the desired result.

8.5 From $\hat{l}_+|l, m\rangle = c_{lm}^+|l, m + 1\rangle$ and Eq. (8.39), we have

$$
\begin{aligned}
\left|c_{lm}^+\right|^2 &= \langle l, m|\hat{l}_-\hat{l}_+|l, m\rangle \\
&= \langle l, m|\left(\hat{\mathbf{l}}^2 - \hat{l}_z^2 - \hbar\hat{l}_z\right)|l, m\rangle \\
&= l(l + 1) - m^2 - m \\
&= [l(l + 1) - m(m + 1)].
\end{aligned}
$$

Therefore, we find $c_{lm}^+ = \sqrt{l(l + 1) - m(m + 1)}$ up to a global phase factor, which can be absorbed into $|l, m + 1\rangle$.

8.6 The first part is straightforward.

$$[\hat{L}_z, \hat{l}_-] = \left[\hat{L}_z, \frac{1}{\hbar}\hat{L}_x - \frac{i}{\hbar}\hat{L}_y\right]$$
$$= \left[\hat{L}_z, \frac{1}{\hbar}\hat{L}_x\right] - i\left[\hat{L}_z, \frac{1}{\hbar}\hat{L}_y\right]$$
$$= i\hat{L}_y - \hat{L}_x$$
$$= -\hbar\hat{l}_-.$$

From

$$\hat{L}_z\hat{l}_-|l, m\rangle = \left(\hat{L}_z\hat{l}_- - \hat{l}_-\hat{L}_z + \hat{l}_-\hat{L}_z\right)|l, m\rangle$$
$$= \left([\hat{L}_z, \hat{l}_-] + \hat{l}_-\hat{L}_z\right)|l, m\rangle$$
$$= \hat{l}_-\hat{L}_z|l, m\rangle - \hbar\hat{l}_-|l, m\rangle$$
$$= m\hbar\hat{l}_-|l, m\rangle - \hbar\hat{l}_-|l, m\rangle$$
$$= (m - 1)\hbar\hat{l}_-|l, m\rangle,$$

it follows that the eigenstates $\hat{l}_-|l, m\rangle$ and $|l, m-1\rangle$ correspond to the same eigenvalue. Hence they have to be proportional to each other.

8.7 The first part is straightforward.

$$\hat{l}_+\hat{l}_- = \left(\hat{l}_x + i\hat{l}_y\right)\left(\hat{l}_x - i\hat{l}_y\right)$$
$$= \hat{l}_x^2 + \hat{l}_y^2 + i\left[\hat{l}_y, \hat{l}_x\right]$$
$$= \hat{l}^2 - \hat{l}_z^2 + \hat{l}_z.$$

From $\hat{l}_-|l, m\rangle = c_{lm}^-|l, m - 1\rangle$ and the above result, we have

$$\left|c_{lm}^-\right|^2 = \langle l, m|\hat{l}_+\hat{l}_-|l, m\rangle$$
$$= \langle l, m|\left(\hat{l}^2 - \hat{l}_z^2 + \hat{l}_z\right)|l, m\rangle$$
$$= l(l + 1) - m^2 + m$$
$$= [l(l + 1) - m(m - 1)].$$

Therefore, we find $c_{lm}^- = \sqrt{l(l + 1) - m(m - 1)}$ up to a global phase factor, which can be absorbed into $|l, m - 1\rangle$.

8.8 From

$$\langle 1, m|\hat{L}_z|1, m'\rangle = m\hbar\delta_{mm'},$$

we have

$$\hat{L}_z = \hbar \begin{bmatrix} 1 & 0 & 0 \\ 0 & 0 & 0 \\ 0 & 0 & -1 \end{bmatrix}.$$

To find the matrix elements of $\hat{L}_x$ and $\hat{L}_y$, we use

$$\frac{1}{\hbar}\hat{L}_x = \frac{1}{2}(\hat{l}_+ + \hat{l}_-), \quad \frac{1}{\hbar}\hat{L}_y = \frac{1}{2i}(\hat{l}_+ - \hat{l}_-).$$

From

$$\langle 1, m|\hat{l}_\pm|1, m'\rangle = \sqrt{2 - m'(m' \pm 1)}\delta_{m, m'\pm 1},$$

we have

$$\hat{l}_+ = \sqrt{2}\begin{bmatrix} 0 & 1 & 0 \\ 0 & 0 & 1 \\ 0 & 0 & 0 \end{bmatrix}, \quad \hat{l}_- = \sqrt{2}\begin{bmatrix} 0 & 0 & 0 \\ 1 & 0 & 0 \\ 0 & 1 & 0 \end{bmatrix}.$$

Hence

$$\hat{L}_x = \frac{\hbar}{\sqrt{2}}\begin{bmatrix} 0 & 1 & 0 \\ 1 & 0 & 1 \\ 0 & 1 & 0 \end{bmatrix}, \quad \hat{L}_y = \frac{\hbar}{\sqrt{2}}\begin{bmatrix} 0 & -i & 0 \\ i & 0 & -i \\ 0 & i & 0 \end{bmatrix}.$$

8.9 Taking into account that

$$\frac{\partial}{\partial\cos\theta} = -\frac{1}{\sin\theta}\frac{\partial}{\partial\theta} \quad \text{and} \quad \frac{\partial}{\partial\tan\phi} = \cos^2\phi\frac{\partial}{\partial\phi},$$

we have

$$\frac{\partial}{\partial x} = \frac{\partial r}{\partial x}\frac{\partial}{\partial r} + \frac{\partial\cos\theta}{\partial x}\frac{\partial}{\partial\cos\theta} + \frac{\partial\tan\phi}{\partial x}\frac{\partial}{\partial\tan\phi}$$

$$= \frac{x}{r}\frac{\partial}{\partial r} + \frac{xz}{r^3}\frac{1}{\sin\theta}\frac{\partial}{\partial\theta} - \frac{y}{x^2}\cos^2\phi\frac{\partial}{\partial\phi}$$

$$= \sin\theta\cos\phi\frac{\partial}{\partial r} + \frac{1}{r}\sin\theta\cos\phi\cos\theta\frac{1}{\sin\theta}\frac{\partial}{\partial\theta}$$

$$- \frac{1}{r}\frac{\sin\theta\sin\phi}{\sin^2\theta\cos^2\phi}\cos^2\phi\frac{\partial}{\partial\phi}$$

$$= \sin\theta\cos\phi\frac{\partial}{\partial r} + \frac{1}{r}\cos\phi\cos\theta\frac{\partial}{\partial\theta} - \frac{1}{r}\frac{\sin\phi}{\sin\theta}\frac{\partial}{\partial\phi}.$$

Similarly, we have

$$\frac{\partial}{\partial y} = \frac{\partial r}{\partial y}\frac{\partial}{\partial r} + \frac{\partial \cos\theta}{\partial y}\frac{\partial}{\partial \cos\theta} + \frac{\partial \tan\phi}{\partial y}\frac{\partial}{\partial \tan\phi}$$

$$= \frac{y}{r}\frac{\partial}{\partial r} + \frac{yz}{r^3}\frac{1}{\sin\theta}\frac{\partial}{\partial\theta} - \frac{1}{x}\cos^2\phi\frac{\partial}{\partial\phi}$$

$$= \sin\theta\sin\phi\frac{\partial}{\partial r} + \frac{1}{r}\sin\theta\sin\phi\cos\theta\frac{1}{\sin\theta}\frac{\partial}{\partial\theta}$$

$$+ \frac{1}{r}\frac{1}{\sin\theta\cos\phi}\cos^2\phi\frac{\partial}{\partial\phi}$$

$$= \sin\theta\sin\phi\frac{\partial}{\partial r} + \frac{1}{r}\sin\phi\cos\theta\frac{\partial}{\partial\theta} + \frac{1}{r}\frac{\cos\phi}{\sin\theta}\frac{\partial}{\partial\phi}$$

and

$$\frac{\partial}{\partial z} = \frac{\partial r}{\partial z}\frac{\partial}{\partial r} + \frac{\partial \cos\theta}{\partial z}\frac{\partial}{\partial \cos\theta} + \frac{\partial \tan\phi}{\partial z}\frac{\partial}{\partial \tan\phi}$$

$$= \frac{z}{r}\frac{\partial}{\partial r} - \left(\frac{1}{r} - \frac{z^2}{r^3}\right)\frac{1}{\sin\theta}\frac{\partial}{\partial\theta}$$

$$= \cos\theta\frac{\partial}{\partial r} - \frac{1}{r}(1 - \cos^2\theta)\frac{1}{\sin\theta}\frac{\partial}{\partial\theta}$$

$$= \cos\theta\frac{\partial}{\partial r} - \frac{1}{r}\sin\theta\frac{\partial}{\partial\theta}.$$

8.10 From Eqs. (8.56), we can express the z component of the orbital angular momentum in spherical coordinates as

$$\hat{L}_z = -i\hbar\left(x\frac{\partial}{\partial y} - y\frac{\partial}{\partial x}\right)$$

$$= -i\hbar\left[r\sin\theta\cos\phi\left(\sin\theta\sin\phi\frac{\partial}{\partial r} + \frac{\cos\theta\sin\phi}{r}\frac{\partial}{\partial\theta} + \frac{\cos\phi}{r\sin\theta}\frac{\partial}{\partial\phi}\right)\right.$$

$$\left. - r\sin\theta\sin\phi\left(\sin\theta\cos\phi\frac{\partial}{\partial r} + \frac{\cos\theta\cos\phi}{r}\frac{\partial}{\partial\theta} - \frac{\sin\phi}{r\sin\theta}\frac{\partial}{\partial\phi}\right)\right]$$

$$= -i\hbar\left(\cos^2\phi + \sin^2\phi\right)\frac{\partial}{\partial\phi}$$

$$= -i\hbar\frac{\partial}{\partial\phi},$$

In other words, $\hat{L}_z$ generates rotations about the z axis, i.e., translations in ϕ. Similarly, for the x component of the orbital angular momentum we have

$$\hat{L}_x = -i\hbar \left(y\frac{\partial}{\partial z} - z\frac{\partial}{\partial y} \right)$$

$$= -i\hbar \left[r\sin\theta\sin\phi \left(\cos\theta\frac{\partial}{\partial r} - \frac{\sin\theta}{r}\frac{\partial}{\partial\theta} \right) \right.$$

$$\left. -r\cos\theta \left(\sin\theta\sin\phi\frac{\partial}{\partial r} + \frac{\cos\theta\sin\phi}{r}\frac{\partial}{\partial\theta} + \frac{\cos\phi}{r\sin\theta}\frac{\partial}{\partial\phi} \right) \right]$$

$$= -i\hbar \left[-(\sin^2\theta + \cos^2\theta)\sin\phi\frac{\partial}{\partial\theta} - \cos\phi\frac{\cos\theta}{\sin\theta}\frac{\partial}{\partial\phi} \right]$$

$$= i\hbar \left(\sin\phi\frac{\partial}{\partial\theta} + \cos\phi\cot\theta\frac{\partial}{\partial\phi} \right),$$

and for the *y* component we have

$$\hat{L}_y = -i\hbar \left(z\frac{\partial}{\partial x} - x\frac{\partial}{\partial z} \right)$$

$$= -i\hbar \left[r\cos\theta \left(\sin\theta\cos\phi\frac{\partial}{\partial r} + \frac{\cos\theta\cos\phi}{r}\frac{\partial}{\partial\theta} - \frac{\sin\phi}{r\sin\theta}\frac{\partial}{\partial\phi} \right) \right.$$

$$\left. -r\sin\theta\cos\phi \left(\cos\theta\frac{\partial}{\partial r} - \frac{\sin\theta}{r}\frac{\partial}{\partial\theta} \right) \right]$$

$$= -i\hbar \left[(\cos^2\theta + \sin^2\theta)\cos\phi\frac{\partial}{\partial\theta} - \sin\phi\frac{\cos\theta}{\sin\theta}\frac{\partial}{\partial\phi} \right]$$

$$= -i\hbar \left(\cos\phi\frac{\partial}{\partial\theta} - \sin\phi\cot\theta\frac{\partial}{\partial\phi} \right).$$

Hence, we obtain

$$\hat{L}_x^2 = -\hbar^2 \left(\sin\phi\frac{\partial}{\partial\theta} + \cos\phi\cot\theta\frac{\partial}{\partial\phi} \right)^2,$$

$$\hat{L}_y^2 = -\hbar^2 \left(\cos\phi\frac{\partial}{\partial\theta} - \sin\phi\cot\theta\frac{\partial}{\partial\phi} \right)^2,$$

$$\hat{L}_z^2 = -\hbar^2\frac{\partial^2}{\partial\phi^2}.$$

Collecting the terms, we find after some straightforward but tedious algebra

$$\hat{\mathbf{L}}^2 = \hat{L}_x^2 + \hat{L}_y^2 + \hat{L}_z^2 = -\hbar^2 \left[\frac{1}{\sin\theta}\frac{\partial}{\partial\theta} \left(\sin\theta\frac{\partial}{\partial\theta} \right) + \frac{1}{\sin^2\theta}\frac{\partial^2}{\partial\phi^2} \right].$$

8.11 Using Eqs. (8.63), (8.64), and the first few Legendre polynomials

$$P_0(x) = 1, \quad P_1(x) = x, \quad P_2(x) = \frac{1}{2}(3x^2 - 1),$$

we obtain the desired results after some straightforward algebra.

8.12 From Eqs. (8.56), we have

$$\frac{\partial^2}{\partial x^2} = \left(\sin\theta \cos\phi \frac{\partial}{\partial r} + \frac{\cos\theta \cos\phi}{r} \frac{\partial}{\partial\theta} - \frac{\sin\phi}{r\sin\theta} \frac{\partial}{\partial\phi} \right)^2,$$

$$\frac{\partial^2}{\partial y^2} = \left(\sin\theta \sin\phi \frac{\partial}{\partial r} + \frac{\cos\theta \sin\phi}{r} \frac{\partial}{\partial\theta} + \frac{\cos\phi}{r\sin\theta} \frac{\partial}{\partial\phi} \right)^2,$$

$$\frac{\partial^2}{\partial z^2} = \left(\cos\theta \frac{\partial}{\partial r} - \frac{\sin\theta}{r} \frac{\partial}{\partial\theta} \right)^2.$$

Collecting the terms, we find after some straightforward but tedious algebra

$$\nabla^2 = \frac{\partial^2}{\partial x^2} + \frac{\partial^2}{\partial y^2} + \frac{\partial^2}{\partial z^2}$$

$$= \frac{1}{r^2} \left[\frac{\partial}{\partial r} \left(r^2 \frac{\partial}{\partial r} \right) + \frac{1}{\sin\theta} \frac{\partial}{\partial\theta} \left(\sin\theta \frac{\partial}{\partial\theta} \right) + \frac{1}{\sin^2\theta} \frac{\partial^2}{\partial\phi^2} \right].$$

8.13 Using Eqs. (8.50) and (8.56), after some straightforward but tedious algebra we can write the gradient operator in spherical coordinates as [see Eq. (6.97)]

$$\nabla = \mathbf{e}_x \frac{\partial}{\partial x} + \mathbf{e}_y \frac{\partial}{\partial y} + \mathbf{e}_z \frac{\partial}{\partial z}$$

$$= \mathbf{e}_r \frac{\partial}{\partial r} + \mathbf{e}_\theta \frac{1}{r} \frac{\partial}{\partial\theta} + \mathbf{e}_\phi \frac{1}{r\sin\theta} \frac{\partial}{\partial\phi},$$

where $\mathbf{e}_r$, $\mathbf{e}_\theta$, and $\mathbf{e}_\phi$ are the spherical unit basis vectors. Therefore, the orbital angular momentum $\hat{\mathbf{L}} = \hat{\mathbf{r}} \times \hat{\mathbf{p}} = -i\hbar\hat{\mathbf{r}} \times \nabla$ in spherical coordinates is given by

$$\hat{\mathbf{L}} = -i\hbar r \mathbf{e}_r \times \left(\mathbf{e}_r \frac{\partial}{\partial r} + \mathbf{e}_\theta \frac{1}{r} \frac{\partial}{\partial\theta} + \mathbf{e}_\phi \frac{1}{r\sin\theta} \frac{\partial}{\partial\phi} \right)$$

$$= i\hbar \left(\mathbf{e}_\theta \frac{1}{\sin\theta} \frac{\partial}{\partial\phi} - \mathbf{e}_\phi \frac{\partial}{\partial\theta} \right),$$

where use has been made of the properties $\mathbf{e}_r \times \mathbf{e}_r = 0$, $\mathbf{e}_r \times \mathbf{e}_\theta = \mathbf{e}_\phi$, and $\mathbf{e}_r \times \mathbf{e}_\phi = -\mathbf{e}_\theta$ [see Eq. (8.52)]. In other words, we have

$$\hat{L}_r = 0, \quad \hat{L}_\theta = i\hbar \frac{1}{\sin\theta}\frac{\partial}{\partial\phi}, \quad \hat{L}_\phi = -i\hbar\frac{\partial}{\partial\theta}.$$

8.14 In order to derive Eq. (8.86), we first write Eq. (8.83) in terms of the constant n as

$$\frac{d^2}{d\tilde{r}^2}\xi(\tilde{r}) + \left[\frac{1}{n^2} - \frac{l(l+1)}{\tilde{r}^2} + \frac{2}{\tilde{r}}\right]\xi(\tilde{r}) = 0.$$

Using the assumption (8.85), we then express the derivatives of $\xi(\tilde{r})$ in terms of those of $W(\tilde{r})$ as

$$\frac{d}{d\tilde{r}}\xi(\tilde{r}) = \left(\tilde{r}^{l+1}e^{-\frac{\tilde{r}}{n}}W\right)' = (l+1)\tilde{r}^l e^{-\frac{\tilde{r}}{n}}W - \frac{1}{n}e^{-\frac{\tilde{r}}{n}}\tilde{r}^{l+1}W + \tilde{r}^{l+1}e^{-\frac{\tilde{r}}{n}}W'$$

$$= \left(\frac{l+1}{\tilde{r}} - \frac{1}{n} + \frac{W'}{W}\right)\tilde{r}^{l+1}e^{-\frac{\tilde{r}}{n}}W = \left(\frac{l+1}{\tilde{r}} - \frac{1}{n} + \frac{W'}{W}\right)\xi,$$

$$\frac{d^2}{d\tilde{r}^2}\xi(\tilde{r}) = \left(\frac{l+1}{\tilde{r}} - \frac{1}{n} + \frac{W'}{W}\right)'\xi + \left(\frac{l+1}{\tilde{r}} - \frac{1}{n} + \frac{W'}{W}\right)\xi'$$

$$= \left[-\frac{l+1}{\tilde{r}^2} + \frac{W''}{W} - \frac{(W')^2}{W^2} + \left(\frac{l+1}{\tilde{r}} - \frac{1}{n} + \frac{W'}{W}\right)^2\right]\xi,$$

where a prime denotes the derivative with respective to $\tilde{r}$ and the dependence on $\tilde{r}$ of W and ξ has been suppressed. Now substituting the latter equations into the first one and making use of the general mathematical formula $(a + b + c)^2 = a^2 + b^2 + c^2 + 2(ab + ac + bc)$, which is true for arbitrary a, b, and c, we obtain

$$\left[\frac{W''}{W} + \left(\frac{2(l+1)}{\tilde{r}} - \frac{2}{n}\right)\frac{W'}{W} + \frac{2}{\tilde{r}}\left(1 - \frac{l+1}{n}\right)\right]\xi = 0,$$

which upon multiplying by $\tilde{r}W$ yields

$$\left[\tilde{r}W''(\tilde{r}) + 2\left(l + 1 - \frac{\tilde{r}}{n}\right)W'(\tilde{r}) + \frac{2(n - l - 1)}{n}W(\tilde{r})\right]\xi(\tilde{r}) = 0,$$

where the dependence on $\tilde{r}$ of W and ξ has been restored. Since $\xi(\tilde{r}) \neq 0$, dividing both sides of the above equation by $\xi(\tilde{r})$ we finally obtain Eq. (8.86).

8.15 This is straightforward.

$$\sum_{l=0}^{n-1}(2l + 1) = 2\sum_{l=0}^{n-1}l + n = 2\frac{n(n-1)}{2} + n = n^2.$$

8.16 From the recurrence relation (8.93), for $n = 2$ and $l = 0$ we have

$$c_1 = \frac{1-2}{2}c_0 = -\frac{1}{2}c_0, \quad c_2 = \frac{2-2}{6}c_1 = 0,$$

with all other $c_j = 0$ for $j > 2$. This means

$$W(\eta) = c_0 \left(1 - \frac{1}{2}\eta \right) = c_0 \left(1 - \frac{\tilde{r}}{2} \right).$$

By taking into account Eqs. (8.85) and (8.87), we have

$$\xi(\tilde{r}) = c_0 \tilde{r} e^{-\frac{\tilde{r}}{2}} \left(1 - \frac{\tilde{r}}{2} \right),$$

which finally gives [see Eqs. (8.73) and (8.81)]

$$f_{20} = \mathcal{N}_{20} \left(1 - \frac{r}{2a_0} \right) e^{-\frac{r}{2a_0}},$$

where $\mathcal{N}_{20}$ is the normalization constant.
For $n = 2$ and $l = 1$, the only non-zero coefficient is c_0. This means $W(\eta) = c_0$ and

$$\xi(\tilde{r}) = c_0 \tilde{r}^2 e^{-\frac{\tilde{r}}{2}}.$$

Hence, we have

$$f_{21} = \mathcal{N}_{21} \left(\frac{r}{a_0} \right) e^{-\frac{r}{2a_0}},$$

where $\mathcal{N}_{21}$ is the normalization constant.

8.17 The states $|\uparrow_z\rangle$ and $|\downarrow_z\rangle$ are eigenstates of $\hat{S}^2$ with eigenvalue $\frac{3\hbar^2}{4}$, hence $\hat{S}^2$ is diagonal in the basis $\{|\uparrow_z\rangle, |\downarrow_z\rangle\}$ and proportional to the identity matrix with the proportional constant being $\frac{3\hbar^2}{4}$. The same result can be obtained from $\langle \frac{1}{2}, m|\hat{S}^2|\frac{1}{2}, m'\rangle = \frac{3\hbar^2}{4}\delta_{mm'}$.

8.18 This is straightforward.

$$\left[\hat{\sigma}_x, \hat{\sigma}_y\right] = \begin{bmatrix} 0 & 1 \\ 1 & 0 \end{bmatrix}\begin{bmatrix} 0 & -i \\ i & 0 \end{bmatrix} - \begin{bmatrix} 0 & -i \\ i & 0 \end{bmatrix}\begin{bmatrix} 0 & 1 \\ 1 & 0 \end{bmatrix}$$

$$= \begin{bmatrix} i & 0 \\ 0 & -i \end{bmatrix} - \begin{bmatrix} -i & 0 \\ 0 & i \end{bmatrix} = 2i\begin{bmatrix} 1 & 0 \\ 0 & -1 \end{bmatrix} = 2i\hat{\sigma}_z,$$

$$\left[\hat{\sigma}_z, \hat{\sigma}_x\right] = \begin{bmatrix} 1 & 0 \\ 0 & -1 \end{bmatrix}\begin{bmatrix} 0 & 1 \\ 1 & 0 \end{bmatrix} - \begin{bmatrix} 0 & 1 \\ 1 & 0 \end{bmatrix}\begin{bmatrix} 1 & 0 \\ 0 & -1 \end{bmatrix}$$

$$= \begin{bmatrix} 0 & 1 \\ -1 & 0 \end{bmatrix} - \begin{bmatrix} 0 & -1 \\ 1 & 0 \end{bmatrix} = 2i\begin{bmatrix} 0 & -i \\ i & 0 \end{bmatrix} = 2i\hat{\sigma}_y,$$

$$\left[\hat{\sigma}_y, \hat{\sigma}_z\right] = \begin{bmatrix} 0 & -i \\ i & 0 \end{bmatrix}\begin{bmatrix} 1 & 0 \\ 0 & -1 \end{bmatrix} - \begin{bmatrix} 1 & 0 \\ 0 & -1 \end{bmatrix}\begin{bmatrix} 0 & -i \\ i & 0 \end{bmatrix}$$

$$= \begin{bmatrix} 0 & i \\ i & 0 \end{bmatrix} - \begin{bmatrix} 0 & -i \\ -i & 0 \end{bmatrix} = 2i\begin{bmatrix} 0 & 1 \\ 1 & 0 \end{bmatrix} = 2i\hat{\sigma}_x.$$

8.19 It suffices to show that

$$\begin{aligned}
\left[\hat{J}_x, \hat{J}_y\right] &= \left[\hat{J}_{1x} + \hat{J}_{2x}, \hat{J}_{1y} + \hat{J}_{2y}\right] \\
&= \left[\hat{J}_{1x}, \hat{J}_{1y}\right] + \left[\hat{J}_{2x} + \hat{J}_{2y}\right] \\
&= i\hbar(\hat{J}_{1z} + \hat{J}_{2z}) \\
&= i\hbar\hat{J}_z.
\end{aligned}$$

Commutation relations for other components can be obtained in a similar manner.

8.20 (a) The particles has a probability of $\frac{1}{2}$ in both states $|\uparrow_z\rangle_1 \otimes |\downarrow_z\rangle_2$ and $|\downarrow_z\rangle_1 \otimes |\uparrow_z\rangle_2$, hence the probability for Alice to obtain $S_{1z} = \frac{\hbar}{2}$ is $\frac{1}{2}$.

(b) If particle 1 is in the state $|\uparrow_z\rangle$, which has a probability of $\frac{1}{2}$, the probability for Alice to obtain $S_{1x} = \frac{\hbar}{2}$ is $|\langle\uparrow_x|\uparrow_z\rangle|^2 = \frac{1}{2}$. If it is in the state $|\downarrow_z\rangle$, which also has a probability of $\frac{1}{2}$, the probability for Alice to obtain $S_{1x} = \frac{\hbar}{2}$ is $|\langle\uparrow_x|\downarrow_z\rangle|^2 = \frac{1}{2}$. Hence the probability for Alice to obtain $S_{1x} = \frac{\hbar}{2}$ is $\frac{1}{2} \times \frac{1}{2} + \frac{1}{2} \times \frac{1}{2} = \frac{1}{2}$.

(c) After Bob's measurement, the particles are in the state $|\downarrow_z\rangle_1 \otimes |\uparrow_z\rangle_2$. Hence the outcome of Alice's measurement is $S_{1z} = -\frac{\hbar}{2}$ with a probability 1.

(d) The outcome of Alice's measurement is $S_{1x} = \frac{\hbar}{2}$ with a probability $|\langle \uparrow_x | \downarrow_z \rangle|^2 = \frac{1}{2}$ and $S_{1x} = -\frac{\hbar}{2}$ with a probability $|\langle \downarrow_x | \downarrow_z \rangle|^2 = \frac{1}{2}$.

Chapter 9

9.1 The fact that the first row of the unitary operator $\hat{U}_{BS}$ has elements of the same sign while the second row has elements of opposite sign means that the lower component of $|\psi\rangle$ is preserved while the upper component is annihilated.

9.2 This is straightforward.

$$\hat{U}'_{BS} |\psi\rangle = \frac{1}{2} \begin{bmatrix} 1 & -1 \\ 1 & 1 \end{bmatrix} \begin{pmatrix} 1 \\ 1 \end{pmatrix} = \frac{1}{2} \begin{pmatrix} 0 \\ 2 \end{pmatrix} = |u\rangle.$$

9.3 The conjugate transpose of the operator $\hat{U}'_{BS}$ is

$$\hat{U}'^{\dagger}_{BS} = \frac{1}{\sqrt{2}} \begin{bmatrix} 1 & 1 \\ -1 & 1 \end{bmatrix},$$

and we have

$$\hat{U}'_{BS} \hat{U}'^{\dagger}_{BS} = \frac{1}{2} \begin{bmatrix} 1 & -1 \\ 1 & 1 \end{bmatrix} \begin{bmatrix} 1 & 1 \\ -1 & 1 \end{bmatrix} = \frac{1}{2} \begin{bmatrix} 2 & 0 \\ 0 & 2 \end{bmatrix} = \hat{I},$$

$$\hat{U}'^{\dagger}_{BS} \hat{U}'_{BS} = \frac{1}{2} \begin{bmatrix} 1 & 1 \\ -1 & 1 \end{bmatrix} \begin{bmatrix} 1 & -1 \\ 1 & 1 \end{bmatrix} = \frac{1}{2} \begin{bmatrix} 2 & 0 \\ 0 & 2 \end{bmatrix} = \hat{I}.$$

9.4 Because it cannot be expressed in terms of a single projector but as a combination of projectors.

9.5 Since we have

$$|o'_k\rangle = \sum_j c_j \langle a'_k | a_j \rangle | o_j \rangle,$$

$$|a'_k\rangle = \sum_j |a_j\rangle \langle a_j | a'_k \rangle,$$

we can write

$$\sum_k |o_k'\rangle|a_k'\rangle = \sum_k \sum_j c_j \langle a_j|a_k'\rangle \langle a_k'|a_j\rangle |o_j\rangle|a_j\rangle$$

$$= \sum_j c_j \langle a_j| \left(\sum_k |a_k'\rangle \langle a_k'| \right) |a_j\rangle|o_j\rangle|a_j\rangle$$

$$= \sum_j c_j \langle a_j|a_j\rangle |o_j\rangle|a_j\rangle$$

$$= \sum_j c_j |o_j\rangle|a_j\rangle.$$

9.6 From the property (9.33) it is conceivable that (based on continuity arguments) we may define $0 \log_b 0 = 0$ for an arbitrary base $b > 0$, then the conclusion follows.

9.7 From Eq. (9.34), the Shannon entropy of the source is given by

$$H = -x \lg x - (1-x) \lg(1-x).$$

The plot of H as a function of x for $0 \le x \le 1$ is depicted in the figure below, which shows clearly that H is maximum at $x = \frac{1}{2}$.

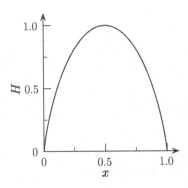

9.8 This is straightforward as the following truth table of $\neg(p \wedge q)$ is identical to that of $\neg p \vee \neg q$ in Table 9.5.

9.9 The rows are the first, third and fifth ones. The relative equation is

$$(r = 0) \wedge \neg(p \wedge q) \quad \text{or} \quad (r = 0) \wedge (\neg p \vee \neg q).$$

Input		Output
p	q	$\neg(p \wedge q)$
1	1	0
1	0	1
0	1	1
0	0	1

Chapter 10

10.2 Using the actions of $\hat{\sigma}_y$ on $|\uparrow_z\rangle$ and $|\downarrow_z\rangle$ tabulated in Table 8.1, we obtain

$$(\hat{\sigma}_{1y} \otimes \hat{\sigma}_{2y})|\Psi_0\rangle = \frac{1}{\sqrt{2}}(\hat{\sigma}_{1y}|\uparrow_z\rangle_1 \otimes \hat{\sigma}_{2y}|\downarrow_z\rangle_2$$
$$- \hat{\sigma}_{1y}|\downarrow_z\rangle_1 \otimes \hat{\sigma}_{2y}|\uparrow_z\rangle_2)$$
$$= \frac{1}{\sqrt{2}}[i|\downarrow_z\rangle_1 \otimes (-i)|\uparrow_z\rangle_2$$
$$- (-i)|\uparrow_z\rangle_1 \otimes i|\downarrow_z\rangle_2]$$
$$= -|\Psi_0\rangle.$$

10.3 Making use of the inverse of Eq. (8.114b),

$$|\uparrow_z\rangle = \frac{1}{\sqrt{2}}(|\uparrow_y\rangle + |\downarrow_y\rangle), \quad |\downarrow_z\rangle = -\frac{i}{\sqrt{2}}(|\uparrow_y\rangle - |\downarrow_y\rangle),$$

we can rewrite $|\Psi_0\rangle$ in terms of $|\uparrow_y\rangle$ and $|\downarrow_y\rangle$ as

$$|\Psi_0\rangle = \frac{1}{\sqrt{2}}(|\uparrow_z\rangle_1 \otimes |\downarrow_z\rangle_2 - |\downarrow_z\rangle_1 \otimes |\uparrow_z\rangle_2)$$
$$= -\frac{i}{2\sqrt{2}}[(|\uparrow_y\rangle_1 + |\downarrow_y\rangle_1) \otimes (|\uparrow_y\rangle_2 - |\downarrow_y\rangle_2)$$
$$- (|\uparrow_y\rangle_1 - |\downarrow_y\rangle_1) \otimes (|\uparrow_y\rangle_2 + |\downarrow_y\rangle_2)]$$
$$= \frac{i}{\sqrt{2}}(|\uparrow_y\rangle_1 \otimes |\downarrow_y\rangle_2 - |\downarrow_y\rangle_1 \otimes |\uparrow_y\rangle_2),$$

which, apart from the global phase factor i, is a singlet state but with the projection of the spin along the y direction.

10.4 This is straightforward.

$$\langle\Psi_0|\hat{\sigma}_{1y}\hat{\sigma}_{2x}|\Psi_0\rangle = \langle\Psi_0|\hat{\sigma}_{1y}\hat{\sigma}_{2x}\left(|\uparrow_z\rangle_1|\downarrow_z\rangle_2 - |\downarrow_z\rangle_1|\uparrow_z\rangle_2\right)$$
$$= \langle\Psi_0|\left[i|\downarrow_z\rangle_1|\uparrow_z\rangle_2 - (-i)|\uparrow_z\rangle_1|\downarrow_z\rangle_2\right]$$
$$= i(\langle\uparrow_z|_1\langle\downarrow_z|_2 - \langle\downarrow_z|_1\langle\uparrow_z|_2)$$
$$\times\left(|\uparrow_z\rangle_1|\downarrow_z\rangle_2 + |\downarrow_z\rangle_1|\uparrow_z\rangle_2\right)$$
$$= 0,$$

$$\langle\Psi_0|\hat{\sigma}_{1y}\hat{\sigma}_{2z}|\Psi_0\rangle = \langle\Psi_0|\hat{\sigma}_{1y}\hat{\sigma}_{2z}\left(|\uparrow_z\rangle_1|\downarrow_z\rangle_2 - |\downarrow_z\rangle_1|\uparrow_z\rangle_2\right)$$
$$= \langle\Psi_0|\left[-i|\downarrow_z\rangle_1|\downarrow_z\rangle_2 - (-i)|\uparrow_z\rangle_1|\uparrow_z\rangle_2\right]$$
$$= i(\langle\uparrow_z|_1\langle\downarrow_z|_2 - \langle\downarrow_z|_1\langle\uparrow_z|_2)$$
$$\times\left(|\uparrow_z\rangle_1|\uparrow_z\rangle_2 - |\downarrow_z\rangle_1|\downarrow_z\rangle_2\right)$$
$$= 0,$$

$$\langle\Psi_0|\hat{\sigma}_{1z}\hat{\sigma}_{2x}|\Psi_0\rangle = \langle\Psi_0|\hat{\sigma}_{1z}\hat{\sigma}_{2x}\left(|\uparrow_z\rangle_1|\downarrow_z\rangle_2 - |\downarrow_z\rangle_1|\uparrow_z\rangle_2\right)$$
$$= \langle\Psi_0|\left[|\uparrow_z\rangle_1|\uparrow_z\rangle_2 - (-1)|\downarrow_z\rangle_1|\downarrow_z\rangle_2\right]$$
$$= (\langle\uparrow_z|_1\langle\downarrow_z|_2 - \langle\downarrow_z|_1\langle\uparrow_z|_2)$$
$$\times\left(|\uparrow_z\rangle_1|\uparrow_z\rangle_2 + |\downarrow_z\rangle_1|\downarrow_z\rangle_2\right)$$
$$= 0,$$

$$\langle\Psi_0|\hat{\sigma}_{1z}\hat{\sigma}_{2y}|\Psi_0\rangle = \langle\Psi_0|\hat{\sigma}_{1z}\hat{\sigma}_{2y}\left(|\uparrow_z\rangle_1|\downarrow_z\rangle_2 - |\downarrow_z\rangle_1|\uparrow_z\rangle_2\right)$$
$$= \langle\Psi_0|\left[-i|\uparrow_z\rangle_1|\uparrow_z\rangle_2 - (-i)|\downarrow_z\rangle_1|\downarrow_z\rangle_2\right]$$
$$= -i(\langle\uparrow_z|_1\langle\downarrow_z|_2 - \langle\downarrow_z|_1\langle\uparrow_z|_2)$$
$$\times\left(|\uparrow_z\rangle_1|\uparrow_z\rangle_2 - |\downarrow_z\rangle_1|\downarrow_z\rangle_2\right)$$
$$= 0,$$

where use has been made of Table 8.1 and Eq. (7.125).

10.5 For coplanar unit vectors **a**, **b**, and **c**, another counterexample could be the case that both **b** and **c** make an angle of $\frac{\pi}{4}$ with **a** (but in the opposite sense) and the angle between **b** and **c** is $\frac{\pi}{2}$. From Eq. (10.26) we have

$$\langle\mathbf{a},\mathbf{b}\rangle_{\Psi_0} = -\mathbf{a}\cdot\mathbf{b} = -\cos\frac{\pi}{4} = -\frac{1}{\sqrt{2}},$$

$$\langle\mathbf{a},\mathbf{c}\rangle_{\Psi_0} = -\mathbf{a}\cdot\mathbf{c} = -\cos\frac{\pi}{4} = -\frac{1}{\sqrt{2}},$$

$$\langle\mathbf{b},\mathbf{c}\rangle_{\Psi_0} = -\mathbf{b}\cdot\mathbf{c} = -\cos\frac{\pi}{2} = 0.$$

Therefore, we obtain

$$\left|\langle \mathbf{a}, \mathbf{b}\rangle_{\Psi_0} + \langle \mathbf{a}, \mathbf{c}\rangle_{\Psi_0}\right| + \langle \mathbf{b}, \mathbf{c}\rangle_{\Psi_0} = \sqrt{2} > 1,$$

which obviously violates the Bell inequality (10.35).

10.6 A straightforward but tedious algebra leads to

$$
\begin{aligned}
|\Psi\rangle &= \frac{1}{2}\Big(|\Psi^+\rangle_{14}|\Psi^+\rangle_{23} - |\Psi^-\rangle_{14}|\Psi^-\rangle_{23} - |\Phi^+\rangle_{14}|\Phi^+\rangle_{23} \\
&\quad + |\Phi^-\rangle_{14}|\Phi^-\rangle_{23}\Big) \\
&= \frac{1}{4}[(|h_1 v_4\rangle + |v_1 h_4\rangle)(|h_2 v_3\rangle + |v_2 h_3\rangle) - (|h_1 v_4\rangle - |v_1 h_4\rangle) \\
&\quad \times (|h_2 v_3\rangle - |v_2 h_3\rangle) \\
&\quad - (|h_1 h_4\rangle + |v_1 v_4\rangle)(|h_2 h_3\rangle + |v_2 v_3\rangle) + (|h_1 h_4\rangle - |v_1 v_4\rangle) \\
&\quad \times (|h_2 h_3\rangle - |v_2 v_3\rangle)] \\
&= \frac{1}{2}(|h_1 v_2\rangle|h_3 v_4\rangle - |h_1 v_2\rangle|v_3 h_4\rangle - |v_1 h_2\rangle|h_3 v_4\rangle \\
&\quad + |v_1 h_2\rangle|v_3 h_4\rangle) \\
&= \frac{1}{2}(|h_1 v_2\rangle - |v_1 h_2\rangle)(|h_3 v_4\rangle - |v_3 h_4\rangle) \\
&= |\psi\rangle_{12} \otimes |\psi\rangle_{34}.
\end{aligned}
$$

where in the intermediate steps we have used the condensed notation for product states, i.e., $|h_1 v_2\rangle = |h\rangle_1 |v\rangle_2$, $|v_3 h_4\rangle = |v\rangle_3 |h\rangle_4$, etc.

10.7 We have

$$
\begin{aligned}
\wp(o_a | o_b, \mathbf{a}, \mathbf{b}) &= \mathrm{Tr}\left(\hat{P}_{o_a, \mathbf{a}}\, \hat{\rho}''\right) \\
&= \frac{\mathrm{Tr}\left(\hat{P}_{o_a, \mathbf{a}} \hat{P}_{o_b, \mathbf{b}}\, \hat{\rho}\, \hat{P}_{o_b, \mathbf{b}}\right)}{\wp(o_b|\mathbf{b})},
\end{aligned}
$$

where

$$\hat{\rho}'' = \frac{\hat{P}_{o_b, \mathbf{b}}\, \hat{\rho}\, \hat{P}_{o_b, \mathbf{b}}}{\wp(o_b|\mathbf{b})} \quad \text{and} \quad \hat{P}_{o_a, \mathbf{a}} = |o_a, \mathbf{a}\rangle\langle o_a, \mathbf{a}|.$$

Then, it follows that

$$
\begin{aligned}
\wp(o_a, o_b | \mathbf{a}, \mathbf{b}) &= \mathrm{Tr}\left(\hat{P}_{o_a, \mathbf{a}} \hat{P}_{o_b, \mathbf{b}}\, \hat{\rho}\, \hat{P}_{o_b, \mathbf{b}}\right) \\
&= \mathrm{Tr}\left(\hat{P}_{o_a, \mathbf{a}} \hat{P}_{o_b, \mathbf{b}}\, \hat{\rho}\right).
\end{aligned}
$$

This allows us to finally obtain

$$\wp(o_a|\mathbf{a}) = \sum_{o_b} \wp(o_a, o_b|\mathbf{a}, \mathbf{b})$$

$$= \sum_{o_b} \text{Tr}\left(\hat{P}_{o_a,\mathbf{a}} \hat{P}_{o_b,\mathbf{b}} \hat{\rho}\right)$$

$$= \text{Tr}\left[\hat{P}_{o_a,\mathbf{a}}\left(\sum_{o_b} \hat{P}_{o_b,\mathbf{b}}\right)\hat{\rho}\right]$$

$$= \text{Tr}\left(\hat{P}_{o_a,\mathbf{a}} \hat{\rho}\right),$$

where use has been made of the property that $\sum_{o_b} \hat{P}_{o_b,\mathbf{b}} = \hat{I}$ for any complete set of orthogonal projectors [see Sec. 3.7].

Chapter 11

11.1 Since two antipodal points on the Bloch sphere have angular coordinates (θ, ϕ) and $(\pi - \theta, \phi + \pi)$, the corresponding two state vectors are given by [see Eq. (11.2)]

$$|\psi_1\rangle = \cos\frac{\theta}{2}|0\rangle + e^{i\phi}\sin\frac{\theta}{2}|1\rangle,$$

$$|\psi_2\rangle = \cos\frac{\pi-\theta}{2}|0\rangle + e^{i(\pi+\phi)}\sin\frac{\pi-\theta}{2}|1\rangle$$

$$= \sin\frac{\theta}{2}|0\rangle - e^{i\phi}\cos\frac{\theta}{2}|1\rangle,$$

where use has been made of Eqs. (2.19) and (2.22). Hence, the states $|\psi_1\rangle$ and $|\psi_2\rangle$ are mutually orthogonal as $\langle\psi_1|\psi_2\rangle = 0$.

11.2 This is straightforward.

$$|1\rangle \xrightarrow{\hat{U}_H} \frac{1}{\sqrt{2}}(|0\rangle - |1\rangle)$$

$$\xrightarrow{\hat{U}_\phi} \frac{1}{\sqrt{2}}(|0\rangle - e^{i\phi}|1\rangle)$$

$$\xrightarrow{\hat{U}_H} \frac{1}{2}[(1 - e^{i\phi})|0\rangle + (1 + e^{i\phi})|1\rangle],$$

where use has been made of Eqs. (11.7) and (11.12).

11.3 The action of the CZ gate on the computational basis states is given by

$$\hat{U}_{CZ}|00\rangle = |00\rangle, \quad \hat{U}_{CZ}|01\rangle = |01\rangle,$$
$$\hat{U}_{CZ}|10\rangle = |10\rangle, \quad \hat{U}_{CZ}|11\rangle = -|11\rangle,$$

where where the first qubit is the control qubit and the second the target qubit, and used has been made of Table 8.1. Hence we have

$$\hat{U}_{CZ} = \begin{bmatrix} 1 & 0 & 0 & 0 \\ 0 & 1 & 0 & 0 \\ 0 & 0 & 1 & 0 \\ 0 & 0 & 0 & -1 \end{bmatrix},$$

which is a unitary matrix.

11.6 Let $|\psi\rangle = a|0\rangle + b|1\rangle$ and $|\phi\rangle = c|0\rangle + d|1\rangle$, then we have

$$|\psi\rangle|\phi\rangle = ac|00\rangle + ad|01\rangle + bc|10\rangle + bd|11\rangle,$$
$$|\phi\rangle|\psi\rangle = ac|00\rangle + bc|01\rangle + ad|10\rangle + bd|11\rangle.$$

Therefore, the action of the swap gate on the computational basis states is given by

$$\hat{U}_{swap}|00\rangle = |00\rangle, \quad \hat{U}_{swap}|01\rangle = |10\rangle,$$
$$\hat{U}_{swap}|10\rangle = |01\rangle, \quad \hat{U}_{swap}|11\rangle = |11\rangle,$$

which implies

$$\hat{U}_{swap} = \begin{bmatrix} 1 & 0 & 0 & 0 \\ 0 & 0 & 1 & 0 \\ 0 & 1 & 0 & 0 \\ 0 & 0 & 0 & 1 \end{bmatrix}.$$

11.7 The state $|\Psi\rangle_{123}$ can be expanded in the Bell basis as

$$|\Psi\rangle = c_-|\Psi^-\rangle + c_+|\Psi^+\rangle + d_-|\Phi^-\rangle + d_+|\Phi^+\rangle,$$

where the subscripts have been suppressed for the sake of notational simplicity. The expansion coefficients are given by

$$c_{\mp} = \langle \Psi^{\mp} | \Psi \rangle$$
$$= \frac{1}{2}(\langle 0|_1 \langle 1|_2 \mp \langle 1|_1 \langle 0|_2)[\alpha(|0\rangle_1 |0\rangle_2 |1\rangle_3 - |0\rangle_1 |1\rangle_2 |0\rangle_3)$$
$$+ \beta(|1\rangle_1 |0\rangle_2 |1\rangle_3 - |1\rangle_1 |1\rangle_2 |0\rangle_3)]$$
$$= -\frac{1}{2}(\alpha|0\rangle_3 \pm \beta|1\rangle_3),$$

$$d_{\mp} = \langle \Phi^{\mp} | \Psi \rangle$$
$$= \frac{1}{2}(\langle 0|_1 \langle 0|_2 \mp \langle 1|_1 \langle 1|_2)[\alpha(|0\rangle_1 |0\rangle_2 |1\rangle_3 - |0\rangle_1 |1\rangle_2 |0\rangle_3)$$
$$+ \beta(|1\rangle_1 |0\rangle_2 |1\rangle_3 - |1\rangle_1 |1\rangle_2 |0\rangle_3)]$$
$$= \frac{1}{2}(\alpha|1\rangle_3 \pm \beta|0\rangle_3),$$

from which we obtain the desired result.

11.8 Each Bell state is as good as the other for implementing quantum teleportation as the four Bell states are related by unitary transformations. The reader can check their equivalence by explicit calculations.

11.9 Let **a**, **b**, and **c** denote respectively the directions given by the B, $\oplus$, and $\otimes$ bases, then from Footnote a in p. 371 and Footnote a in p. 374 it follows that

$$\mathbf{a} = \frac{1}{\sqrt{2}}(\mathbf{e}_z + \mathbf{e}_x), \quad \mathbf{b} = \mathbf{e}_z, \quad \mathbf{c} = \mathbf{e}_x.$$

Since $|\Psi^-\rangle$ and $|\Psi_0\rangle$ represent the same Bell state, using Eq. (10.26) we have

$$\langle \mathbf{a}, \mathbf{b} \rangle_{\psi-} = -\mathbf{a} \cdot \mathbf{b} = -\frac{1}{\sqrt{2}},$$

$$\langle \mathbf{a}, \mathbf{c} \rangle_{\psi-} = -\mathbf{a} \cdot \mathbf{c} = -\frac{1}{\sqrt{2}},$$

$$\langle \mathbf{b}, \mathbf{c} \rangle_{\psi-} = -\mathbf{b} \cdot \mathbf{c} = 0.$$

Therefore, we obtain

$$|\langle \mathbf{a}, \mathbf{b} \rangle_{\psi-} + \langle \mathbf{a}, \mathbf{c} \rangle_{\psi-}| + \langle \mathbf{b}, \mathbf{c} \rangle_{\psi-} = \sqrt{2} > 1,$$

which indeed violates the Bell inequality (10.35).

11.10 It suffices to consider the definition (11.58).

11.11 From the given information

$$\wp(J = 1, K = 1) = 0, \quad \wp(J = 2, K = 1) = \frac{3}{4},$$

$$\wp(J = 1, K = 2) = \frac{1}{8}, \quad \wp(J = 2, K = 2) = \frac{1}{8}.$$

it follows that

$$\wp(J = 1) = \frac{1}{8} + 0 = \frac{1}{8}, \quad \wp(J = 2) = \frac{3}{4} + \frac{1}{8} = \frac{7}{8}$$

$$\wp(K = 1) = 0 + \frac{3}{4} = \frac{3}{4}, \quad \wp(K = 2) = \frac{1}{8} + \frac{1}{8} = \frac{1}{4}.$$

Hence, we have

$$H(J, K) = -\frac{3}{4}\lg\frac{3}{4} - \frac{1}{8}\lg\frac{1}{8} - \frac{1}{8}\lg\frac{1}{8}$$

$$\simeq 1.061 \text{ bits,}$$

$$H(J) = -\frac{1}{8}\lg\frac{1}{8} - \frac{7}{8}\lg\frac{7}{8}$$

$$\simeq 0.544 \text{ bits,}$$

$$H(K) = -\frac{3}{4}\lg\frac{3}{4} - \frac{1}{4}\lg\frac{1}{4}$$

$$\simeq 0.811 \text{ bits,}$$

from which we obtain

$$I(J : K) = H(J) + H(K) - H(J, K)$$

$$= -\frac{7}{8}\lg\frac{7}{8} - \frac{1}{4}\lg\frac{1}{4} + \frac{1}{8}\lg\frac{1}{8}$$

$$\simeq 0.294 \text{ bits,}$$

$$H(J|K) = H(J) - I(J : K)$$

$$= \frac{1}{4}\lg\frac{1}{4} - \frac{1}{8}\lg\frac{1}{8} - \frac{1}{8}\lg\frac{1}{8}$$

$$= \frac{1}{4} \text{ bits,}$$

$$H(K|J) = H(K) - I(J : K)$$

$$= -\frac{3}{4}\lg\frac{3}{4} + \frac{7}{8}\lg\frac{7}{8} - \frac{1}{8}\lg\frac{1}{8}$$

$$\simeq 0.518 \text{ bits.}$$

Chapter 12

12.1 From the definition of $\hat{E}_j$, we have

$$
\begin{aligned}
\hat{E}_j^\dagger &= \text{Tr}_S\left[\left(\hat{U}_t^\dagger \hat{P}_{a_j} \hat{U}_t \hat{\rho}_S\right)^\dagger\right] \\
&= \text{Tr}_S\left(\hat{\rho}_S \hat{U}_t^\dagger \hat{P}_{a_j} \hat{U}_t\right) \\
&= \text{Tr}_S\left(\hat{U}_t^\dagger \hat{P}_{a_j} \hat{U}_t \hat{\rho}_S\right) \\
&= \hat{E}_j,
\end{aligned}
$$

where use has been made of the fact that the density matrix and projector are Hermitian, and the cyclic property of the trace (7.115c). Alternatively, from Eq. (12.20) we find

$$
\hat{E}_j^\dagger = \hat{\vartheta}_j^\dagger \hat{\vartheta}_j = \hat{E}_j.
$$

Let $|\phi_A\rangle$ be an arbitrary state in $\mathcal{H}_A$, then again from Eq. (12.20) we have

$$
\langle \hat{E}_j \rangle_{\phi_A} = \langle \phi_A | \hat{\vartheta}_j^\dagger \hat{\vartheta}_j | \phi_A \rangle = \left\| \hat{\vartheta}_j | \phi_A \rangle \right\|^2 \geq 0.
$$

12.2 This is straightforward but tedious.

$$
\begin{aligned}
\hat{\mathcal{A}} &= \frac{1}{\sqrt{2}}\left[\hat{O}_a^2 + \frac{(\hat{O}_b + \hat{O}_{b'})^2}{2} - 2\frac{\hat{O}_a(\hat{O}_b + \hat{O}_{b'})}{\sqrt{2}} + \hat{O}_{a'}^2 \right. \\
&\quad \left. + \frac{(\hat{O}_b - \hat{O}_{b'})^2}{2} - 2\frac{\hat{O}_{a'}(\hat{O}_b - \hat{O}_{b'})}{\sqrt{2}} \right] \\
&= \frac{1}{\sqrt{2}}\frac{1}{2\sqrt{2}}\left[2\sqrt{2}\hat{O}_a^2 + \sqrt{2}\hat{O}_b^2 + \sqrt{2}\hat{O}_{b'}^2 + \sqrt{2}(\hat{O}_b\hat{O}_{b'} + \hat{O}_{b'}\hat{O}_b) \right. \\
&\quad - 4\hat{O}_a\hat{O}_b - 4\hat{O}_a\hat{O}_{b'} + 2\sqrt{2}\hat{O}_{a'}^2 + \sqrt{2}\hat{O}_b^2 + \sqrt{2}\hat{O}_{b'}^2 \\
&\quad \left. - \sqrt{2}(\hat{O}_b\hat{O}_{b'} + \hat{O}_{b'}\hat{O}_b) - 4\hat{O}_{a'}\hat{O}_b + 4\hat{O}_{a'}\hat{O}_{b'} \right] \\
&= \frac{1}{4}\left[2\sqrt{2}(\hat{O}_a^2 + \hat{O}_{a'}^2 + \hat{O}_b^2 + \hat{O}_{b'}^2) \right. \\
&\quad \left. - 4(\hat{O}_a\hat{O}_b + \hat{O}_{a'}\hat{O}_b + \hat{O}_a\hat{O}_{b'} - \hat{O}_{a'}\hat{O}_{b'}) \right] \\
&= 2\sqrt{2}\hat{I} - \hat{\mathcal{B}},
\end{aligned}
$$

wherein the second equality it is noted that $\hat{O}_b\hat{O}_{b'} + \hat{O}_{b'}\hat{O}_b \neq 2\hat{O}_b\hat{O}_{b'}$ since $\hat{O}_b$ and $\hat{O}_{b'}$ do not commute.

12.3 Consider the Hermitian operator

$$\hat{A}' = 2\sqrt{2}\hat{I} + \hat{B}$$

$$= \frac{1}{\sqrt{2}}\left(\hat{O}_a^2 + \hat{O}_{a'}^2 + \hat{O}_b^2 + \hat{O}_{b'}^2\right) + \hat{B},$$

which can be written as [see Problem 12.2]

$$\hat{A}' = \frac{1}{\sqrt{2}}\left[\left(\hat{O}_a + \frac{\hat{O}_b + \hat{O}_{b'}}{\sqrt{2}}\right)^2 + \left(\hat{O}_{a'} + \frac{\hat{O}_b - \hat{O}_{b'}}{\sqrt{2}}\right)^2\right].$$

Since $\hat{A}'$ is in the form of the sum of squares of Hermitian operators, it is positive semidefinite and has a non-negative expectation value, that is,

$$\langle\hat{A}'\rangle = 2\sqrt{2} + \langle\hat{B}\rangle \geq 0.$$

Hence, we have

$$\langle\hat{B}\rangle \geq -2\sqrt{2},$$

which completes the proof.

12.4 The squares of the observables $\hat{O}$ and $\hat{O}'$ are given by

$$\hat{O}^2 = (|1\rangle\langle1| - |2\rangle\langle2|)(|1\rangle\langle1| - |2\rangle\langle2|) = |1\rangle\langle1| + |2\rangle\langle2|,$$

$$\hat{O}'^2 = (|3\rangle\langle3| - |4\rangle\langle4|)(|3\rangle\langle3| - |4\rangle\langle4|) = |3\rangle\langle3| + |4\rangle\langle4|,$$

which yields

$$\hat{A} = \frac{1}{\sqrt{2}}(|1\rangle\langle1| + |2\rangle\langle2| + |3\rangle\langle3| + |4\rangle\langle4|)$$

$$= \sqrt{2}(|1\rangle\langle1| + |2\rangle\langle2|).$$

In obtaining the last equality, we have made use of the relations

$$|3\rangle\langle3| = \frac{1}{2}(|1\rangle\langle1| + |2\rangle\langle2| + |1\rangle\langle2| + |2\rangle\langle1|),$$

$$|4\rangle\langle4| = \frac{1}{2}(|1\rangle\langle1| + |2\rangle\langle2| - |1\rangle\langle2| - |2\rangle\langle1|),$$

The main lesson here is that the square of a Hermitian operator turns out to be a sum of only positive semidefinite operators. Let

$$|\psi\rangle = c_1|1\rangle + c_2|2\rangle,$$

where c_1 and c_2 are probability amplitudes satisfying the normalization condition $|c_1|^2 + |c_2|^2 = 1$, then $\langle \hat{A} \rangle_\psi$ is given by

$$\langle \hat{A} \rangle_\psi = \sqrt{2}(\langle \psi | 1 \rangle \langle 1 | \psi \rangle + \langle \psi | 2 \rangle \langle 2 | \psi \rangle)$$
$$= \sqrt{2}(|c_1|^2 + |c_2|^2)$$
$$= \sqrt{2} \geq 0.$$

It is noted that the equality holds if and only if $|\psi\rangle = 0$.

Printed in the United States
by Baker & Taylor Publisher Services